Biochemistry of Dioxygen

BIOCHEMISTRY OF THE ELEMENTS

Series Editor: **Earl Frieden**
Florida State University
Tallahassee, Florida

A Continuation Order Plan is available for this series. A continuation order will bring delivery of each new volume immediately upon publication. Volumes are billed only upon actual shipment. For further information please contact the publisher.

Biochemistry of Dioxygen

Lloyd L. Ingraham

and

Damon L. Meyer

University of California, Davis
Davis, California

PLENUM PRESS • NEW YORK AND LONDON

Library of Congress Cataloging in Publication Data

Ingraham, Lloyd, L.
  Biochemistry of dioxygen.

  (Biochemistry of the elements; v. 4)
  Includes bibliographies and index.
  1. Oxygen—Metabolism. I. Meyer, Damon, L. II. Title. III. Series.
QP535.O1I46   1985   599'.019214   85-12055
ISBN 0-306-41948-3

©1985 Plenum Press, New York
A Division of Plenum Publishing Corporation
233 Spring Street, New York, N.Y. 10013

Printed in the United States of America

Preface

This book is written for the research biochemist who needs to know more about the particular field of dioxygen metabolism, whether this be for designing lectures for a graduate level course or for his or her own research needs. We hope researchers in a given area of dioxygen metabolism will gain knowledge of related fields of dioxygen metabolism.

We have decided to use the term *dioxygen* to distinguish molecular oxygen from divalent oxygen in water and organic compounds, dioxygen being a simpler term than molecular oxygen. We do not intend to review the metabolism of all compounds that contain oxygen, since this would include all of biochemistry.

An understanding of dioxygen chemistry is essential to the discussion of the biochemistry of dioxygen. While this statement could be made about any biochemical constituent, the chemistry of dioxygen is so unusual that interpretations without detailed chemical background are futile. Prediction of dioxygen reaction products by analogy with other oxidants is impossible.

The partial reduction products of dioxygen, superoxide ion and peroxides, develop naturally in the chemistry of dioxygen. It would be difficult to discuss dioxygen biochemistry without first discussing these partial reduction products.

The first chapters stress the chemistry of dioxygen. These are followed by chapters that delve into the intricacies of dioxygen metabolism. This procedure has tended to divide the book into two sections. The sections are purposely not well defined because certain biological reactions of dioxygen and the reduction products of dioxygen are better discussed in the early chemical chapter.

After an introductory chapter, the discussion is focused on the chemistry and physical properties of ground-state triplet dioxygen. An important aspect of dioxygen chemistry is the kinetic barrier to oxidations by dioxygen. Because of this kinetic barrier, there is a need for a mechanism to activate dioxygen before it can oxidize most substrates.

The activation of dioxygen in biological systems is usually accomplished by a metalloenzyme. These form metal–dioxygen complexes that are commonly diamagnetic. These (singlet-state) metal–dioxygen complexes perform the great majority of the biological oxidations. Singlet metal dioxygen complexes react

v

much more like singlet dioxygen than like ground-state dioxygen, so that knowledge of singlet-dioxygen chemistry facilitates an understanding and appreciation of dioxygen biochemistry. The reactions of singlet dioxygen that may have relevance to biological dioxygen metabolism are discussed thoroughly in Chapter 3. There are also many reactions of singlet dioxygen that have no relevance to biological reactions at this time, but the potential importance of these reactions to dioxygen metabolism dictates that they be included. These reactions are included as an addendum designated Chapter 21.

The discussion of triplet and singlet dioxygen is followed by a discussion of the chemistry of the reduction products of dioxygen, superoxide ion and peroxides. Superoxide dismutase and prostaglandins very naturally fit into these chapters. The assumption is made that the reader is aware of the physiological function of prostaglandins but wishes to know more about the chemistry and biosynthesis. Discussions of catalases and peroxidases logically follow the chemistry and biochemistry of peroxides. A familiarity with the enzyme intermediates in catalase and peroxidase is helpful in the understanding of enzyme intermediates in dioxygen reactions.

The discussion returns to dioxygen again to focus on the important problem of activation. Activation by metals makes an easy transition to metal carriers for dioxygen, both models and actual biological carriers.

Once dioxygen is activated, it can react in three different ways: (1) as a one-, two-, or four-electron acceptor from the substrate; (2) as a two-electron acceptor plus an oxygen atom donor; or (3) as a donor of two oxygen atoms. These are discussed in order beginning with the four-electron oxidases. The two-electron-plus-oxygen-atom donors are called monooxygenases. These are classified for discussion with respect to the cofactor involved: flavins, pterins, copper, and iron. The monooxygenases are followed by the enzymes that donate two oxygen atoms, the dioxygenases. The next two chapters discuss the special topics of bioluminescence and toxicity, followed by the remainder of the singlet-dioxygen chemistry in Chapter 21.

It is hoped that the reader will have time to read the whole book in order to appreciate the complete story of dioxygen metabolism. However, this book should be useful as a reference book when the reader desires information on a given aspect of dioxygen chemistry or metabolism.

Lloyd L. Ingraham
Damon L. Meyer

Contents

Introduction

1

1.1 Chemical Reactions of Dioxygen

Dioxygen is a fascinating molecule for biologists, chemists, and physicists. Its utilization is necessary for all higher animals, and it is either produced or utilized, or both, by much of the rest of the living world. To the layman, oxygen utilization is almost synonymous with life. Dioxygen has several characteristics that make it ideal for a terminal oxidant in a biological system: its high oxidizing potential, its barrier to oxidation, and the fact that it forms innocuous products.

The high oxidizing potential is valuable for the simple reason that oxidations by dioxygen make large amounts of energy available to the organism. A poor oxidizing agent would make less efficient use of foods.

The kinetic barrier to reaction is also an important factor because without it, foodstuffs and the organism itself would be oxidized indiscriminately by the terminal oxidant. It is difficult to picture life in an atmosphere of chlorine.

The production of innocuous products is an important but seldom appreciated factor. The products water and carbon dioxide are neutral and unreactive. A terminal oxidase system that produced hydrochloric or hydrobromic acid would pose some difficult problems for any conceivable organism.

However, there are also problems connected with the use of dioxygen as a terminal oxidant in biological systems. The desirable high oxidizing potential of dioxygen also means that dioxygen and some of its partial reduction products are potentially hazardous to the cell. Certain compounds that react with dioxygen without requiring prior enzymatic activation of the dioxygen can be toxic to the cell. The one- and two-electron reduction products of dioxygen pose a particular problem because they are often reactive. Thus, the cell must have enzymes that reduce the level of these partially reduced products of dioxygen that may be toxic to the cell. The two notable enzymes in this category are catalase and superoxide dismutase.

The kinetic barrier that is a benefit to the organism when storing dioxygen presents a problem when the organism needs to use the dioxygen to produce energy. The forms of catalysis that reduce this barrier for living systems illustrate nature's novelty at its highest level.

Nature faces another problem with dioxygen: It has a relatively low solubility. Water at 20°C will dissolve only 3.1% of its volume of dioxygen (STP). Dioxygen is more soluble in organic solvents. For example, ethanol will dissolve 41.5% of its volume in dioxygen. Nature not only needs enzymes to activate dioxygen, but also must have carriers to get the dioxygen to the point of reaction. The problems in designing a dioxygen carrier are an important facet of the metabolism of dioxygen.

The kinetic barrier to oxidations by dioxygen is the result of two properties of dioxygen. One is that ground-state dioxygen is a triplet state. The other is that the lowest orbital available to accept an electron is an antibonding orbital.

Ground-state dioxygen does not react as a normal double bond, but as a diradical because of the triplet ground state. To maintain spin conservation during the reaction, dioxygen must either react with another molecule with unpaired electrons or produce a triplet-state product. Stable triplet states are unusual, so this places a kinetic restriction on reactions of dioxygen. Electron transfer requires that the first electron be placed in a partially filled antibonding π-orbital. Again, this places a barrier to electron-transfer reactions.

1.2   Biological Reactions of Dioxygen

In the previous section, we saw the need for activation of ground-state dioxygen for reaction. This activation is commonly accomplished in biological systems by forming a complex with a transition metal. The complex eliminates both of the aforementioned barriers. The complex is a singlet state instead of a triplet state, and back-bonding from the metal to the dioxygen adds electrons that make the antibonding π-orbitals more like electron pairs to allow for electron acceptance without a barrier.

The reactions of the metal–dioxygen complexes are much different from those of ground-state triplet dioxygen. Ground-state dioxygen commonly reacts in chain reactions. Chain reactions are difficult to control and consequently would tend to burn up more of the cell than simply the intended energy source. Singlet-dioxygen and biological dioxygen reactions tend to be reactions in which only one substrate is oxidized and the reaction stops. Metal–dioxygen complexes are not singlet dioxygen, but the reactions are similar.

1.3   Reduction Products of Dioxygen

An integral part of dioxygen chemistry is the one- and two-electron reduction products. The one-electron product, superoxide ion, and its conjugate acid, HO_2, have fascinating properties. Hydrogen superoxide is thermodynamically unstable with respect to dihydrogen and dioxygen. The reduction of dioxygen to superoxide ion has a negative potential, but the one-electron potential at superoxide ion is highly positive. Dioxygen becomes a stronger one-electron oxidant by adding an electron. Despite the positive potential, superoxide ion tends not to react as an oxidant. This is because the addition of one more electron is repelled by the existing negative charge on the superoxide ion. Superoxide can act as a base because the favorable disproportion reaction consumes hydrogen ions:

$$2H^+ + 2O_2^- \longrightarrow H_2O_2 + O_2$$

The real danger of superoxide ion appears to be in the metal-ion-catalyzed Haber–Weiss reaction, which produces the very reactive, and consequently toxic, hydroxyl radical:

$$O_2^- + H_2O_2 \xrightarrow{\text{Metal ion}} OH^- + \cdot OH + O_2$$

Very little of the dioxygen consumed in a cell produces superoxide, but there are extraneous reactions that are not part of a normal dioxygen metabolism that do produce superoxide in the cell. Superoxide is kept at a low concentration in the cell by the disproportionation reaction catalyzed by the enzyme superoxide dismutase.

The two-electron reduction product of dioxygen, hydrogen peroxide, is formed by both enzymatic and nonenzymatic reactions. Hydrogen peroxide is a reactive compound, so it must be kept at a low concentration in the cell. This is accomplished by the enzyme catalase, which decomposes hydrogen peroxide to water and dioxygen. It is interesting to note that the structure of the intermediate in hydrogen peroxide decomposition is probably the same as that of the intermediate in certain hydroxylation reactions.

1.4   Dioxygen Enzymes

The main reactions of dioxygen metabolism can be classified into three broad groups. The first group includes those reactions in which oxygen acts by merely accepting electrons from the substrate. The second group is comprised of reactions in which an oxygen atom is donated to a substrate and two electrons are accepted from one of the substrates to form water. These reactions appear to be quite diverse in mechanism and utilize several types of cofactors including copper, iron, flavins, and pterins. The third group of reactions includes those in which two oxygen atoms are added to the substrate in one step. These reactions are reminiscent of those of singlet dioxygen.

Finally, the high potential of dioxygen as an oxidizing agent automatically causes it to be a dangerous chemical despite the kinetic barriers. The toxicity of dioxygen is an important subject.

Dioxygen is the oxidant in bioluminescence reactions. The dioxygen actually acts as in a dioxygenase reaction, but the uniqueness of the reactions gives reason for grouping them in a separate category.

1.5   Oxygen-17 Nuclear Magnetic Resonance Spectroscopy of Oxygen

To study dioxygen metabolism, it is necessary to be able to trace the pathway of the oxygen atom. In early work, this was usually performed by mass spectroscopy of ^{18}O. Many newer studies utilize the NMR detection of ^{17}O. Some of the applications of ^{17}O NMR spectroscopy to oxygen metabolism are reviewed here.

Oxygen exists as three natural isotopes, ^{16}O, ^{17}O, and ^{18}O, as 99.759, 0.037, and 1.204% of natural oxygen. The ^{17}O isotope is rather expensive because of its low natural abundance, but the cost is compensated for by a nuclear spin of $^{5}/_{2}$ that allows it to be detected by NMR. The ^{18}O isotope is more readily available than ^{17}O, but can be detected only by mass spectrometry. There is a radioactive isotope of oxygen, ^{15}O, but its half-life of 124 sec precludes it from having much use in tracer chemistry.

Oxygen-17 NMR spectroscopy is a promising analytical technique for the study of oxygen environments and reaction mechanisms. The technique has not been used extensively in the past due to the low natural abundance of ^{17}O and to the fact that the nucleus is a quadrapole. In a large number of cases, these disadvantages have been offset by the advent of availability of Fourier-transform NMR apparatus and ^{17}O enriched H_2O and O_2. Signal-to-noise ratios are raised experimentally by maximizing the rate of molecular tumbling, i.e., by raising the temperature or lowering the solution viscosity.

Oxygen-17 NMR has proved very informative in those cases in which the difficulties have been overcome. The resonances cover a very wide range of chemical shift even within, for instance, the category of oxygen atoms bonded to carbon (Sugiwara *et al.*, 1979). These values range from about -30 to $+600$ ppm from H_2O. In this series, and also in those compounds with nitrogen–oxygen bonds, there is a good correlation between bond order and chemical shift, so that alcohols appear at the -30 ppm end of the scale and aldehydes in the neighborhood of $+600$ ppm. Ethers, esters, carbonates, anhydrides, acids, acyl halides, amides, and other compounds fall in various regions between these extremes, depending on resonance and hydrogen bonding. Nitrogen-bonded oxygens absorb as far as 915 ppm downfield. The chemical shift for alcohols can be correlated with the O–H stretching frequency (Takasuka, 1981).

In other types of compounds, the position of absorption is more difficult to predict. Ozone, for example, has two resonances—one at 1598 ppm (terminal oxygen) and the other at 1032 ppm. Oxygen atoms bound to NMR-active nuclei show split patterns so long as solvent exchange is slow compared with the observation time scale.

Block et al. (1980) (see also Dyer et al., 1982) observed anomalies in the ^{17}O NMR spectra of four-membered ring sulfoxides and sulfones that parallel anomalies in the ^{13}C spectra of the compounds. The values obtained for both types of signals did not follow the pattern established for decreasing ring size from six-membered to three-membered. They also observed that oxygens on asymmetrical sulfones give distinguishable signals that are very sensitive to remote substitutents. Of particular interest to biochemists are the ^{17}O NMR studies of adenine nucleotides (Gerlt et al., 1982; Huang and Tsai, 1982).

Many of the experiments that have utilized ^{17}O NMR have been studies of rates of hydration and exchange in inorganic complexes. In some cases, these measurements have been made on compounds of biological interest. Mn and Mg complexes of ATP, for example, have been explored (Zetter et al., 1973), as have Ni complexes of EDTA (Grant et al., 1971). For cases in which solvent exchange is very rapid, the equilibrium causes broadening of the signal, which can be related to kinetic parameters. Slower reactions are observed by using isotopically enriched samples of solvent and measuring the rate of appearance or disappearance of a characteristic signal.

The cyclic phosphate esters A and B (Figure 1-1) can be distinguished (Coderre et al., 1981) by their ^{17}O NMR spectra. The absolute configurations were first established by ^{17}O perturbations of the phosphorus chemical shift of derivatives of these compounds. It was then observed that the ^{17}O NMR spectra of the diastereomers differed in chemical shift, ^{31}P–^{17}O coupling constant, and line width. In phosphorus-decoupled spectra of a diastereomeric mixture, the signals were resolved. The spectra were recorded at 95°C to minimize line width.

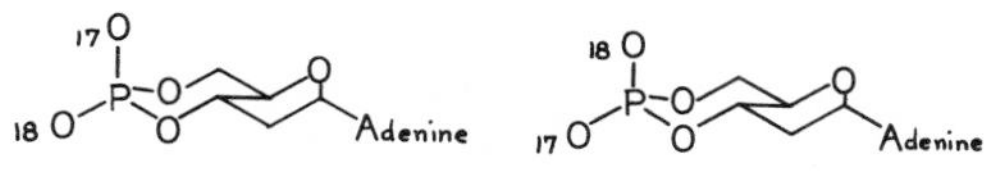

Figure 1-1. Structure of two deoxyadenosine cyclic phosphates as determined by NMR.

The limitation of ^{17}O NMR makes its feasibility for enzymatic studies remote. The tumbling rates of large molecules are quite slow, since they are related, ideally at least, to the third power of the radius. Furthermore, increasing the signal-to-noise ratio by raising the concentration also increases the viscosity, which slows the tumbling. Significant temperature elevation is also impossible.

However, model studies with this technique could be very instructive. Attempts to observe metal–dioxygen complexes have failed (Lapidot and Irving, 1972), but it has been pointed out that these procedures definitely could be improved (Klemperer, 1978). The benefits of doing so are clear. The differences between several possible metal–dioxygen complexes of biological complexes have shown (Klemperer, 1978) that the chemical shifts are related to the degree of back-bonding. Oxygen-17 NMR would clearly show whether or not the two oxygen atoms were equivalent.

References

Block, E., Bazzi, A. A., Lambert, J. B., Wharry, S. M., Anderson, K. K., Dittmer, D. C., Patwardhan, B. H., and Smith, D. J. H., 1980. Resonance studies of organosulfur compounds: The four-membered-ring sulfone effect, *J. Org. Chem.* 45:4807–4810.

Codderre, J. A., Mehdi, S., Demou, P. C., Weber, R., Traficante, D. D., and Gerth, J. A., 1981. Oxygen chiral phosphodiesters. 3. Use of ^{17}O NMR spectroscopy to demonstrate configurational differences in diastereomers of cyclic 2'-deoxyadenosine 3',5'-([17]O,[18]O) monophosphate, *J. Am. Chem. Soc.* 103:1870–1872.

Dyer, J. C., Harris, D. L., and Evans, S. A., Jr., 1982. Oxygen-17 nuclear magnetic resonance spectroscopy of sulfoxides and sulfones: Alkyl substituent induced chemical shift effects, *J. Org. Chem.* 47:3660–3664.

Gerlt, J. A., Demov, P. C., and Mehdi, S., 1982. ^{17}O NMR spectral properties of simple phosphate esters and adenine nucleotides, *J. Am. Chem. Soc.* 104:2848–2856.

Grant, M. W., Dodgen, H. W., and Hunt, J. P., 1971. An oxygen-17 nuclear magnetic resonance study of nickel(II) ethylene-diaminetetraacetate complexes in aqueous solution, *J. Am. Chem. Soc.* 93:6828–6831.

Huang, S. L., and Tsai, M.-D., 1982. Does the magnesium(II) ion interact with the α-phosphate of adenosine triphosphate? An investigation by oxygen-17 nuclear magnetic resonance, *Biochemistry* 21:951–959.

Klemperer, W. G., 1978. ^{17}O-NMR spectroscopy as a structural probe, *Angew. Chemie Int. Ed. Engl.* 17:246–254.

Lapidot, A., and Irving, C. S., 1972, Oxygen-17 nuclear magnetic resonance spectroscopy and iridium and rhodium molecular oxygen complexes, *J. Chem. Soc. Dalton Trans.* 1972:668–670.

Sugiwara, T., Kawada, Y., Katch, M., and Iwamura, H., 1979, Oxygen-17 nuclear magnetic resonance. III. Oxygen with coordination number of two, *Bull. Chem. Soc. Jpn.* 52:3391–3396.

Takasuka, M., 1981. Relationship of the [17]O chemical shift to the stretching frequency of the hydroxy-group in saturated alcohols, *J. Chem. Soc. Perkin Trans.* 2:1558–1561.

Zetter, M. S., Dodgen, H. W., and Hunt, J. P., 1973. Measurement of the water exchange rate of bound water in the manganese II–adenosine triphosphate complex by oxygen-17 nuclear magnetic resonance, *Biochemistry* 12:778–782.

Ground-State Dioxygen

2

2.1 Properties of Dioxygen (Table 2-1)

The critical characteristic of ground state dioxygen is that it is a triplet instead of a singlet state. This fact contributes to the kinetic barrier in reactions with ground-state dioxygen and also influences the type of reactions that do occur.

The unique electronic structure of dioxygen and its resulting unique chemistry can be better understood if we consider the individual orbitals in dioxygen (see Table 2-1). The second valence shell of atomic oxygen has 6 electrons, so that dioxygen has 12 electrons in the second shell. Two of these electrons are in each 2s orbital to form lone pairs.* If we define the Z direction to be the direction of the bond between the two nuclei, we can combine the $2p_z$ orbitals to form a sigma bond with 2 electrons. The p_x and $2p_y$ are thus perpendicular to the sigma bond and form π-bonds that we will label π_x and π_y. Two electrons are placed in each of these orbitals, making a total of 10 electrons. However, we still have 2 more electrons. The $2p_x$ and $2p_y$ also form antibonding π-orbitals, π_x^* and π_y^*, which are at a lower energy than the σ^* orbital. The question is, how do we place these 2 electrons in the π_x^* and π_y^* orbitals? Placing 1 in each antibonding orbital will produce much less electron–electron repulsion than placing them both in the same antibonding orbital. The 2 electrons are in separate orbitals, as shown in Table 2-2, and as a result are unpaired, since unpairing also reduces the interaction between the electrons (Coulson, 1947). The most stable configuration of dioxygen is a triplet state, since 2 unpaired electrons form a triplet state.

Ground-state dioxygen is usually designated as $^3\Sigma_g$, indicating a triplet state with zero resultant orbital momentum. Note that in the π-orbitals of dioxygen,

---

* The orbitals are actually hybridized so that the lone pair is an sp^2 hybrid. Nevertheless, for simplicity, we shall add electrons to pure s and p orbitals.

7

Table 2-1. Properties of Dioxygen

Pale blue liquid; colorless gas			
Solubilities (cm³/liter at 1 atm)	10°C	20°C	25°C
Water	38.15	31.36	28.6
Seawater	—	5.10	4.73
Chloroform	219.5	—	—

Bond energy, 118 kcal;
bond distance, 1.21 Å

Isotopes

	Percent in air	Stability
^{15}O	—	Half-life 122.6-sec
^{16}O	99.7587	Stable
^{17}O	0.0374	Stable; spin $= \frac{3}{2}$
^{18}O	0.2039	Stable

Boiling point: 16_{O_2}, 90.19°K; 18_{O_2}, 90.30°K

Thermodynamics

$$2H_2 \text{ (g)} + O_2 \text{ (g)} \rightarrow 2H_2O \text{ (}\ell\text{)} \qquad \Delta H = -136.634 \text{ kcal/mole}$$

$$O_2 + 4H^+ + 4e^- \rightarrow 2H_2O \qquad E° = 1.229 \text{ V}$$

$$O_2 + 2H_2O + 4e^- \rightarrow 4OH^- \qquad E° = 0.401 \text{ V}$$

$$O_2 + 4H^+ (10^{-7} \text{ M}) + 4e^- \rightarrow 2H_2O \qquad E° = 0.82 \text{ V}$$

$$O_2 + e^- \rightarrow O_2- \qquad E° = -0.33 \text{ V}$$

O–O frequency $\qquad$ 1556 cm^{-1a}

[a] Sawyer and Gibian (1979).

Table 2-2. Structure of Ground-State Dioxygen

$2s$:	$2s$:		
$2p_x$	$2p_x$	π_x:	$\pi_x{}^* \cdot \uparrow$
$2p_y$	$2p_y$	π_y:	$\pi_y{}^* \cdot \uparrow$
$2p_z$	$2p_z$	π_z:	

there are six bonding electrons and two antibonding electrons, leaving an excess of, effectively, four bonding electrons. Four bonding electrons correspond to a double bond between the oxygen atoms. As dioxygen is reduced, the effective number of bonding electrons is reduced, and correspondingly the bond distance is increased. The one-electron reduction product, O_2^-, called superoxide ion, has effectively three bonding electrons, and the bond distance is increased from 1.21 Å in dioxygen to 1.28 Å in superoxide ions. Reduction by one more electron to form peroxide dianion reduces the number of effective bonding electrons to two, and the bond distance increases to 1.49 Å. Reduction also results in a lower bond-dissociation energy. The 118 kcal/mole for dioxygen reduces to 35 kcal in hydrogen peroxide.

It was noted in Chapter 1 that oxygen has a kinetic barrier to the oxidation of most substrates. The rate of reaction of ground-state $^3\Sigma_g$ dioxygen is limited by two factors: One is that dioxygen is difficult to reduce by one electron; the other results from the triplet nature of dioxygen. One-electron reduction of dioxygen is difficult because the electron must be added in an antibonding orbital that is already partially occupied. There is not only the energy of an antibonding orbital to contend with, but also the problem of electron repulsion. Thus, reactions of dioxygen involving electron transfer are generally slow. The potentials of the reduction of dioxygen are summarized in Figure 2-1. Despite dioxygen's being a very powerful oxidant [1.23 V in acid solution and 0.82 V in neutral solution (George, 1964)], it is very difficult to add the first electron. The first electron to form O_2^- from dioxygen is added at a potential of only -0.33 V* (Wood, 1974). This potential should be compared with that of the Co^{2+}–Co couple of -0.277 V or the Ni^{2+}–Ni couple of -0.250 V (Latimer, 1938). Dioxygen in one-electron oxidations is a less powerful oxidant than Ni^{2+} and Co^{2+}. It is interesting to note that superoxide ion is a mild reducing agent.

After one electron is added, the O–O bond distance increases, thus decreasing the bonding between the oxygens and making the p-orbitals more like lone pairs. The next electrons are added relatively easily if protons are available because they are added to essentially empty lone-pair orbitals. If protons are not available, the electron must be added in repulsion to the existing negative charge on the superoxide ion. Superoxide ion has the potential of a good oxidizing agent even at pH 7 (Wood, 1974):

$$O_2^- + 2H^+ + e^- \rightarrow H_2O_2 \qquad E_o' = +0.94 \text{ V at pH } 7.0$$

The next two electrons are added (George, 1964) at 0.80 and 2.74 volts in acid solution or at 0.38 and 2.33 volts in neutral solution. The intermediates in dioxygen reduction are all powerful oxidizing agents.

---

* This potential is also sometimes reported as -0.16 V. The value of -0.33 V is for 1 atm dioxygen and the value -0.16 V for molar dioxygen.

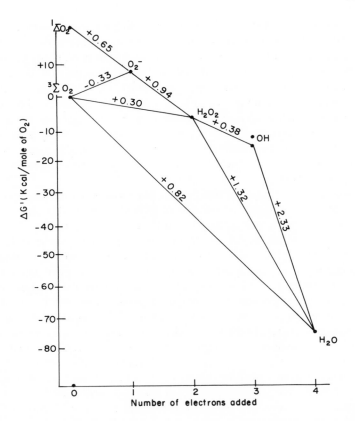

Figure 2-1. Reduction of ground-state dioxygen at pH 7 and 298°C. Note that after one uphill reduction of-ground state ($^3\Sigma_g$) dioxygen to O_2^-, all the rest are downhill. Reduction potentials are placed on the reaction lines. There is no barrier on the one-electron pathway of excited dioxygen ($^1\Delta$). Excited ($^1\Delta$) dioxygen is discussed in Chapter 2.

The reduction of dioxygen at a metal electrode is not always what one would expect. On a clean electrode, the potential is independent of pH and electrode material. However, on an oxidized electrode, the potential is pH-dependent and occurs at the potential of the metal oxide of the electrode material (Wilshire and Sawyer, 1978).

$$M(OH)_2 + 2H^+ + 2e^- \rightarrow M + 2H_2O$$

$$M(OH)_2 + 2e^- \rightarrow M + 2OH^-$$

$$2M + O_2 + 2H_2O \rightarrow 2M\,(OH)_2$$

The triplet character of $^3\Sigma_g$ dioxygen also reduces its rate of reaction. The unpaired electrons on $^3\Sigma_g$ dioxygen tend either to stay unpaired or to pair with an electron on the other reactant to form a new bond. If the other reactant does not have an unpaired electron, the product will also be a triplet. The reaction of dioxygen with a diamagnetic reactant will proceed only if the product has a low-energy triplet state.

2.2 Reactions of Dioxygen

Triplet dioxygen will undergo three general types of reactions:

1. $^3O_2 + R \rightarrow {}^3RO_2$

2. $^3O_2 + R \rightarrow O_2^- + R\cdot^+$

3. $^3O_2 + R\cdot \rightarrow RO_2\cdot$

Reaction 1 is an allowed reaction that depends on whether or not RO_2 has a low-energy triplet state. In general, this reaction is limited by the few low-energy triplet states. In the most common examples of reaction 1, RO_2 is a molecular complex between R and O_2. The low-energy triplet-state complex between R and O_2 collapses to produce a singlet RO_2 product.

Reaction 2 is also allowed, but is limited by the thermodynamics of the reaction. When triplet dioxygen does react in electron-transfer reactions, it does so by a one-electron transfer (Sawyer and Seo, 1977). The one-electron oxidizing potential of dioxygen is very poor, however, so the reactant has to be a very easily oxidizable compound for this reaction to occur.

Reaction 3 of triplet dioxygen with a radical is an allowed reaction and a common reaction. The initiation step limits this reaction. After the first radical is formed, the oxidation occurs in a chain reaction in which the $RO_2\cdot$ removes a hydrogen atom from the reactant to form a new radical. The reaction of triplet oxygen with free radicals can be extremely rapid (Barber *et al.*, 1963). In these reactions, one unpaired electron of the dioxygen forms a bond with the unpaired electron on the radical (Clyne and Thrush, 1963). These reactions are comparable to radical dimerization reactions. Even so, dioxygen as a triplet reacts much slower than simple doublet radicals with other radicals (Taube, 1965).

A reaction that is probably an example of Type 1, forming a triplet molecular complex, is shown in Figure 2-2. Tetrakis-dimethylaminoethylene will react with ground-state dioxygen to form tetramethyl urea plus light (Paris, 1965). The olefin in this case has an exceptionally high electron density and the light emission is from an excited state of the starting material (tetrakis-dimethylaminoethylene),

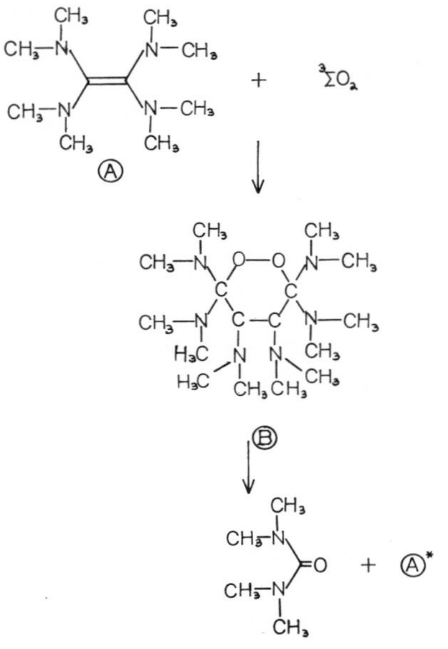

Figure 2-2. Reaction of tetrakis-dimethylamino-ethylene (A) with ground-state dioxygen to form tetramethyl urea (B).

not product (Fletcher and Heller, 1965). Kinetics require the participation of two reactant molecules. These data suggest that the reaction proceeds by break-down of intermediate 2-2B.

Another example of a reaction that can form an oxygen substrate that has a sufficiently low triplet state to allow the reaction to proceed is the reaction of oxygen with hexaphenylethane (Methoff and Branch, 1930). A type 1 reaction of interest to biochemists is the reaction of ground-state dioxygen with reduced flavins. Triplet dioxygen reacts with the anion of reduced flavin to form a triplet complex. The triplet complex then decays to a singlet complex that reacts to give the 4a-hydroperoxide (Kemal et al., 1977).

Ground-state dioxygen oxidizes sulfite ion slowly. The type 1 reaction of dioxygen with sulfite ion to produce triplet peroxy sulfate does not occur (Alyea and Backstrom, 1929) because the energy of the triplet state is not sufficiently low. Instead, the reaction occurs by a type 2 reaction proceeding through free radicals.

The reaction of dioxygen with carbanions to form an alkyl hydroperoxide is exothermic by about 30 kcal. Because the triplet state of the product is too high for reaction 1 to occur, the reaction proceeds instead by an electron transfer (Russell et al., 1962) in reaction 2.

2.3 Chain Reactions of Dioxygen

Most oxidations of organic compounds by $^3\Sigma_g$ dioxygen proceed by chain reactions in which the dioxygen reacts with a radical and the resulting peroxy radical reacts with another substrate (Walsh, 1946; Mayo, 1968; Ingold, 1969). Any formation of a radical will initiate the process, and most bimolecular reactions between radicals will terminate the chain. Combination of two peroxy radicals, however, leads to unstable molecules that disintegrate to form new radicals and continue the chain.

Initiation

4. $I + RH \rightarrow IH + R\cdot$

Chain reaction

5. $R\cdot + O_2 \rightarrow RO_2\cdot$

6. $RO_2\cdot + RH \rightarrow RO_2H + R$

Chain termination

7. $RO_2\cdot + RO_2\cdot \rightarrow$ many products

As discussed above, reaction 5 can be extremely rapid. The rate of oxidation depends on reaction 6. Although this is unfavorable for the usual unactivated hydrocarbon with a C–H bond energy of 98.7 kcal, it is favorable for many compounds. The O–H bond energy in ROOH is approximately 90 kcal/mole, which is comparable to the C–H bond in a tertiary hydrocarbon of 90–92 kcal/mole. Many compounds can be easily oxidized; for example, the C–H aldehydic bond is only 86 kcal/mole. Many other bonds of hydrogen to other atoms are even lower. Tertiary hydrogens are more susceptible to attack than secondary, and secondary hydrogens are more susceptible than primary. Reaction 7 is much more important in chain-termination reactions of secondary peroxy radicals than in those of tertiary peroxy radicals (Hiatt et al., 1968a,b; Howard and Ingold, 1968a,b). A general reaction mechanism for those cases in which ROO· is a secondary peroxy radical, suggested by Russell (1957), is shown in Figure 2-3. The products, dioxygen, ketone, and alcohol, are formed directly from the tetroxide intermediate. This mechanism is supported by the observation that the decomposition of secondary peroxy radicals exhibits a deuterium isotope effect. The electrons are all paired in the tetroxide and would therefore be expected to remain all paired in the product. The dioxygen is therefore produced in the (excited) singlet state.

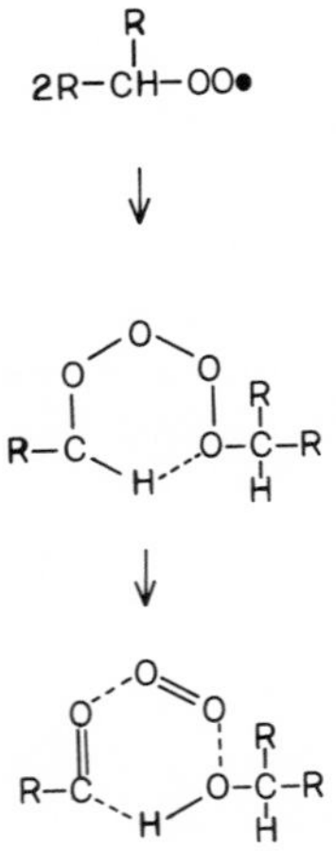

$$2R{-}\underset{\underset{\textstyle H}{|}}{\overset{\overset{\textstyle R}{|}}{C}}{-}OO\bullet$$

↓

↓

Figure 2-3. Thermal decomposition of secondary hydroperoxy radicals.

Howard and Ingold (1968c) found a small production of singlet oxygen from secondary butyl peroxy radical. The singlet oxygen was detected by the reaction with 9,10-diphenylanthracene to form a transannular peroxide.* On the other hand, *t*-butyl peroxy radical gave no transannular peroxide with 9,10-diphenylanthracene. The *t*-butyl peroxy radical probably decomposes directly into triplet oxygen and *t*-butoxy radical:

$$2t\text{-BuOO·} \rightarrow O_2\ (^3\Sigma_g) + 2t\text{-BuO·}$$

2.4 Examples of Dioxygen Chain Reactions

The oxidation of hydrocarbon fuels at flame temperatures proceeds by the same free-radical mechanism discussed previously for the formation of hydroperoxides (Walsh, 1946). Again, the tertiary hydrogens are removed more easily than the secondary and the secondary more easily than the primary hydrogens. However, in a flame, the hydroperoxides decompose, often with explosive force. Tertiary hydroperoxides cleave as shown in Figure 2-4 to form alkoxy radicals and hydroxyl radicals, and the alkoxy radicals cleave at the weakest bond to the α-carbon to form a ketone and another alkyl radical. Secondary peroxides lose water to form ketones as shown in Figure 2-5. Secondary peroxides (Figure 2-

---

* The subject of detection of singlet dioxygen is discussed in Chapter 3.

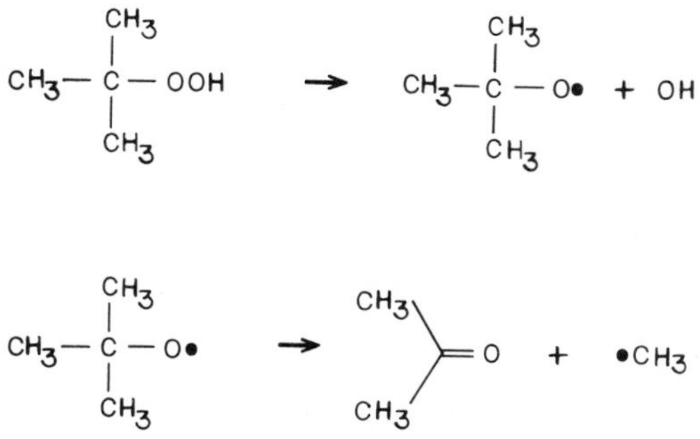

Figure 2-4. Thermal decomposition of a tertiary hydroperoxide at flame temperatures.

6) can also decompose explosively by the same mechanism described for tertiary peroxides. Primary peroxides will undergo the same reactions as the secondary peroxides. The products are now aldehydes instead of ketones. The second reaction can be explosive.

$$CH_3\text{-}CH_2OOH \rightarrow CH_3CHO + H_2O$$

$$CH_3\text{-}CH_2OOH \rightarrow CH_3CH_2O\cdot + \cdot OH$$

$$CH_3\text{-}CH_2O\cdot \rightarrow \cdot CH_3 + CH_2O$$

The oxidation of isooctane (2,2,4-trimethylpentane), the ideal constituent of gasoline, can be written (Figure 2-7) with a knowledge of the reactions discussed above. At higher flame temperatures, the aldehydes are excited sufficiently to decarbonylate (Walsh, 1946):

$$RCHO \rightarrow RH + CO$$

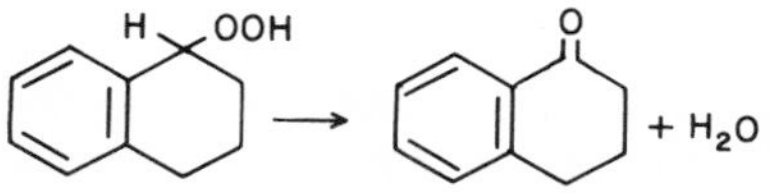

Figure 2-5. Thermal decomposition of a secondary hydroperoxide to a ketone at flame temperatures.

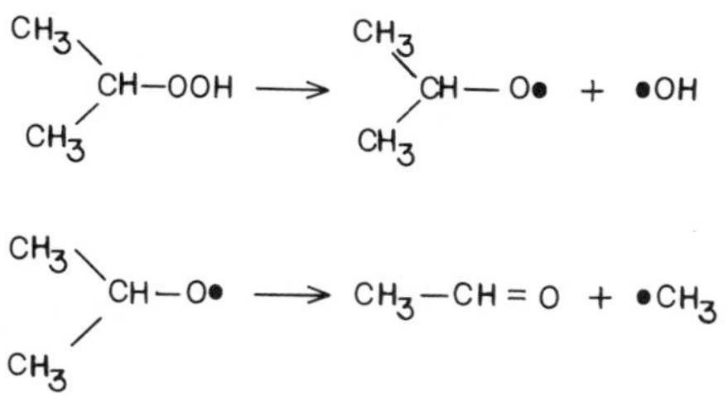

Figure 2-6. Thermal decomposition of a secondary hydroperoxide to an aldehyde at flame temperatures.

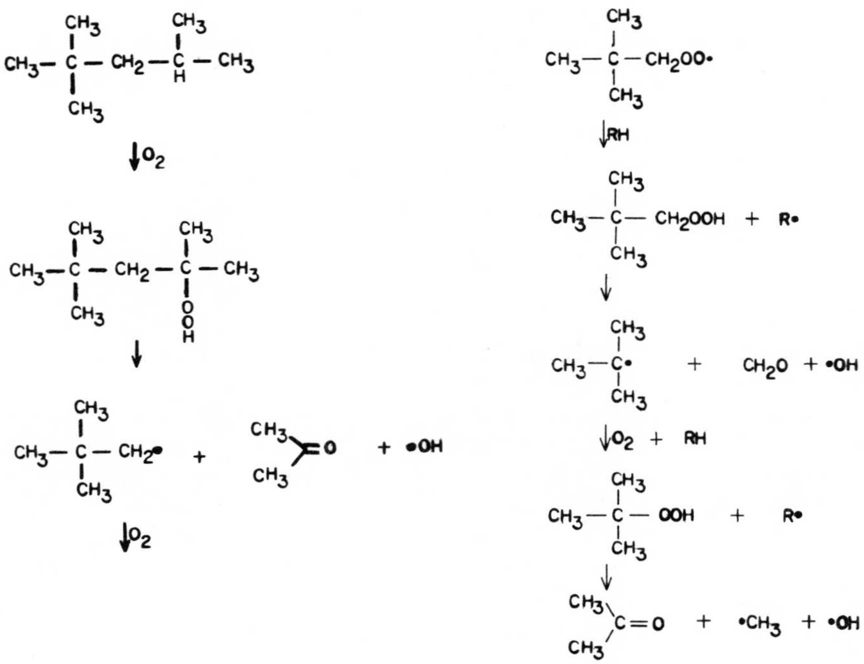

Figure 2-7. Reaction of 2,2,4-trimethylpentane (gasoline) with dioxygen.

For example, the burning of methane produces formaldehyde as an early product that decomposes to hydrogen and carbon monoxide:

$$\cdot CH_3 + O_2 \rightarrow CH_3O_2\cdot$$

$$CH_3O_2\cdot \rightarrow H_2CO + \cdot OH$$

$$H_2CO \rightarrow H_2 + CO$$

$$\cdot OH + CH_4 \rightarrow \cdot CH_3 + H_2O$$

The oxidation of lipids by ground-state dioxygen is a thoroughly studied area of research [Porter *et al.* (1981) and references listed therein]. The mechanism proceeds as described earlier, by a lipid radical reacting with dioxygen to form a peroxy radical that in turn reacts with another lipid to form a new lipid radical and a lipid hydroperoxide. If there is a double bond β to the peroxy radical, it can close to form a cyclic peroxide (Porter *et al.*, 1978) or a series of cyclic peroxides in a polyolefin (Porter *et al.*, 1980b). The use of liquid-chromatography techniques allowed Porter *et al.* (1980a) to study the stereo-chemistry of *cis,cis* diolefinic lipid autoxidation (Figure 2-8). The reaction of linoleic acid with molecular oxygen results in four hydroperoxides, two with *trans,trans,* and two with *trans,cis* geometries. The hydroperoxides do not equi-librate with one another, since the product ratios do not change with the extent of oxidation (total oxidation ≤2%) over a period of 8 hr.

The product distribution is dependent on temperature and fatty acid con-centration. The ratio of *trans,cis/trans,trans* products varies linearly with fatty acid concentration from about 0.1 to 0.4 at 10°C and from 0.1 to about 0.5 at 50°C. Product distribution is independent of dioxygen concentration, as is the rate of autoxidation. If the mechanism required interconversion of carbon rad-icals, then one would expect to find a higher percentage of the product derived from the initial radical at high dioxygen concentration than at low dioxygen concentration. Hence, the stereochemical distribution does not depend on iso-merization of the pentadienyl radical.

The observations require a mechanism in which the peroxy radicals stereo-isomerize (rotation A and β-scission") to achieve the *cis–trans* isomerized rad-icals. In this case, the product ratio is independent of dioxygen concentration because the isomerization process is not stopped by reaction with dioxygen. Hence, pentadienyl radical equilibration (Figure 2-9) may occur, but does not necessarily occur, according to these data.

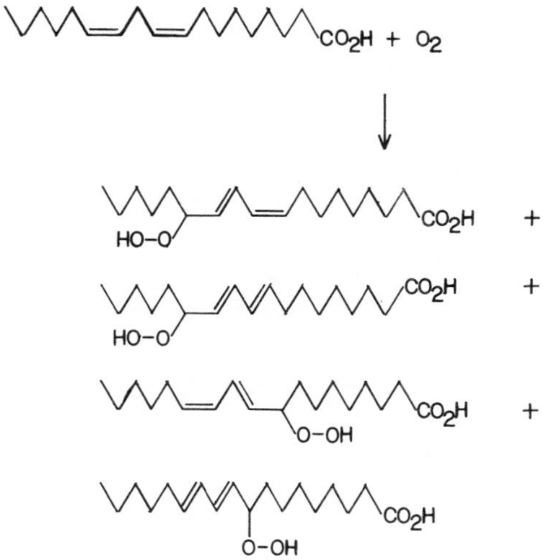

Figure 2-8. Reaction of linoleic acid with ground-state dioxygen.

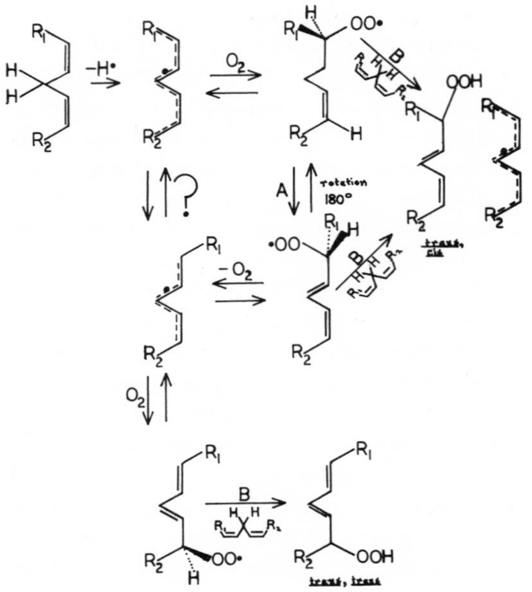

Figure 2-9. A partial mechanism for autoxidation of a *cis,cis* fatty acid, showing the reactions necessary to obtain *trans,cis* and *trans,trans* products. Carbon-radical isomerization, if it does occur, does not affect the product ratio.

However, since the hydroperoxides do not equilibrate, the product-forming reaction, B, does affect the product distribution. And, since the rate of B depends on fatty acid concentration, the product distribution also depends on the fatty acid concentration. Thus, the rate of formation of *trans,cis* product depends on the concentration of fatty acid and is independent of the rate of rotation A (since either conformer can form *trans,cis* product), while the rate of formation of *trans,trans* product depends on the rate of rotation, the rate of β-scission, and the fatty acid concentration. Rotation A and β-scission, then, are responsible for the temperature dependence of the product distribution, while product-forming reaction B explains the dependence on reactant concentration.

Kinetic analysis shows the *trans,cis*/*trans,trans* ratio to be linearly dependent on fatty acid concentration. It also shows the activation enthalpy for β-scission to be 10.5 kcal/mole higher than the activation enthalpy for reaction B.

Autoxidation inhibitors decrease the rate of reaction B by increasing the concentration of H· donors. The *trans,cis*/*trans,trans* ratio is therefore also related to *p*-methoxyphenol concentration. Consequently, the effectiveness of a prospective antioxidant could be measured by the product ratio it produces.

Autoxidation of phospholipid emulsions and lecithin bilayers is very similar to solution oxidation.

Tovrog *et al.* (1980) have shown that ground-state dioxygen may be incorporated into organic molecules via the combination of a cobalt–nitro complex and a palladium–olefin complex. The metals function catalytically. The product is acetaldehyde (from ethylene) and acetone (from propylene) in quantitative yield. No epoxide intermediate is involved.

These reactions occur by oxygen transfer from the nitro ligand, resulting in reduction to a nitrosyl ligand. This species is reoxidized to a nitro group by dioxygen. For gaseous substrates (e.g., ethylene), the reaction works best in an oxygen-poor atmosphere. Under these conditions, oxidation of the nitrosyl ligand is the rate-limiting reaction.

The nitro group in this reaction acts as a very weak oxygen-centered nucleophil. Hence, the presence of other nucleophils (including acetic acid) abolishes the oxygenation reaction by reacting more rapidly with the palladium-bound olefin.

References

Alyea, H. M., and Backstrom, H. J. J., 1929. The inhibitive effect of alcohols on the oxidation of sodium sulfite, *J. Am. Chem. Soc.* 51:90–109.

Barber, M., Faven, J., and Linnet, J. W., 1963. The mass spectrophotometric study of the reaction of methyl radicals with oxygen, *Proc. R. Soc. London Ser. A* 274:306–318.

Clyne, M. A. A., and Thrush, B. A., 1963. Rates of elementary processes in the chain reaction between hydrogen and oxygen, *Proc. R. Soc. London Ser. A* 275:559–574.

Coulson, C. A., 1947. Representation of simple molecules by molecular orbitals, *Q. Rev. Chem. Soc.* 1:144–178.

Fletcher, A. N. and Heller, C. A., 1965. Chemiluminescence quenching terms, *Photochem and Photobiol.* 4:1051–1058.

George, P., 1964. The fitness of oxygen, in *Oxidases and Related Redox Systems,* T. E. King, H. S. Mason, and M. Morrison (eds.), John Wiley, New York, pp. 1–36.

Hiatt, R., Mill, T., and Mayo, F. R., 1968a. Homolytic decompositions of hydroperoxides. I. Summary and implications for autoxidation, *J. Org. Chem.* 33:1416–1420.

Hiatt, R., Mill, T., Irvin, K. C., and Castleman, J. H., 1968b. Homolytic decompositions of hydroperoxides. III. Radical-induced decompositions of primary and secondary hydroperoxides, *J. Org. Chem.* 33:1428–1430.

Howard, J. A., and Ingold, K. U., 1968a. Absolute rate constants for hydrocarbon oxidation. XI. The reactions of tertiary peroxy radicals, *Can. J. Chem.* 46:2655–2660.

Howard, J. A., and Ingold, K. U., 1968b. Absolute rate constants for hydrocarbon oxidation. XII. The reactions of secondary peroxy radicals, *Can. J. Chem.* 46:2661–2666.

Howard, J. A., and Ingold, K. U., 1968c. The self-reaction of sec-butyl-peroxy-radicals: Confirmation of the Russell mechanism, *J. Am. Chem. Soc.* 90:1056–1058.

Ingold, K. V., 1969. Peroxy radicals, *Acc. Chem. Res.* 2:1–9.

Kemal, C., Chan, T. W., and Bruice, T. C., 1977. Reaction of $^3O_2$ with dihydroflavins. 1. $N^{3,5}$-Dimethyl-1,5-dihydrolumiflavin and 1,5-dihydroisoalloxazines, *J. Am. Chem. Soc.* 99:7272–7286.

Latimer, W. M. (ed.), 1938. *The Oxidation States of the Elements and Their Potentials in Aqueous Solutions,* Prentice Hall, New York.

Mayo, F. R., 1968. Free radical autoxidation of hydrocarbons, *Acc. Chem. Res.* 1:193–201.

Methoff, R. C., and Branch, G. E. K., 1930. The kinetics of the reaction of hexaphenylethylene with oxygen, *J. Am. Chem. Soc.* 52:255–268.

Paris, D. P., 1965. Chemiluminescence of tetrakis-(dimethylamino)-ethylene, *Photochem and Photobiol.* 4:1059–1065.

Porter, N. A., Dixon, J., and Ramdas, I., 1978. Cyclic peroxides and the thiobarbiturate assay, *Biochim. Biophys. Acta* 441:506–512.

Porter, N. A., Weber, B. A., Weenen, H., and Khan, J. A., 1980a. Autoxidation of polyunsaturated lipids: Factors controlling the stereochemistry of product hydroperoxides, *J. Am. Chem. Soc.* 102:5597–5601.

Porter, N. A., Roe, N. A., and McPhail, A. T., 1980b. Serial cyclization of peroxy free radicals: Models for polyolefin oxidation, *J. Am. Chem. Soc.* 102:7574–7576.

Porter, N. A., Lehman, L. S., Weber, B. A., and Smith, K. J., 1981. Unified mechanism for polyunsaturated fatty acid autoxidation: Composition of peroxy radical hydrogen atom abstraction, β-scission and cyclization, *J. Am. Chem. Soc.* 103:6447–6455.

Russell, G. A., 1957. Deuterium-isotope effects in the autoxidation of aralkyl hydrocarbons: Mechanism of the interaction of peroxy radicals, *J. Am. Chem. Soc.* 79:3871–3877.

Russell, G. H., Moye, A. J., and Nagpal, K. L., 1962. Effect of structure on the rate of reaction of carbanions with molecular oxygen, *J. Am. Chem. Soc.* 84:4154–4155.

Sawyer, D. T., and Gibian, M. J., 1979. The chemistry of superoxide ion, *Tetrahedron* 35:1471–1481.

Sawyer, D. T., and Seo, E. T., 1977. One electron mechanism for the electrondonical reduction of molecular oxygen, *Inorg. Chem.* 16:499–501.

Taube, H., 1965. Mechanisms of oxidation with oxygen, *J. Gen. Physiol.* 49:Part 2, 29–50.

Tovrog, B. S., Mares, F., and Diamond, S. E. E., 1980. Cobalt–nitro complexes as oxygen transfer agents: Oxidation of olefins, *J. Am. Chem. Soc.* 102:6616–6618.

Walsh, A. D., 1946. Processes in the oxidation of hydrocarbon fuels II, *Trans. Faraday Soc.* 43:297–304.

Wilshire, J., and Sawyer, D. T., 1978. Redox chemistry of dioxygen species, *Acc. Chem. Res.* 12:105–110.

Wood, P. M., 1974. The redox potential of the system oxygen–superoxide, *FEBS Lett.* 44:22–24.

Singlet Dioxygen

3.1 Physical Properties and Generation (Table 3-1)

The two lowest excited states of dioxygen are singlet states. The lower-energy state of the two has both electrons in one of the antibonding π^* orbitals. The other state has the electrons paired, but in separate π^* orbitals. The lower state is 23.4 kcal and the higher state 37.5 kcal above the triplet ground state (Kasha and Khan, 1970). An orbital momentum of 0 gives a Σ state; an orbital momentum of 1, a π state; and an orbital momentum of 2, a Δ state. The lower singlet state has an orbital momentum of 2 and is therefore a Δ state; the other has an orbital moment of 0 and is therefore a Σ state.

Excitation to the $^1\Sigma$ state of dioxygen requires 37.5 kcal, and the lifetime of the state is only 10^{-11} sec (Kearns, 1971; Khan, 1976). Hence, $^1\Sigma$ dioxygen contributes little to the chemistry of dioxygen except by virtue of its ability to decay to $^1\Delta$ dioxygen.

On the other hand, excitation to the $^1\Delta$ state requires only 23.4 kcal (relative to the naturally abundant $^3\Sigma_g$ state), and the state has a half-life of 2 μsec in water (Kearns, 1971; Khan, 1976). This is ample time for many reactions.

The relatively low excitation energy and the sufficiently long half-life allow $^1\Delta$ dioxygen to contribute to the overall chemistry of dioxygen. Since no transition dipole is associated with the excitation conversion, the $^1\Delta$ state of dioxygen cannot be obtained directly by light without the intermediary function of a dye molecule. This is another important factor in the low reactivity of molecular oxygen that is extremely beneficial to life.

The lifetime of singlet dioxygen is highly solvent-dependent (Ogilby and Foote, 1983). Surprisingly, the lifetime of $^1O_2$ increases from 2 μsec in normal water to 20 μsec in deuterated water (Merkel *et al.*, 1972). Ogilby and Foote (1981) have obtained rate constants for the decay of singlet oxygen in four

Table 3-1. Properties of Singlet Dioxygen

1_{Σ_g} state 37.5 kcal above ground state; 1_{Δ_g} state 23.4 kcal above ground state	
O–O bond distance	1.216 Å[a]
Bond energy	94.7 kcal[a]
O–O frequency	1483.5 cm^{-1a}

[a] Jones *et al.* (1979).

common solvents (acetone, acetonitrile, benzene, chloroform) and their deuterated analogues. In each case, the lifetime is much longer in the deuterated than in the protonated solvent. Kinetic analysis showed clearly that the observed effect is not due to differential quenching of the sensitizer triplet excited state. However, no explanation was offered for the unexpected solvent deuterium isotope effect.

Singlet Δ dioxygen has slightly different physical properties than ground-state dioxygen (Jones *et al.*, 1979). The O–O distance is 1.216 Å, the bond energy 94.7 kcal, and the O–O frequency 1483.5 cm^{-1} compared with 1.207 Å, 117.2 kcal, and 1554.7 cm^{-1} for ground-state dioxygen. All the parameters for singlet dioxygen are characteristic of a weaker O–O bond.

Chemically, excitation to the $^1\Delta$ state removes both barriers that slow the reaction between ground-state dioxygen and most organic molecules, i.e., the problem of matching electron spin and the problem of adding an electron to an occupied orbital. As one would expect, the resulting molecule is strongly electrophilic and short-lived.

While singlet dioxygen is not believed to be involved in most biological reactions* of dioxygen, some iron–oxygen complexes undergo reactions very similar to those of singlet dioxygen. An understanding of singlet dioxygen reactions may aid in our understanding of biological reactants.

Singlet oxygen can be generated either by radiation in the presence of a dye or by chemical reaction. Singlet dioxygen can also be formed by electrical discharges (Foner and Hudson, 1956; Herron and Schiff, 1958; Elias *et al.*, 1959), but this method is not useful to solution chemists.

The oxidation of a compound by dioxygen is often catalyzed by a dye in

---

* Singlet dioxygen has been implicated in phytocytosis by myeloperoxidase (Krinsky, 1977; Rosen and Klebanoff, 1977) and as a product or by-product in adrenoxin reductase, xanthine oxidase, lipoxygenase, quercitin dioxygenase, prostaglandin, and oxygenated cytochrome P-450 [Krinsky (1977) and references therein] (see also Chapter 18). However, most implications depend on trapping experiments, which may be suspect and several of which have been shown to be erroneous (Held and Hurst, 1978; Smith and Stroud, 1978).

the presence of light. This process is called photooxidation. The mechanism for photooxidation was first proposed by Kautsky in 1931 (Kautsky and deBruijn, 1931). The dye molecule is excited and singlet dioxygen is produced through energy transfer between the excited dye molecule and ground-state triplet dioxygen. The actual oxidation of the substrate is performed by the singlet oxygen. This type of mechanism occurs with many dyes, one of the most common being methylene blue (Kautsky and deBruijn, 1931; Kautsky, 1937). Flavins may also act in this manner. Simple generators have been designed to produce singlet dioxygen photochemically (Midden and Wang, 1983).

There are several methods to form singlet dioxygen chemically, the most common of which utilize hydrogen peroxide. Singlet dioxygen is formed when hydrogen peroxide is oxidized by a two-electron process (McKeown and Waters, 1966). Hydrogen peroxide is in the singlet state, so if no electrons are unpaired in the oxidation process, the product must be singlet dioxygen. The usual oxidizing agent is hypochlorite. The reaction of hydrogen peroxide with hypochlorite probably proceeds through the intermediate HOOCl, which decomposes with conservation of spin to form the singlet dioxygen. The reaction is exothermic enough to leave dioxygen in the $^1\Delta$ state (Foote and Wexler, 1964a).

The emission of red light resulting from decay of singlet dioxygen to ground-state dioxygen has been used as an indication of $^1\Delta$ dioxygen production (Krinsky, 1977). It has been known for some time that the reaction of Cl_2 with H_2O_2 in basic solution will produce red light (Gattow and Schneider, 1954; Khan and Kasha, 1963; Brown and Ogryzlo, 1964). Certain other reactions involving one-electron oxidation of superoxide ion by ferric ion and by ceric ion will also emit red light (Stauff and Lohmann, 1964):

$$Fe^{3+} + HO_2 \rightarrow Fe^{2+} + O_2 + H^+$$

$$Ce^{4+} + HO_2 \rightarrow Ce^{3+} + O_2 + H^+$$

Singlet dioxygen may also be formed by the decomposition of perchromate ion in aqueous solution (Peters et al., 1972).

Singlet dioxygen may be produced by treatment of certain carbonic acid derivatives with hydrogen peroxide (Rebek et al., 1979) forming percarbamate derivatives (Figure 3-1) that subsequently decompose to produce singlet dioxygen. The parent percarbonate is unavailable by direct reaction of hydrogen peroxide and carbon dioxide. If an olefin is present, these reagents are efficient epoxidizing agents; otherwise, they produce singlet dioxygen that can be detected by diphenylanthracene in very high yield.

The actual singlet-dioxygen-producing species in these experiments could not be determined, since it was the mixture of dehydrating agent (benzoylisocyanate or carbonylditriazole) and nonaqueous hydrogen peroxide that caused

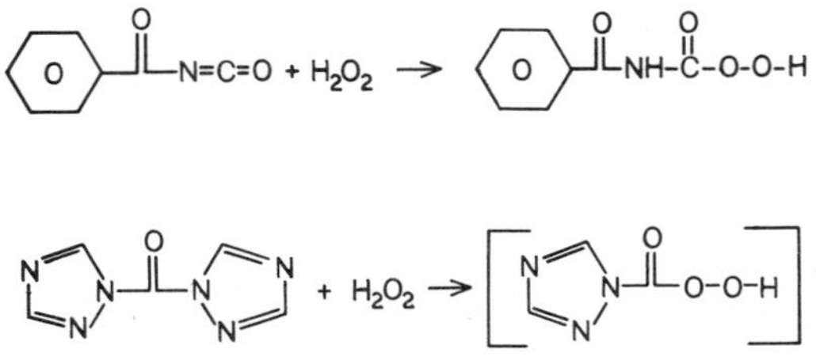

Figure 3-1. Preparation of percarbamates. Compounds A and B are both good epoxidizing agents if olefin is present and good singlet-dioxygen generators if it is not. Compound A can be prepared in crystalline form, but B can be prepared only *in situ*.

singlet-dioxygen evolution. Also, compound A in Figure 3-1 produced singlet dioxygen only on mixture with nonaqueous hydrogen peroxide. Other dehydrating agents such as carbodiimides had similar effects on treatment with H_2O_2, but both epoxidation and singlet-dioxygen formation were smaller percentages of the total reaction. This was ascribed partly to side reactions and partly to unique hydrogen-bonding geometries of reagents A and B.

Singlet dioxygen can be stored in polymers of methyl-substituted vinyl-naphthalene (Saito *et al.*, 1981). The polymer absorbs the singlet dioxygen in dichloromethane solution at 0°C to form a 1,4-naphthalene endoperoxide (Figure 3-2). The yield of endoperoxide is about 90% per naphthalene group. When the polymer is warmed to 30°C, the singlet dioxygen is released.

Singlet dioxygen is also formed as an unwanted product in biological systems. For example, the oxidation of NADPH by liver microsomes produces singlet dioxygen, which is believed to result from the breakdown of lipid peroxides (King *et al.*, 1975).

Many substances are quenchers for singlet dioxygen. The most-studied is

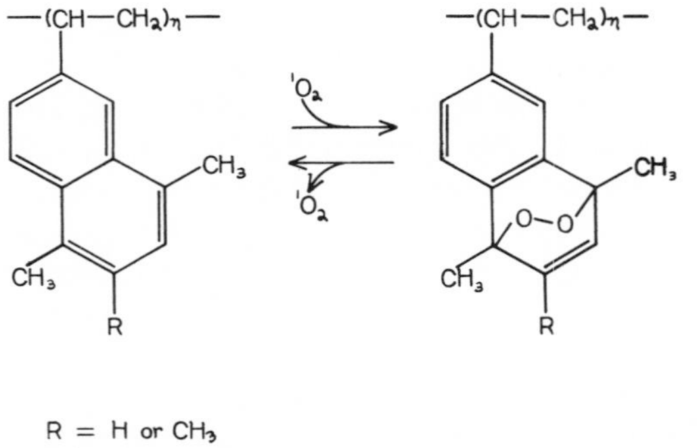

R = H or CH₃

Figure 3-2. A reversible carrier for singlet dioxygen.

β-carotene. The mechanism of quenching by β-carotene appears to be a transfer of energy to form a triplet state of β-carotene. There is little loss of β-carotene by oxidation. One β-carotene molecule can quench as many as 250 singlet dioxygen molecules. This may be an important function of β-carotene in chloroplasts (Foote and Denny, 1968).

Taube has suggested that $^1\Delta$ O_2 could combine with a base. Note in the $^1\Delta$ structure that one of the π^* orbitals is empty and conceivably able to accept two electrons from a base (Taube, 1965). Interestingly, amines are known to catalyze the conversion of $^1\Delta$ dioxygen to $^3\Sigma_g$ dioxygen. This could proceed through an intermediate in which the electron pair of the amine is coordinated with the dioxygen.

Singlet dioxygen will oxidize nitrogen to nitrate ion (Anbar, 1966). Fortunately, our atmosphere does not contain singlet dioxygen. Among various miscellaneous reactions of singlet dioxygen is the formation of ozonides from a diazo compound and an aldehyde (Higley and Murray, 1974). Singlet dioxygen reacts quite readily with sulfides, forming sulfoxides and other products (Monroe, 1979; Kacher and Foote, 1979).

3.2 Detection of Singlet Dioxygen

That singlet dioxygen has been suggested at one time or another as an intermediate in many mechanisms of enzymatic dioxygen reaction (Chan, 1971; Finazzi-Agro *et al.*, 1973; King *et al.*, 1975; Krinsky, 1979), but at present is not a strong contender in most of these mechanisms, is testimony to the uncertainty of common methods for its detection. Detection is most difficult in aqueous solution at low concentration, and these are exactly the conditions that obtain in enzyme-mechanism studies. The uncertainties derive not from reaction of the trapping agents with singlet dioxygen, but from possible reactions of the trapping agents with reactive species other than singlet dioxygen (Smith and Stroud, 1978).

The reactions commonly used for this purpose are compounds that can react with singlet dioxygen via either a Diels–Alder-type reaction (dienes) or an allylic hydroperoxide reaction (olefins). The most reactive dienes are furans (Wilson, 1966) such as 2,5-dimethylfuran or 1,3-diphenylisobenzofuran (DPBF). Both these reagents have some disadvantages. 2,5-Dimethylfuran is soluble in water, but must be monitored at a relatively short wavelength of 215 nm. DPBF has the disadvantage of a low solubility in water. The reaction of 2,5-diphenylfuran with singlet dioxygen is also used. The reaction can be followed by the loss of absorbance at 324 nm. However, this method is somewhat hazardous, since other oxidative molecules commonly present can also cause the bleaching effect (Krinsky, 1977; Smith and Stroud, 1978). Polynuclear aromatic hydrocarbons such as 9,10-dimethylanthracene or other cyclic dienes such as tetracyclone have also been used.

As Smith and Stroud (1978) and Borg *et al.* (1978) and others have persuasively argued, these reagents are also not as specific as one would wish, and many erroneous conclusions have been drawn from their indications. These compounds are generally fairly labile to any of several types of oxidation, so that even if the reaction with singlet dioxygen is the fastest among them, one cannot tell qualitatively whether or not singlet dioxygen is involved. Foote (1979) lists peroxy radicals, hydrogen peroxide, and halogens as reagents that will convert conjugated furans to diketones at an appreciable rate. In some cases, the products vary with the type of oxidation, so that product analysis decreases the ambiguity. Particularly inconclusive, however, are studies in which only decrease in optical absorption at a particular wavelength is taken as evidence of reaction with singlet dioxygen.

The photochemistry of DPBF, one of the most common trapping reagents, was thoroughly investigated rather long after its use as a singlet-dioxygen-detecting agent became popular (Singh *et al.*, 1978). It was shown that several reactions do occur besides the formation of the endoperoxide from singlet dioxygen and ground-state DPBF. For instance, DPBF probably can sensitize the formation of singlet dioxygen. In addition endoperoxide formation reaction is reversible, so that quantitative or kinetic observations are complex, at best. Also, visible-light photolysis of the endoperoxide to give radicals probably occurs, as does photobleaching of DPBF, by mechanisms other than reaction with singlet dioxygen. It is interesting to note that the same product is obtained with the sulfur and nitrogen analogues of DPBF on irradiation of an aerated solution (Figure 3-3). However, it has been shown by a number of techniques (Olmstead, 1980) that DPBF does not physically quench singlet dioxygen. Many similar

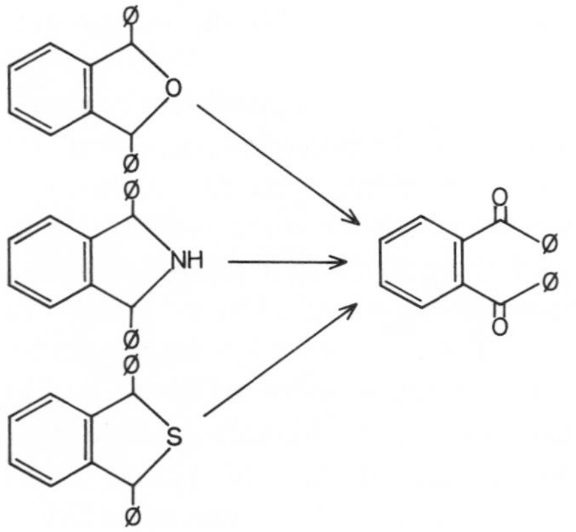

Figure 3-3. Product formed from the reaction of DPBF and certain analogues with singlet dioxygen.

complications probably occur with other related trapping agents. Some of the polynuclear aromatic hydrocarbons, for instance, are known as moderate singlet-dioxygen sensitizers.

In systems for which it is known that reactive radicals such as superoxide or hydroxyl radical are not present, oxidation of reactive dienes becomes a somewhat more definitive test. However, this information is not generally available for enzymatic systems.

It has been recommended that trapping agents be used that give a variety of characteristic products, i.e., a "fingerprint," duplication of which by a different mechanism is unlikely (Foote, 1978). This has been accomplished by the use of olefins that react to form characteristic allylic hydroperoxides. These are often different products, in the case of many complex olefins, from the allylic hydroperoxides formed via autoxidation. Generally, either mechanism will produce one or two allylic hydroperoxides unique to that mechanism and other allylic peroxides common to both; the products common to both, however, appear in different ratios for one mechanism than for the other. The allylic hydroperoxide reaction is generally slower than the Diels–Alder-type reaction, but the multiplicity of products is advantageous in that the degree of certainty in the conclusion can be assessed by how closely the observed product distribution resembles that for the control case.

Optically active olefins are attractive for this purpose, since racemization generally occurs during free-radical reactions but not during singlet-dioxygen reactions. (+)Limonene has been suggested for some specific applications, and its oxidation products have been analyzed (Rawls and Estes, 1978). Studies with other diolefins indicate that they could also be useful if solubility problems were overcome (Tanielian and Chaineaux, 1978).

Teng and Smith (1973) proposed cholesterol for this purpose, since it is already present in many biological systems. Singlet dioxygen reacts to form 5-hydroxycholesterol, but free-radical oxidation gives the 7-hydroxycholesterol, after reduction, and other oxidation procedures give other distinguishable products. Furthermore, cholesterol does not react with superoxide (Smith et al., 1977), so that oxidized product indicates participation of some species other than an Fe^{3+}–superoxide complex. Hence, cholesterol is one of the most specific trapping agents to have been thoroughly tested.

The 5-hydroperoxycholesterol has been observed in photolytically lysed erythrocytes. Formation of the hydroperoxy derivative apparently has a strong destabilizing influence on the erythrocyte membrane (Foote, 1976). However, cholesterol is very large relative to most enzyme active sites, so that lack of an oxidation product when testing an enzymatic system for singlet-dioxygen participation is an indeterminate result. Also, dissolution of cholesterol in an aqueous system requires the use of detergents, which may complicate the reaction. Foote et al. (1980) have avoided this problem by using cholesterol either attached to polystyrene "beadlets" or in mineral oil droplets. The reaction of cholesterol is also very inefficient.

A singlet-dioxygen trap that may prove useful for biochemists is the sodium salt of anthracene-9,10-bisethanesulfonic acid (Botsivali and Evans, 1979). This reagent is also too large to fit into the active center of most enzymes, so it will measure the singlet dioxygen in solution. It is quite soluble in water, whereas many of the other trapping agents are not. It may be monitored by an absorption at 399 nm. The product is cleanly the 9,10-endoperoxide, which has no absorption above 320 nm. Anthracene-9,10-bisethanesulfonic acid may be prepared from 9,10-dibromoanthracene.

For systems in which it is known that singlet dioxygen will be produced, much less selective conditions have been used to advantage. For instance, *p*-nitroso-*N,N*-dimethylaniline has been shown to be bleached by the endoperoxide formed by reaction of singlet dioxygen with imidazole derivatives (Kraljic and Mohsni, 1978; Ryang and Foote, 1979). This reagent has been proposed and tested for use in the study of "photodynamic action," in which it is fairly well established that cellular damage is caused by singlet dioxygen formed by electronic excitation via a sensitizing dye. However, this same reagent has been used as a "specific" trapping agent for both hydroxyl radical and superoxide.

The traditional methods of studying the photodynamic effect have relied primarily on quenching of the phenomenon by azide or other physical quenching agent such as diazabicyclo[2.2.2]octane (DABCO) (Figure 3-4) (Ouannes and Wilson, 1968) and enhancement of the effect by D_2O (Ito, 1978; Kobayashi, 1978). For instance, different types of cellular damage have been correlated with different sensitizing dyes and the location of the dye within the cell, using these techniques. Nonsensitizing dyes produce little damage, in proportion to their sensitizing efficiencies. Dyes that can intercalate with nucleic acids cause genetic damage, dyes that can cross membranes but not intercalate cause cellular organelle damage, dyes that cannot cross membranes, but are efficient sensitizers, cause cell-wall damage, and so on.

It should be emphasized, however, that reaction quenching with azide and

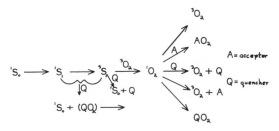

Figure 3-4. Sensitization and quenching of singlet dioxygen.

enhancement with D_2O are only very rudimentary evidence of participation of singlet dioxygen and should not be taken as proof of such (Hasty *et al.*, 1972). Azide does react with free radicals and cannot be expected to be universally innocuous in enzyme preparations. Also, the quenching effects observed with azide, when quantitated, vary over several orders of magnitude (Hasty *et al.*, 1972). D_2O rate effects are, of course, expected any time proton removal appears in any rate expression, so that change of reaction rate due to use of D_2O, even under optimal conditions, can be used only as confirmation of other results.

Superoxide ion is also an efficient quencher for singlet dioxygen. In aprotic solvents, bromide ion, iodide ion, ferrocyanate ion, thiosulfate ion, and adeno-cyanate ion act as quenchers. The mechanism is believed to be through a singlet charge-transfer complex that decays to a triplet state and dissociates (Bellus, 1979):

$$I^- + {}^1O_2 \xrightarrow[\text{H}^+]{\text{No}} I^- + {}^3O_2$$

$$3I^- + 2{}^1O_2 \xrightarrow[\text{MeOH}]{\text{H}_2\text{O or}} I_3^- + 2O_2^-$$

In aqueous solution, the charge transfer can result in electron transfer.

It is interesting to note that the electron-transfer reaction for superoxide ion gives the same products as quenching. This may be why superoxide ion is such an efficient quencher (Guir and Foote, 1976).

$$O_2^- + {}^1O_2 \longrightarrow {}^3O_2 + O_2^-$$

The physical quenching of singlet dioxygen is a complex process that has been reviewed (Bellus, 1979). Especially for cases in which the excited state is generated by sensitization, a large number of reactions are possible that complicate both quenching studies and D_2O rate studies (Figure 3-4). Carotenoids are generally the most efficient quenchers, with rates up to about 50% of the diffusion limit of 3×10^{10} M^{-1} sec^{-1}. Since chlorophyll is among the most efficient sensitizers of singlet dioxygen, it is believed that protection from photooxidation is a primary role of carotenoids in chloroplasts (Foote, 1976). Carotenoids can act by reacting with singlet dioxygen or by quenching the chlorophyll triplet state, which excites dioxygen, in addition to the singlet-dioxygen quenching mode, which is believed to result in triplet carotenoid and triplet dioxygen. Hence, when used to indicate the presence of singlet dioxygen, carotenoid concentration should be low enough that quenching of the sensitizer is not a factor. Also, quantitation of carotenoid quenching is very difficult because of carotene's low aqueous solubility and consequent unpredictable aggregation (Foote, 1979).

The interaction of singlet oxygen with carotenoids to produce ground-state dioxygen is a somewhat mysterious process. Considerable emphasis has been placed on the reaction of singlet oxygen with ground-state carotene to produce triplet dioxygen and triplet carotene (Bellus, 1979):

$$^1O_2 + {}^1\beta\text{-carotene} \rightarrow {}^3O_2 + {}^3\beta\text{-carotene}$$

Energetically, this reaction seems feasible, since the ground-state–triplet-state energy gap for carotene is between 17 and 25 kcal/mole. Also, the reaction offers one explanation for the rapid decrease in quenching ability for polyolefins between nine and seven bonds in conjugation. However, mechanistically it is very difficult (and unusual) to postulate a reaction between two singlet species that results directly in the formation of two triplet species. (Initial formation of singlet excited carotene that decays to triplet excited carotene is apparently discountable on the basis of energetics.) Also, one would expect isomerization and radical reactions to compete with decay of triplet carotene to the ground state, rendering it a much less effective quenching agent. Furthermore, there are other possible quenching mechanisms that are not excluded by the data. Quenching could be the result of spin–orbit coupling, which would yield a thermally excited (but singlet) carotenoid, along with ground-state dioxygen. This mechanism does not immediately explain the increase in efficiency for olefins with more than eight π-bonds in conjugation, but neither is it incompatible with these data. (Spin–orbit-coupled quenching should generally increase with length of the conjugated system due to increasing polarizability.) Or an electron-transfer mechanism could be involved that would require some minimum chain length to have low enough oxidation potential to allow the electron transfer to occur.

We do not believe that the available evidence is sufficient to confirm or eliminate any of these possibilities. However, if the reaction above is invoked, some further explanation of it is required, since it is quantum-mechanically forbidden and mechanistically unprecedented.

Amines are also often used as a detection method via quenching of singlet dioxygen. The rate constants for the best aliphatic amines are around 10^7 M^{-1} sec^{-1}—much slower than carotenoids, but still useful. The rate constant correlates inversely with ionization potential of the amine (this also holds true for inorganic anions), with aromatic amines showing a much wider range of quenching rates. The correlation with ionization potential has been used as evidence that amines and inorganic anions quench by means of a partial charge-transfer complex and that intersystem crossing is the result of spin–orbit coupling between the singlet and triplet states of this intermediate. This is also the basis for quenching by superoxide. Photooxidation of amines occurs via the same charge-transfer intermediate if the nitrogen atom can form a double bond with the adjacent carbon atom (i.e., if the nitrogen atom is not at a bridgehead position and the adjacent carbon atom is bound to a proton). The fact that this complicating reaction cannot occur is one of the reasons DABCO is an effective experimental reagent (other factors include water solubility and transparency at wavelengths greater than 300 nm). It should be mentioned that quite a high concentration of DABCO is required to quench significant levels of singlet oxygen (Foote, 1979).

One mechanism for quenching by tertiary amines has been studied by Saito *et al.* (1981). Advantage was taken of the retro-Diels–Alder reaction of endoperoxide (Figure 3-5) and its water solubility to study the electron-transfer reaction (Figure 3-6) between the electron-rich quencher *N,N*-dimethyl-*p*-anisidine (3-5) and singlet oxygen to produce superoxide. Superoxide was detected in 1.1% yield by its reaction with nitroblue tetrazolium. This reaction could be inhibited by superoxide dismutase and azide. The reaction was carried out at pH 7.5, where superoxide disproportionates rapidly, so the yield is not a reflection of the amount of superoxide formed. However, superoxide formation dropped drastically with increasing oxidation potential in a series of amine quenchers. It was concluded that electron transfer is important only for quenching agents with oxidation potentials less than 0.5 vs. Standard Calomel Electrode in water. However, this certainly represents a complicating factor in the study of a reaction that might involve singlet dioxygen intermediates.

Another method for detection of singlet dioxygen lies in observing the energy it emits in returning to the ground state. Detection of chemiluminescence is extremely sensitive, so that even very low-intensity emission can be observed. By the same token, however, the sensitivity is so great that when detecting reaction intermediates, it is often impossible to tell whether or not the observed luminescence represents the primary pathway or is the result of some contaminating influence or minor pathway (Kasha, 1978).

There are five emission bands for singlet dioxygen at 1268, 762, 634, 476, and 381 nm due to singlet delta, singlet sigma, singlet delta pair, the singlet sigma–singlet delta pair, and the singlet sigma–singlet sigma pair, respectively. Theoretically, detection of each of these at the correct relative intensities would be a good indication of the presence of singlet dioxygen (though not necessarily of its participation in the main reaction pathway). Unfortunately, observation of all five is generally impossible due to the large number of contaminants that

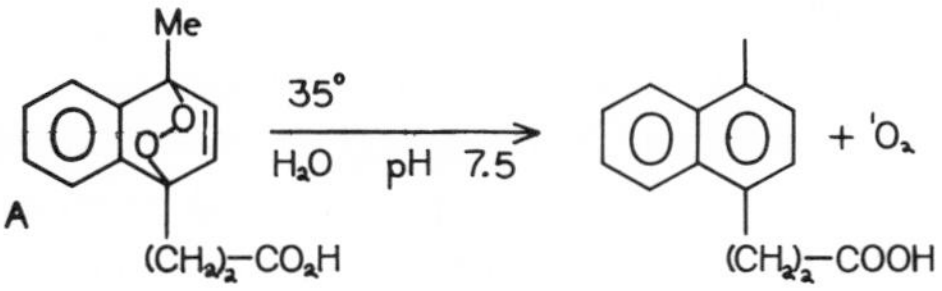

Figure 3-5. Preparation of singlet dioxygen from an endoperoxide. (A) *N,N*-Dimethyl-*p*-anisidine.

$$^1O_2 + MeO\!-\!\!\left\langle\!\!\bigcirc\!\!\right\rangle\!\!-\!N(Me)_2 \;\longrightarrow\; O_2^{\bar{\cdot}} + MeO\!-\!\!\left\langle\!\!\bigcirc\!\!\right\rangle\!\!-\!\overset{+\cdot}{N}(Me)_2$$

Figure 3-6. Use of a reagent for the detection of singlet dioxygen.

have fluorescence in this range. Also, singlet-dioxygen luminescence is so inefficient (Foote, 1979) that determination of an emission-wavelength maximum is difficult.

The photosensitization of singlet dioxygen by dyes also works in reverse, resulting in very intense emission from any sensitizing dye present, and it has been shown (Brabham and Kasha, 1974) that this can be used to detect singlet dioxygen at concentrations below those at which direct observation is possible. Such an experiment assumes, of course, that the state of singlet dioxygen produced during the reaction is the correct one to sensitize the dye being used. The fact that many of these dyes require the energy of two molecules of singlet dioxygen makes it likely that there is a weak interaction between molecules of singlet dioxygen.

The decay of two $^1\Delta$ dioxygen molecules will cause Zn tetraphenylporphine to fluoresce, and the simultaneous decay of one molecule of $^1\Delta$ dioxygen and one of $^1\Sigma$ dioxygen has enough energy to excite eosin and cause fluorescence. Two molecules of $^1\Sigma$ dioxygen give enough energy to excite anthracene (Khan and Kasha, 1966).

Spin trapping of singlet dioxygen has also been suggested as a detection method (Moan and Wold, 1979; Lion *et al.*, 1976), since singlet dioxygen reacts with several cyclic amines to produce nitroxyl radicals. The reaction is pH-dependent, and useful amines are therefore restricted to those with a *pK* near neutrality such as 2,2,6,6-tetramethyl-4-piperidone. The sensitivity of this method is sufficient for biochemically produced singlet dioxygen, but hydrogen peroxide and, presumably, superoxide and hydroxyl radicals would produce identical products. Hence, the utility of this method is probably very limited under normal circumstances. Foote (1979) and co-workers have elucidated a general photochemical oxidation that does not involve singlet dioxygen. This reaction complicates the search for singlet oxygen in biological systems because it produces many of the same products as singlet-oxygen reactions and because it provides another pathway by which photochemical transformations can occur. An electron-deficient dye when excited to a singlet state may have sufficient oxidation potential to oxidize the reaction substrate by one electron. The substrate cation radical may then react with ground-state dioxygen to produce a substrate peroxy cation radical. This radical is then reduced by the sensitizer radical anion to produce dioxygenated product and ground-state sensitizer. Conversely, the initially formed sensitizer anion radical may reduce oxygen to superoxide that would combine with the substrate cation radical to form dioxygenated product. Formation of both endoperoxides and allylic hydroperoxides has been observed by this mechanism. In addition, olefin cleavage (as in dioxetane cleavage) has been observed with certain olefins that do not react with singlet oxygen. Besides pointing out another complication in the study of oxygen biochemistry, this mechanism suggests another possibility for heretofore mysterious dioxygenase reactions (Foote, 1979).

Foote (1979) has also provided a general method for calculating kinetic results of singlet-oxygen experiments. The constraints on this analysis illustrate the difficulty of analysis in a heterogeneous (i.e., biological) system. Taking into account a trap (reaction), a known quenching exogenous rate constant or an endogenous rate constant measured in a control experiment, and decay in the solvent used (Figure 3-7), a rate expression can be obtained that predicts that the D_2O rate effect will be substantial only if decay is the predominant pathway for removal of singlet oxygen.

3.3 Organic Reactions of Singlet Dioxygen

When singlet dioxygen oxidizes molecules, it usually forms products that contain both atoms of oxygen. The reactions can be of value in organic synthesis (Wasserman and Ives, 1981). The most common reactions with olefins tend to be of two types: (1) endoperoxide-forming reactions and (2) allylic-hydroperoxide-forming reactions. The better understood of the two is the first reaction, in which

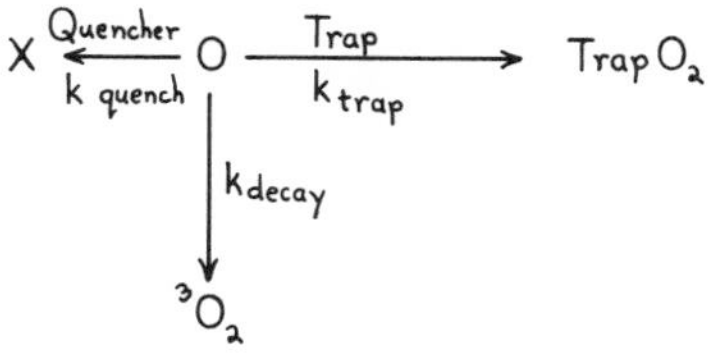

$$X \xleftarrow[k_{quench}]{Quencher} O \xrightarrow[k_{trap}]{Trap} Trap\, O_2$$

$$\downarrow k_{decay}$$

$$^3O_2$$

$$\text{Yield } {}^1O_2 = \text{Yield Trap } O_2 \left(1 + \frac{k_{decay} + k_{quench}[Quencher]}{k_{trap}\,[Trap]} \right)$$

$$\frac{\text{rate in } D_2O}{\text{rate in } H_2O} = \frac{k_{trap}[Trap] + k_{quench}[Quencher] + k_{decay}(H_2O)}{k_{trap}[Trap] + k_{quench}[Quencher] + k_{decay}(D_2O)}$$

Figure 3-7. General kinetics of singlet-dioxygen reactions.

singlet dioxygen cyclizes with a diene in a concerted electrocyclic (4 + 2) reaction (Figure 3-8A) to form an endoperoxide (Foote and Wexler, 1964b). The reaction is facile (Kearns, 1969) and obeys established selection rules (Woodward and Hoffman, 1970; Turro, 1978). The reaction may also occur with aromatic organics (Figure 3-8B) and appears to be entirely analogous to the Diels–Alder reaction (House, 1972). However, some important differences do exist.

The mechanism of the (4 + 2) cycloaddition reaction of singlet dioxygen has been studied with regard to stereochemistry (Pacquette *et al.*, 1980a,b) and compared to the analogous reaction with *N*-methyltriazolinedione (Figure 3-9) as dienophile (Pacquette *et al.*, 1980c). The reactions are related in that both dienophiles are very reactive so that both reactions have very low activation energies [$E_a = 0$ (Gorman *et al.*, 1979)]. More typical Diels–Alder reactions have activation energies in the range from 10 to 18 kcal/mol. The minimal activation energy for the singlet dioxygen [and *N*-methyltriazolinedione (Figure 3-9)] reaction has the effect of minimizing the energy difference of approach to the diene from opposite sides. However, the high reactivity also has the effect of advancing the transition state to a very early stage in the reaction, which allegedly magnifies whatever energetic differences exist.

In the case of norbornyl-fused naphthalenes, σ–π interactions of the second-highest π-orbital are invoked to establish a difference in repulsive electron density between top and bottom sides of the olefin. The influence of the second-highest occupied π-orbital is rationalized by the fact that the singlet-oxygen orbitals with which the repulsive interaction occurs are very low in energy compared to the olefin π-orbitals and by the fact that interaction of the second-highest π-orbital

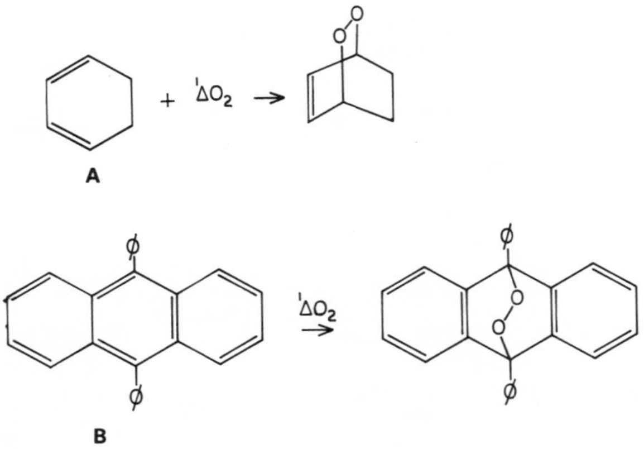

Figure 3-8. Reaction of singlet dioxygen with dienes.

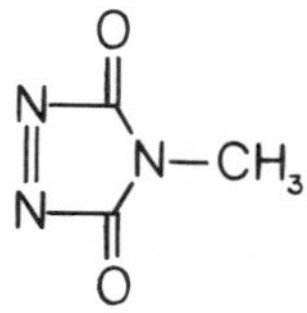

Figure 3-9. *N*-Methyltriazolinedione, a reactant for singlet dioxygen.

with the σ-orbital of the same symmetry gives an electron distribution that correctly explains the observed products.

The reverse of this reaction produces both singlet and triplet dioxygen in certain proportions depending on the exact endoperoxide involved. Interestingly, this singlet/triplet ratio can be influenced, in certain cases, by the presence of a strong magnetic field during the reaction (Turro and Chow, 1979).

The Diels–Alder type of addition of dioxygen to a 1,3-diene is not common in enzymatic systems. One interesting example that appears similar to this re- action is the prostaglandin synthetase reaction (see Chapter 18). It has been shown that the reaction is not, in fact, concerted, but the resemblance is striking. Also, certain types of (but not most) cellular photooxidation damage can be ascribed to this reaction. Guanosine, in particular, is the nucleoside that reacts fastest with singlet dioxygen, probably via a concerted cycloaddition, and this may be the reaction that is responsible for genetic damage resulting from light exposure in the presence of intercalating dyes.

The other of the two characteristic reactions of singlet dioxygen is also mimicked in biological systems. It is the reaction with olefins to form allylic hydroperoxides. In a typical reaction, the double bond migrates to the next carbon and the previously unsaturated carbon forms a hydroperoxide (Figure 3-10)

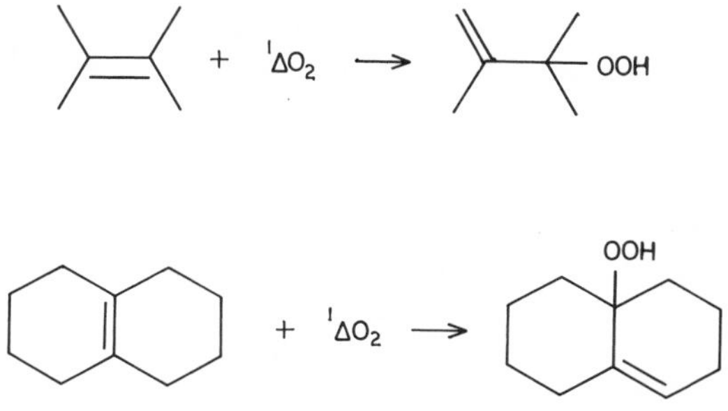

Figure 3-10. Reaction of singlet dioxygen with olefins.

(Foote *et al.*, 1968). This reaction is often accompanied by a side product of carbonyl compounds, the proportion of which depends on how electron-rich the olefin is. Enamines give entirely these carbonyl products.

Frimer (1979) has reviewed and condensed the confusing rearrangement reactions of the products of singlet-dioxygen reactions: dioxetanes, endoperoxides, and allylic hydroperoxides (see Chapter 21 for a detailed examination of these reactions).

3.4   Singlet Dioxygen–Tocopherol Reaction

A biologically interesting example of a related singlet-dioxygen reaction is the oxidation of tocopherol (Figure 3-11) (Clough *et al.*, 1979). Using low-temperature techniques, the authors determined a hydroperoxide to be the primary product of the reaction. This appears to be a vinylogous variation of the standard reaction with enols to form hydroperoxides, in which the α oxygen participates only minimally. In this case, though, the hydrogen source is the phenolic functionality, rather than an allylic hydrogen. This source could, of course, supply either radicals or protons (Figure 3-12). However, the fact that the reaction is slower, relative to quenching (which tocopherol also does) in aprotic solvents, may be very indirect evidence for a charged intermediate. Also, it is interesting to note that a concerted pathway does not appear to be necessary in this case, since the reaction does proceed in aprotic solvents even though the oxygen bridge is clearly too short to reach the hydrogen source.

3.5   Metathesis

Section 3.3 discussed the allowed addition of singlet dioxygen in concerted $(4 + 2)$ electrocyclic reactions. The addition of singlet dioxygen to an olefin in a concerted $(2 + 2)$ electrocyclic reaction is forbidden by symmetry (Woodward and Hoffman, 1970). Because many of the dioxygenase enzymes contain transition-state metals, the matter of $(2 + 2)$ cycloadditions must be reconsidered.

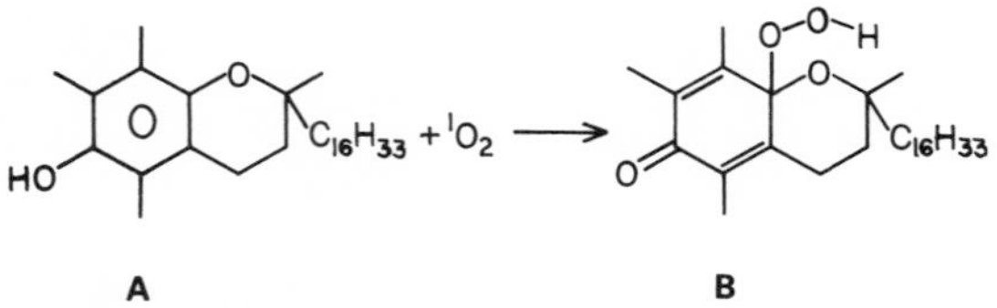

Figure 3-11. Reaction of singlet dioxygen with tocopherol (A) giving a hydroperoxide (B) as the primary product.

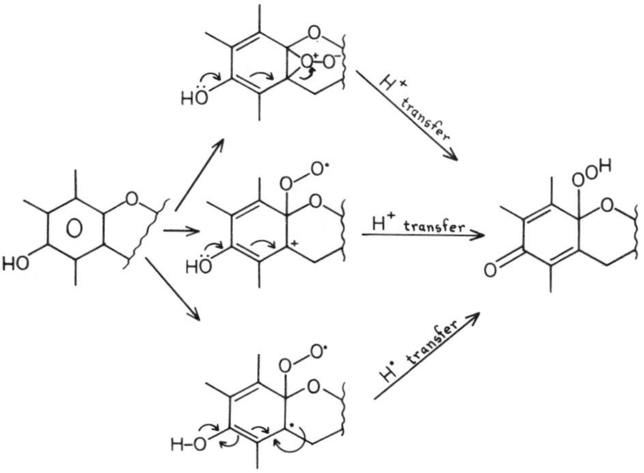

Figure 3-12. Possible mechanisms of the singlet-dioxygen reaction with tocopherol.

Transition metals can drastically modify the symmetry restrictions that prevent (2 + 2) additions of olefins, by allowing electron flow between orthogonal d-orbitals (Figure 3-13) (Mango and Schachtschneider, 1971). Without metal catalysis, the reaction could proceed through the ground state only via a *supra antara* mode (Figure 3-14). Since steric hindrance prevents this (Woodward and Hoffman, 1970), the reaction does not proceed except in cases in which the excited state of the products is more stable than the ground state of the reactants.

In the presence of a transition metal, the restrictions disappear. The olefins may first bond with a d-orbital of a certain symmetry with respect to the reaction pathway. These electrons may then simply switch to the orthoganol d-orbital of

Figure 3-13. Mango theory of how orthogonal d-orbitals can eliminate the symmetry restrictions in (2 + 2) cycloaddition reactions.

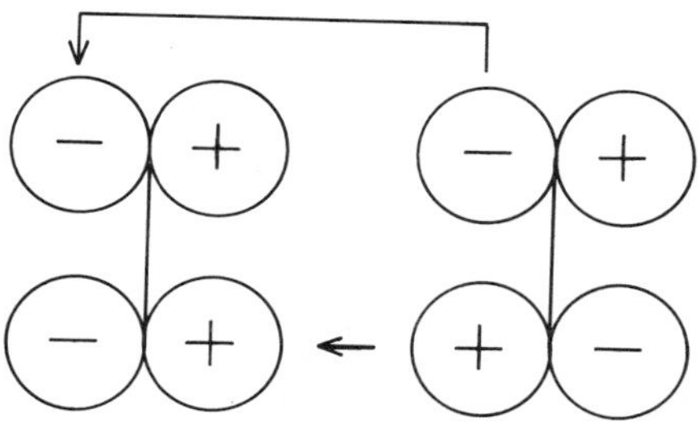

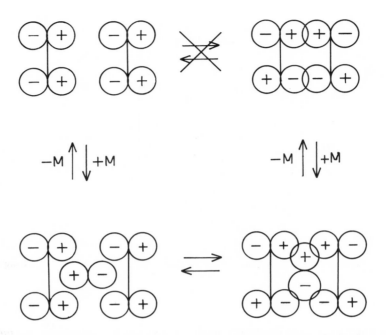

Figure 3-14. Allowed *supra antara* mode of a (2 + 2) cycloaddition reaction in the absence of metals.

opposite symmetry with respect to the reaction pathway, which gives the correct symmetry for ground-state bonding in the product. Theoretically, the cyclobutane could then dissociate from the metal. In practice, dissociation of the cyclic product has not been observed; in fact, cyclobutane does not react at all. Instead, the adduct appears to be an intermediate or activated complex in the equilibrium of olefin interchange ("metathesis," as drawn in Figure 3-14). With the proper metals, this reaction occurs very quickly at room temperature. The factors that control the rate of the reaction and determine which metals may act as catalysts are the ligand geometry and the oxidation state. Clearly, if the required d-orbitals (the d_{zx} and d_{zy}) are not degenerate, then interchanging electrons between them is an energetic process.

3.6   Phototaxis

One biological phenomenon for which singlet dioxygen may be responsible is the phototactic response of plants to blue light. This conclusion was reached (Galston and Baker, 1950) before the intermediacy in "photooxidation" of singlet dioxygen was firmly established. However, it was known that riboflavin is an effective mediator of photooxidation, that indoles are readily degraded by photo-oxidation, and that indoleacetic acid is the primary plant growth hormone (auxin).

It was shown that growth of tissue in simple media containing indoleacetic acid could be inhibited by exposure to white light and that auxin removed from seedlings into agar blocks could be inactivated by riboflavin-mediated photo-oxidation. A phototropic action spectrum was measured for the *Avena* coleoptile and shown to be very similar to the visible spectrum of riboflavin. Unfortunately, a clear conclusion could not be drawn because the spectrum of carotene is very similar to that of riboflavin. However, the phototropic response of the *Avena* base was much closer (Haig, 1935) to the riboflavin spectrum than to the carotene spectrum, suggesting that the bending that occurs at the base of the stem is mediated by riboflavin.

Analogous results have been obtained by Lagarias and Poff (1985) in *Dictyostelium discoideum*. Both ameba and pseudoplasmodia show phototactic behavior. The action spectrum of the ameba movement closely resembles the spectrum of protoporphyrin (protoporphyrin has been isolated from the ameba and is known to be an excellent sensitizing agent). Also, the light-initiated morphogenetic change from ameba to pseudoplasmodium exhibits an action spectrum very similar to the protoporphyrin spectrum.

Amebas grown in intense light have much less protoporphyrin than amebas grown in the dark. This could be due to photooxidation of the pigment. The direction of phototaxis (toward or away from the light) is affected by the level of protoporphyrin in the amebas. Dark-grown amebas (more protoporphyrin) change from moving toward the light to away from the light at a much lower light intensity than do light-grown amebas.

The photodynamic effects of protoporphyrin are well known. The authors suggest that a similar mechanism involving singlet dioxygen may be responsible for ameba phototaxis.

The photodynamic effect itself has been studied in detail and appears to be primarily a singlet-dioxygen phenomenon. A comprehensive list of photo-sensitizing dyes is available (Santamaria and Prino, 1972). Also, the reactions of singlet dioxygen with individual amino acids have been studied to explain photodynamic effects on proteins (Foote, 1976). The amino acids primarily involved are tryptophan, methionine, and histidine. Nucleic acids are also affected.

3.7 Singlet Sigma Dioxygen

All the reactions discussed above are probably $^1\Delta$ O_2 reactions. $^1\Sigma$ O_2 can be formed by laser excitation, and it would be interesting to know what would be formed by the reaction of $^1\Sigma\ O_2$ with an olefin. The reaction of 1,1-dimethoxy-2-methyl propene with laser-generated singlet dioxygen gives products derived from a dioxetane in addition to the hydroperoxide. The singlet dioxygen is a mixture of $^1\Sigma\ O_2$ and $^1\Delta\ O_2$ because $^1\Sigma\ O_2$ decays to $^1\Delta\ O_2$. One would expect a higher activation energy for dioxetane formation so that $^1\Sigma\ O_2$ may be giving this product. However, the addition of ammonia, which quenches $^1\Sigma\ O_2$, has no effect on the product ratio (Hammond, 1979). The conclusion is that $^1\Sigma\ O_2$ and $^1\Delta\ O_2$ have the same reactivities in this reaction and possibly in other reactions also.

Summary

This discussion includes the most useful material on singlet dioxygen for a biochemist. Most reactions of singlet dioxygen add both atoms of oxygen to the product, and the most important reactions are endoperoxide formation from dienes and allylic hydroperoxides from olefins. Those who wish to learn some further details of singlet-dioxygen chemistry are referred to Chapter 21.

References

Anbar, M., 1966. Oxidation of molecular nitrogen by excited singlet oxygen molecules in aqueous solution, *J. Am. Chem. Soc.* 88:5924–5926.

Bellus, D., 1979. Physical quenchers of singlet molecular oxygen, in *Advances in Photochemistry*, Vol. 11, J. M. Pitts, Jr., G. S. Hammond, K. Gollnick, and D. Grosjean (eds.), John Wiley, New York, pp. 105–205.

Borg, D. C., Schach, K. M., Elmore, J. J., Jr., and Bell, J. A., 1978. Cytotoxic reactions of free radical species of oxygen, *Photochem. Photobiol.* 28:887–908.

Botsivali, M., and Evans, D. F., 1979. A new trap for singlet oxygen in aqueous solution, *J. Chem. Soc. Chem. Commun.* 1979:1114–1116.

Brabham, O. E., and Kasha, M., 1974. Novel sensitized chemiluminescence: Infra-red emissions from thiazine dyes sensitized by singlet oxygen, *Chem. Phys. Lett.* 29:159–162.

Brown, R. J., and Ogryzlo, E. A., 1964. Chemiluminescence from the reaction of chlorine with aqueous hydrogen peroxide, *Proc. Chem. Soc.* 1964:117.

Chan, H. W.-S., 1971. Singlet oxygen analogs in biological systems: Coupled oxygenation of 1,3-dienes by soybean lipoxidase, *J. Am. Chem. Soc.* 93:2357–2358.

Clough, R. L., Yee, B. G., and Foote, C. S., 1979. Chemistry of singlet oxygen. 30. The unstable primary product of tocopherol photooxidation, *J. Am. Chem. Soc.* 101:683–686.

Elias, E., Ogryzlo, E. A., and Schiff, H. I., 1959. The study of electrically discharged O_2 by means of an isothermal colorimetric detector, *Can. J. Chem.* 37:1680–1689.

Finazzi-Agro, A., Avigliano, L., Veldnik, G. A., Vliegenthart, J. F. G., and Boldingh, J., 1973. The influence of oxygen on the fluorescence of lipoxygenase, *Biochim. Biophys. Acta* 326:462–470.

Foner, S. N., and Hudson, R. L., 1956. Metastable oxygen molecules produced by electrical discharge, *J. Chem. Phys.* 25:601–602.

Foote, C. S., 1976. Photosensitized oxidation and singlet oxygen: Consequences in biological systems, in *Free Radicals in Biology,* Vol. 2, W. A. Pryor (ed.), Academic Press, New York, pp. 85–133.

Foote, C. S., 1978. *Photochem. Photobiol.* 28:922 (in the Discussion).

Foote, C. S., 1979. Detection of singlet oxygen in complex systems: A critique, in *Biochemical and Clinical Aspects of Oxygen,* W. S. Caughey (ed.), Academic Press, New York, pp. 603–626.

Foote, C. S., and Denny, R. W., 1968. Chemistry of singlet oxygen. VII. Quenching by β-carotene, *J. Am. Chem. Soc.* 90:6233–6235.

Foote, C. S., and Wexler, S., 1964a. Olefin oxidations with excited singlet molecular oxygen, *J. Am. Chem. Soc.* 86:3879–3880.

Foote, C. S., and Wexler, S., 1964b. Singlet oxygen: A probable intermediate in photosensitized autoxidations, *J. Am. Chem. Soc.* 86:3880–3881.

Foote, C. S., Wexler, S., Ando, W., and Higgins, R., 1968. Chemistry of singlet oxygen. IV. Oxygenations with hypochlorite–hydrogen peroxide, *J. Am. Chem. Soc.* 90:975–981.

Foote, C. S., Shook, F. C., and Abakerli, R. A., 1980. Chemistry of superoxide ion. 4. Singlet oxygen is not a major product of dismutation, *J. Am. Chem. Soc.* 80:2503–2504.

Frimer, A. A., 1979. The reaction of singlet oxygen with olefins: The question of mechanism, *Chem. Rev.* 79:359–387.

Galston, A. W., and Baker, R. S., 1950. Studies on the physiology of light action. II. The photodynamic action of riboflavin, *Am. J. Bot.* 36:773–780.

Gattow, G., and Schneider, A., 1954. Eine anorganische Lumineszensreaktion, *Naturwissenschaften* 41:116.

Gorman, A. A., Lovering, G., and Rogers, M. A. J., 1979. The entropy-controlled reactivity of singlet oxygen ($^1\Delta_g$) toward furans and indoles in toluene: A variable-temperature study by pulse radiolysis, *J. Am. Chem. Soc.* 101:3050–3055.

Guir, H. J., and Foote, C. S., 1976. Chemistry of superoxide ion III quenching of singlet oxygen, *J. Am. Chem. Soc.* 98:1984–1986.

Haig, C., 1935. The phototropic responses of *Avena* in relation to intensity and wavelength, *Biol. Bull.* 69:305–324.

Hammond, W. B., 1979. On the relative reactivity of 'E *vs.*' O_2: Reactions of laser-generated singlet oxygen with methoxy-olefins, *Tetra. Lett.* 1979:2309–2312.

Hasty, N., Merkel, P. B., Radlick, P., and Kearns, D. R., 1972. Role of azide in singlet oxygen reactions: Reaction of azide with singlet oxygen, *Tetrahedron. Lett.* 1:49–52.

Held, A. M., and Hurst, J. K., 1978. Ambiguity associated with the use of singlet oxygen trapping agents in myeloperoxidase-catalyzed reactions, *Biochem. Biophys. Res. Commun.* 81:878–885.

Herron, I. T., and Schiff, H. I., 1958. A mass spectrophotometric study of normal oxygen and oxygen subjected to electrical discharge, *Can. J. Chem.* 36:1159–1170.

Higley, D. P., and Murray, R. W., 1974. Oxidation of diazo compounds with singlet oxygen: Formation of ozonides, *J. Am. Chem. Soc.* 96:3330–3332.

House, H. O., 1972. *Modern Synthetic Reactions,* Benjamin/Cummings, Menlo Park, California, p. 337.

Ito, T., 1978. Cellular and subcellular mechanisms of photodynamic action: The $^1O_2$ hypothesis as a driving force in recent research, *Photochem. Photobiol.* 28:493–508.

Jones, R. D., Summerville, D. A., and Basolo, F., 1979. Synthetic oxygen carriers related to biological systems, *Chem. Rev.* 79:139–179.

Kacher, M. L., and Foote, C. S., 1979. Chemistry of singlet oxygen. XXVIII. Steric and electronic effects on the reactivity of sulfides with singlet oxygen, *Photochem. Photobiol.* 29:765–769.

Kasha, M., 1978. *Photochem. Photobiol.* 28:391 (in the Discussion).

Kasha, M., and Khan, A. U., 1970. The physics, chemistry, and biology of singlet molecular oxygen, *Ann. N. Y. Acad. Sci.* 171:5–23.

Kautsky, H., 1937. Die Wechselwirkung zwischen sensibilisataen und Sauerstoff im Licht, *Biochemistry* 291:271–284.

Kautsky, H., and deBruijn, H., 1931. The explanation of the inhibition of photoluminescence of fluorescent systems by oxygen: The formation of active diffusing oxygen molecules by sensitization, *Naturwissenschaften* 19:1043.

Kearns, D. R., 1969. Selection rules for singlet-oxygen reactions: Concerted addition reactions, *J. Am. Chem. Soc.* 91:6554–6563.

Kearns, D. R., 1971. Physical and chemical properties of singlet molecular oxygen, *Chem. Rev.* 71:395–427.

Khan, A. V., 1976. Singlet molecular oxygen: A new kind of oxygen, *J. Phys. Chem.* 80:2219–2228.

Khan, A. J., and Kasha, M., 1963. Red chemiluminescence of molecular oxygen in aqueous solution, *J. Chem. Phys.* 39:2105–2106.

Khan, A. V., and Kasha, M., 1966. Physical theory of chemiluminescence in systems evolving molecular oxygen, *J. Am. Chem. Soc.* 88:1574–1576.

King, M. M., Lai, E. K., and McKay, P. B., 1975. Singlet oxygen production associated with enzyme-catalyzed lipid peroxidation in liver microsomes, *J. Biol. Chem.* 250:6496–6502.

Kobayashi, K., 1978. Effect of sodium azide on photodynamic induction of genetic changes in yeast, *Photochem. Photobiol.* 28:535–538.

Kraljic, I., and Mohsni, S. El., 1978. A new method for the detection of singlet dioxygen in aqueous solutions, *Photochem. Photobiol.* 28:577–581.

Krinsky, N. I., 1977. Singlet oxygen in biological systems, *Trends Biochem. Sci.* 2(2):35–38.

Krinsky, N. I., 1979. Biological roles of singlet oxygen, *Org. Chem. Ser. Monogr. 40: Singlet Oxygen*, Chapt. 12, pp. 597–641.

Lagarias, J. C. and Poff, K. L., 1985. To be published.

Lion, Y., Delmelle, M., and van de Vorst, A., 1976. New method of detecting singlet oxygen production, *Nature (London)* 263:442–443.

Mango, F. D., and Schachtschneider, J. H., 1971. Orbital symmetry restraints to transition metal catalyzed [2 + 2] cycloaddition reactions, *J. Am. Chem. Soc.* 93:1123–1130.

McKeown, E., and Waters, W. A., 1966. The oxidation of organic compounds by "singlet" oxygen, *J. Chem. Soc. B* 1966:1040–1046.

Merkel, P. B., Nilsson, R., and Kearns, D. R., 1972. Deuterium effects on singlet oxygen lifetimes in solutions: A new test of singlet oxygen reactions, *J. Am. Chem. Soc.* 94:1030–1031.

Midden, W. R., and Wang, S. Y., 1983. Singlet oxygen generation of solution kinetics: Clean and simple, *J. Am. Chem. Soc.* 105:4129–4135.

Moan, J., and Wold, E., 1979. Detection of singlet oxygen production by ESR, *Nature (London)* 279:450–451.

Monroe, B. M., 1979. Rates of reaction of singlet oxygen with sulfides, *Photochem. Photobiol.* 29:761–764.

Ogilby, P. R., and Foote, C. S., 1981. Chemistry of singlet oxygen. 34. Unexpected solvent deuterium isotope effects on the lifetime of singlet molecular oxygen ($^1\Delta_g$), *J. Am. Chem. Soc.* 103:1219–1221.

Ogilby, P. R., and Foote, C. S., 1983. Chemistry of singlet oxygen. 42. Effect of solvent, solvent isotopic substitution, and temperature on the lifetime of singlet molecular oxygen ($^1\Delta_g$), *J. Am. Chem. Soc.* 105:3423–3430.

Olmstead, J., III, 1980. Photocalorimetric studies of singlet oxygen reactions, *J. Am. Chem. Soc.* 102:66–71.

Ouannes, C., and Wilson, T., 1968. Quenching of singlet oxygen by tertiary aliphatic amines: Effect of DABCO, *J. Am. Chem. Soc.* 90:6527–6528.

Pacquette, L. A., Bellamy, F., Böhm, M. C., and Gleiter, R., 1980a. Electronic control of stereoselectivity. 6. Directionality of singlet oxygen addition to 1,4-dimethoxynaphthalenes laterally fused to bridged bicyclic systems, *J. Org. Chem.* 45:4913–4921.

Pacquette, L. A., Carr, R. V. C., Arnold, E., and Clardy, J., 1980b. Electronic control of stereo-selectivity. 5. Stereochemistry of singlet oxygen capture by cyclopentadiene rings fused to norbornyl and norbornenyl frameworks, *J. Org. Chem.* 45:4907–4913.

Pacquette, L. A., Carr, R. V. C., Charumilind, P., and Blount, J. F., 1980c. Electronic control of stereoselectivity. 7. Stereospecificity of *N*-methyltriazolinedione cycloaddition to tricyclo-[5.2.1.O^{2,6}]deca-2,5-diene, tricyclo[5.2.1.O^{2,6}]deca-2,5,8-triene, and tricyclo[5.2.2.O^{2,6}]undeca-2,5,8-triene, *J. Org. Chem.* 45:4922–4926.

Peters, J. W., Pitts, J. N., Rosenthal, I., and Fuhr, H., 1972. A new and unique chemical source of singlet oxygen: Potassium perchromate, *J. Am. Chem. Soc.* 94:4348–4350.

Rawls, H. R., and Estes, F. L., 1978. A preliminary investigation of the use of limonene for detecting singlet oxygen as a component of polluted air, *Photochem. Photobiol.* 28:465–468.

Rebek, J., McCready, R., Wolf, S., and Mossman, A., 1979. New oxidizing agents from the dehydration of hydrogen peroxide, *J. Org. Chem.* 44:1485–1493.

Rosen, H., and Klebanoff, S. J., 1977. Formation of singlet oxygen by the myeloperoxidase-mediated antimicrobial systems, *J. Biol. Chem.* 252:4803–4810.

Ryang, H. S., and Foote, C. S., 1979. Chemistry of singlet oxygen. 31. Low temperature studies of dye-sensitized photooxygenation of imidazoles: Direct observation of unstable 2,5-endoper-oxide intermediates, *J. Am. Chem. Soc.* 101:6683–6687.

Saito, I., Nagata, R., and Matsuura, T., 1981. On the use of methyl substituted poly (vinyl naph-thalene) as a reversible singlet oxygen carrier, *Tetrahedron Lett.* 22(42):4231–4234.

Santamaria, L., and Prino, G., 1972. List of the photodynamic substances, *Res. Prog. Org.–Biol. Med. Chem.* 3(I):XI–XXXV.

Singh, A., McIntyre, N. R., and Koroll, G. W., 1978. Photochemical formation of metastable species from 1,3-diphenylisobenzofuran, *Photochem. Photobiol.* 28:595–601.

Smith, L. L., and Stroud, J. P., 1978. Sterol metabolism. XLII. On the interception of singlet molecular oxygen by sterols, *Photochem. Photobiol.* 28:479–485.

Smith, L. L., Kulig, M. J., and Teng, J. I., 1977. Sterol metabolism XI: On the failure of superoxide radical anion to react with cholesterol, *Chem. Phys. Lipids* 20:211–215.

Stauff, J., and Lohmann, F., 1964. Spectren der Chemilumineszenz einiger anorganischer Oxydation reaktion, *Z. Phys. Chem. N.F.* 40:123–126.

Tanielian, C., and Chaineaux, J., 1978. Singlet oxygen reaction with 2,7-dimethyl-2,6-octadiene, *Photochem. Photobiol.* 28:487–492.

Taube, H., 1965. Mechanisms of oxidation with oxygen, *J. Gen. Physiol.* 49:29–50.

Teng, J. I., and Smith, L. L., 1973. Steroid metabolism. XXIV. On the unlikely participation of singlet molecular oxygen in several enzyme oxygenations, *J. Am. Chem. Soc.* 95:4060–4061.

Turro, N. J., 1978. *Modern Molecular Photochemistry,* Benjamin/Cummings, Menlo Park, California, Chapter 14.

Turro, N. J., and Chow, M.-F., 1979. Magnetic field effects on the thermolysis of endoperoxides of aromatic compounds: Correlations with singlet oxygen yield and activation entropies, *J. Am. Chem. Soc.* 101:3701–3703.

Wasserman, H. H., and Ives, J. L., 1981. Singlet oxygen in organic synthesis, *Tetrahedron* 37:1825–1852.

Wilson, T., 1966. Excited singlet molecular oxygen in photooxidation, *J. Am. Chem. Soc.* 88:2898–2902.

Woodward, R. B., and Hoffman, R., 1970. *The Conservation of Orbital Symmetry,* Academic Press, Weinheim, Germany, pp. 43 and 67–70.

Superoxide Ion

4

4.1 Properties, Sources, and Stability (Table 4-1)

One-electron reductions of dioxygen form superoxide ion, O_2^- (Sawyer and Valentine, 1981). Although triplet dioxygen is a rather poor one-electron oxidant, superoxide ion is a strong one-electron oxidant if protons are available.

$$E°(pH\ 7)$$

$$O_2 + e^- \rightarrow O_2^- \cdot \qquad\qquad -0.33$$

$$O_2^- \cdot + e^- + 2H^+ \rightarrow H_2O_2 \qquad +0.89$$

If protons are not available, the product is the highly unstable peroxydianion O_2^{2-} so that superoxide does not act as an oxidant in the absence of protons. It is interesting to note that superoxide ion is also a mild reducing agent. As discussed in section 4.5, superoxide ion often acts on oxidized organic compounds to form radical anions.

The conjugate acid of superoxide ion, perhydroxyl radical, is one of those few compounds with a positive free energy of formation. It is unstable with respect to H_2 and O_2 by 2500 $\pm$ 600 calories (Howard, 1980). Superoxide anion has a strong tendency to disproportionate to hydrogen peroxide and dioxygen:

$$2\ O_2^- + 2H^+ \rightleftharpoons H_2O_2 + O_2 \quad K\ (pH\ 7) = 4 \times 10^{20}$$

This reaction causes superoxide anion to act as a strong base in many reactions. Although the pK of HO_2 is only 4.69, the strong tendency for dis-

Table 4-1. Properties of Superoxide

Property		Reference
O–O bond distance 1.33 Å[a]		Sawyer and Gibian (1979)
O–O frequency		
$\quad$ HO_2	1101 cm^{-1}	Milligan and Jacox (1963)
$\quad$ LiO_2	1097 cm^{-1}	Andrews (1968)
$\quad$ "Free" O_2-	1090 cm^{-1}	Andrews (1968)
$^3\Sigma_g O_2 + e^- \rightarrow O_2-$	$E° = -0.33$ V	Ilan $et\ al.$ (1974)
$\Delta O_2 + e^- \rightarrow O_2-$	$E° = +0.65$ V	Ilan $et\ al.$ (1974)
$O_2- + e^- + 2H^+ (10^{-7}$ M$) \rightarrow H_2O_2$	$E° = +0.59$ V	Ilan $et\ al.$ (1974)
$O_2- + e^- + 2H^+ (1$ M$) \rightarrow H_2O_2$	$E° = +0.94$ V	Ilan $et\ al.$ (1974)
$HO_2 \rightarrow O_2 + {}^1/_2\ O_2$	$\Delta G° = -2.50$ kcal	Ilan $et\ al.$ (1974)
$HO_2 \rightarrow H^+ + O_2-$	$pK = 4.69 \pm 0.08$	Bielski (1978)
$^2O_2- + 2H^+ (10^{-7}) \rightleftharpoons H_2O_2 + O_2$	$pK = 4 \times 10^{20}$	Sawyer and Gibian (1979)
$^1O_1 + H_2O \rightleftharpoons O_2 + HO_2- + OH^-$	$pK = 2.5 \times 10^8$	Sawyer and Gibian (1979)
In aqueous solution		Bielski (1978)
$\quad$ $O_2-\ \lambda_{max} = 245$ nm, $\varepsilon = 2350 \pm 120$ M^{-1} cm^{-1}		
$\quad$ $HO_2\ \lambda_{max} = 225$ nm, $\varepsilon = 1400 \pm 80$ M^{-1} cm^{-1}		
In acetonitrile + 0.1 M tetraphenylammonium perchlorate		Sawyer and Gibian (1979)
$\quad$ $O_2-\ \lambda_{max} = 225$ nm, $\varepsilon = 1460$ M^{-1} cm^{-1}		

proportionation causes superoxide to act like the conjugate base of an acid with a pK of approximately 24 (Sawyer and Valentine, 1981).

$\quad$ Organic chemists, who need reagent quantities of superoxide ion, commonly obtain it from KO_2, whereas biochemists, who need only trace amounts, commonly generate superoxide ion in solution by the xanthine oxidase–xanthine reaction or by pulse radiolysis of oxygenated aqueous solutions containing formate ion by several reactions:

$$H_2O \xrightarrow{hv} (OH + H\cdot + H^+ + e^-)$$

$$H\cdot + O_2 \rightarrow HO_2$$

$$e^- + O_2 \rightarrow O_2^-$$

$$\cdot OH + HCO_2^- \rightarrow H_2O + CO_2^-$$

$$CO_2^- + O_2 \rightarrow CO_2 + O_2^-$$

$$H^+ + HCO_2^- \rightarrow HCOOH$$

Potassium superoxide is produced commercially by oxidation of potassium, a process first described in 1811 by Guy Lussac (Lee-Ruff, 1977).

A convenient well-defined source of superoxide ion is by electrolysis of oxygenated solutions of dimethylsulfoxide (DMSO) or of acetonitrile (Fee and Hildebrand, 1974; Ozawa *et al.*, 1977). These solutions can be diluted in water.

A crucial obstacle in both the early use and understanding of superoxide reactions was the reagent's negligible solubility in organic and aprotic solvents. Two methods have been found to circumvent this problem (Foote, 1978). Foote and his group have found that while potassium superoxide is insoluble in organic solvents, tetramethylammonium superoxide is "a very nice reagent . . . , soluble in all organic solvents" (Peters and Foote, 1976; Guiraud and Foote, 1976). The method for preparation of this salt was reported by McElroy and Hashman (1964). It involves a solid-phase reaction, *in vacuo,* between potassium superoxide and tetramethylammonium hydroxide pentahydrate. The product is isolated by extraction with liquid ammonia.

Others have used crown ethers to increase the solubility of potassium superoxide in organic solvents, particularly DMSO (Valentine and Curtis, 1975). Choice of solvent is crucial in oxidation reactions because entirely different products can be obtained (see below). The reactivity of superoxide toward a wide variety of solvents has been tested (Sanka and Martinsons, 1968). Use of crown ethers in solubilization of the anion in nonpolar solvents has the advantage that since it is known that the mechanism of solubilization is via complexation of the cation (Pedersen, 1967; Liotta and Harris, 1974), it can be inferred for mechanistic purposes that the reactive species is "naked" superoxide anion, as opposed to KO_2. It has been observed in at least one case that the potassium superoxide reaction can proceed with only a catalytic concentration of the ether (Corey *et al.*, 1975). Generally, however, at least one equivalent of macrocyclic ether is added to maximize the reaction rate and simplify the kinetics.

Potassium superoxide is quite soluble in water. However, it decomposes very rapidly in aqueous solution so that further study in this solvent is limited. The decomposition of superoxide catalyzed by water is a complex process that has been the subject of some experimentation. At alkaline pH, superoxide has a relatively long half-life (about 50 sec at pH 14), which decreases with pH (1.2 sec at pH 10). In this range, decomposition depends critically on hydration rather than on protonation (Khan, 1978). The rate of alkaline decomposition is first-order in superoxide ion, rather than second-order (Bielski and Saito, 1971). The rate constant for reaction of two superoxide anion radicals is essentially zero (Bielski and Allen, 1977). The alkaline decomposition is analogous to the decomposition in aprotic solvents because the hydration spheres in both solutions are considered to be the limiting factor in decomposition of superoxide ion. The half-life of superoxide (crown ether) in dry pyridine, for instance, is about 30 min.

In neutral to acidic pH aqueous solution, decomposition is due to bimolecular dismutation of either two protonated molecules or one protonated and one unprotonated molecule (Behar *et al.*, 1970). The observed rate constant follows the equation

$$k_{\text{obs.}} = \frac{k_1 + k_2 x}{(1 + x)^2}$$

where k_1 is the rate constant for the reaction of two perhydroxyl radicals, k_2 is the rate constant for reaction of superoxide anion and perhydroxyl radical, and x is the ratio of the acidity constant to the H^+ concentration in aqueous solution.

$$2\ HO_2\cdot \xrightarrow{k_1} H_2O_2 + O_2$$

$$H_2O + HO_2 + O_2^- \xrightarrow{k_2} H_2O_2 + O_2 + OH^-$$

$$HO_2\cdot \underset{}{\overset{K}{\rightleftharpoons}} O_2^-\cdot + H^+ \qquad x = \frac{K}{[H^+]}$$

The values for the elementary rate constants have been measured as $k_1 = 8.60 \pm 0.62 \times 10^5\ M^{-1}\ sec^{-1}$ and $k_2 = 1.02 \pm 0.49 \times 10^8\ M^{-1}\ sec^{-1}$ (Bielski, 1978).

A pH-dependent deuterium isotope effect has also been measured (Bielski and Saito, 1971). Products of the reaction are O_2 and either H_2O_2 or OOH^- depending on the conditions. Although theoretical arguments (Koppenol and Butler, 1977) may be made that the overlap of partially filled orbitals in O_2^- and $HO_2\cdot$ leading to singlet dioxygen is more likely than the overlap of a partially filled with a filled orbital leading to triplet dioxygen, the experimental observation is that the dioxygen molecule produced in the nonaqueous or high-pH reaction is triplet (ground-state) dioxygen (Poupko and Rosenthal, 1973; Nilsson and Kearns, 1974). However, whether the disproportionation reaction in water (i.e., involving HOO· + O_2^- or HOO·) results in singlet dioxygen has been controversial (Khan, 1978). Central to the question is the theoretical consideration that superoxide is expected to be an extremely efficient quencher of singlet dioxygen. Hence, trapping or direct observation of the excited state may be unlikely even if it is formed initially.

Foote *et al.* (1980a,b) have reported good evidence that singlet dioxygen is not formed (in significant amounts) in the aqueous dismutation of superoxide. They performed this demonstration by first adding superoxide to a solution in which singlet dioxygen was being produced at a known rate. (Detection was by cholesterol on an inert support.) This showed the efficiency with which super-oxide quenched the singlet dioxygen produced. They then repeated the experi-ment, but without singlet-dioxygen generation. Knowing the system's detection efficiency and the superoxide quenching efficiency under identical conditions, they were able to calculate that no more than a few tenths of a percent of singlet dioxygen was produced (relative to superoxide added) in the pH range of 4–8.

Aubry *et al.* (1981) have reached the same conclusion. Their system in-volved use of the potassium salt of rubrene-2,3,8,9-tetracarboxylic acid as a water-soluble trap that on reaction with singlet oxygen forms an endoperoxide that can be isolated by high-performance liquid chromatography, but is inert to dioxygen, superoxide, and peroxides. Superoxide was generated at a low steady-state level by radiolysis of a formate solution, minimizing the problem of singlet-oxygen quenching by superoxide. No rubrene endoperoxide was detected, and all degradation of the trap molecule could be accounted for by the action of hydrated electrons generated in the radiolysis process.

The observation of the formation of only triplet dioxygen in the dismutation reaction can be rationalized. The dismutation reaction could proceed by either singlet or triplet transition state according to whether the two radicals combine with electron spins paired or unpaired. The stability of triplet dioxygen over singlet dioxygen causes only the triplet transition state to break down to give products (Sawyer and Valentine, 1981):

$$HO_2\cdot + O_2^-\cdot \rightleftharpoons [\cdot O_2H\,O_2\cdot]_T^- \rightarrow HO_2^- + {}^3O_2$$

$$HO_2\cdot + O_2^- \rightleftharpoons [\cdot O_2\,HO_2\cdot]_S^- \rightarrow HO_2^- + {}^1O_2$$

Alcohols can furnish the proton needed in the dismutation reaction:

$$ROH + 2O_2^-\cdot \rightarrow RO^- + HO^- + O_2$$

Thus, the disproportionation forms bases much stronger than superoxide (Lee-Ruff, 1977; Sawyer *et al.*, 1978).

The disproportionation reaction of superoxide ion has been useful in the regeneration of breathing atmospheres in a closed environment. A chemical is needed that will remove carbon dioxide and moisture from the air and replace it with oxygen. Potassium superoxide suits this purpose very well. Moisture in the used air causes it to disporportionate to dioxygen, the monoanion of hydrogen peroxide, and hydroxide ion. Under these conditions, carbon dioxide is removed as carbonate, which in turn keeps the system from reacting further, until more moisture is absorbed, by removing any excess water. For this and other purposes, a quantitative titration of superoxide has been developed (Seyb and Kleinberg, 1951). The salt is quantitatively decomposed by addition of acetic acid in the presence of diethyl phthalate. The dioxygen liberated is quantitated in a gas burette, and the diester serves to trap the remaining alkaline peroxide. Some of the peroxide oxygen can then be liberated with ferric chloride and hyrochloric acid and the dioxygen again measured in a gas burrette as a check on the first measurement. Results are within 2%.

4.2 Detection of Superoxide Ion

Analytical methods for superoxide ion have been reviewed by Bors *et al.* (1978). The method of choice depends on the system in which it is to be detected. In nonaqueous systems, either the method of Hirata and Hayaishi (1975) or that of Seyb and Kleinberg (1951) can be used, both of which utilize the dismutation reaction catalyzed by protons.

In aqueous systems, the methods commonly utilize the oxidation or reduction of a reagent. Examples of the oxidation of a reagent to a colored compound are the oxidation of adrenaline to adrenochrome and the oxidation of Tiron (1,2-dihydroxy-3,5-disulfonic acid). Other reagents oxidized by superoxide ion are hydroquinone or phenol derivatives, sulfite, hydroxylamine, and nitroxides. All these reagents are fairly labile and therefore nonspecific. The observation of chemiluminescence in the oxidation of luminol is a very sensitive measure of superoxide ion. Reagents that trap by being oxidized are generally oxidized by other mild oxidizing agents, autoxidize at an appreciable rate, and often give the same products on reaction with hydroxyl radicals as with superoxide. The autoxidation process acting on hydroquinones, at least, also produces superoxide, so a fair amount of complication can be expected.

Other methods for the detection of superoxide ion utilize its reducing power. A wide range of such reagents have been used (Bors *et al.*, 1978). These include reagents that are reduced by superoxide, such as tetrazolium salts (Bielski and Richter, 1977), tetranitromethane, and native or acetylated ferricytochrome. Acetylated ferricytochrome *c* is a relatively specific reagent for superoxide ion (Azzi *et al.*, 1975). Acetylated ferricytochrome is reduced by superoxide ion, but not by mitochrondrial nonmicrosomal reductases. However, other common reducing agents such as dithionite can also reduce acetylated ferricytochrome *c*.

Many of the assays for superoxide ion also gave positive responses to hydroxyl radical. Generally, specificity is incorporated into the assay by inhibiting the trapping reaction with superoxide dismutase (SOD). Hence, it is a decrease in effect that is often measured. If the superoxide is formed at the active center of an enzyme, SOD may not be able to act on the superoxide. In this case, a copper salt (cupric salicylate) has been used as SOD substitute (Mayer *et al.*, 1980). This dismutation reaction is similar in rate to the enzymatic reaction.

4.3 Haber–Weiss Reaction

Superoxide ion may react by the "Haber–Weiss" reaction. This is the reaction of superoxide (protonated or anionic) and hydrogen peroxide to form dioxygen, hydroxide ions, and the very reactive hydroxyl radicals. The reaction was first proposed as one of the steps in the ferrous iron decomposition of hydrogen peroxide (Haber and Weiss, 1934):

$$Fe(II) + H_2O_2 \rightarrow Fe(III) + \cdot OH + OH^-$$

$$\cdot OH + H_2O_2 \rightarrow HO_2\cdot + H_2O$$

$$HO_2\cdot + H_2O_2 \rightarrow \cdot OH + H_2O + O_2$$

$$Fe(II) + \cdot OH \rightarrow Fe(III) + OH^-$$

The dioxygen appears to be formed in the singlet state because the Haber–Weiss products will add 1,3-diphenylisobenzofuran (DPBF) (Kobayashi and Ando, 1979). Beauchamp and Fridovich (1970) noticed that removal of either superoxide (by SOD) or hydrogen peroxide (by catalase) from the xanthine oxidase system inhibited the formation of ethylene from methional. They hypothesized that superoxide itself was not a potent enough oxidizing agent to cause the reaction, but that hydroxyl radicals, formed via the Haber–Weiss process, could perform the reaction:

$$O_2^-\cdot + H_2O_2 \rightarrow O_2 + {}^-OH + \cdot OH$$

$$\cdot OH + methional \rightarrow ethylene + other \ products$$

However, radicals less active than hydroxyl radicals can cause the ethylene formation observed by Fridovich (Pryor and Tang, 1978).

The reaction of superoxide ion with hydrogen peroxide has been found to be negligibly slow in the absence of a catalyst (Czapski and Ilan, 1978; Koppenol et al., 1978; Ferradini et al., 1978; Weinstein and Bielski, 1979). The bimolecular rate constant is 2.25 ± 0.2 M^{-1} sec^{-1} (Ferradini et al., 1978).

Several plausible mechanisms have been suggested, and tested to a certain extent, by which hydroxyl radicals could be generated in the case originally observed by Fridovich and others. For instance, two-step iron catalysis is a highly plausible mechanism for this reaction (Koppenol et al., 1978; Ferradini et al., 1978; McCord and Day, 1978):

$$O_2^- \cdot + Fe(III) \rightarrow O_2 + Fe(II)$$

$$Fe(II) + H_2O_2 \rightarrow Fe(III) + OH^- + OH\cdot$$

Alternatively, superoxide could simply complex with a transition metal, and the stable complex could react with H_2O_2 to form the hydroxyl radicals.

The two-step iron-catalysis mechanism is, of course, not restricted to superoxide. Many commonly available (*in vivo*) reducing agents could presumably carry out the first step of the process. This being the case, any contact of a ferrous enzyme with hydrogen peroxide would be extremely deleterious to the cell.

Other interesting facets of this reaction are that alkyl (such as lipid) hydroperoxides in the presence of catalysts may react with superoxide much faster than does hydrogen peroxide (Thomas *et al.*, 1976; Peters and Foote, 1978) in certain solutions:

$$O_2^- \cdot + ROOH \rightarrow O_2 + RO\cdot + OH^-$$

However, in very clean solutions, there is probably no reaction of superoxide with organic hydroperoxides (Bors *et al.*, 1979; Stanley, 1980).

Hydrogen peroxide will disproportionate under alkaline aprotic conditions:

$$HO_2^- + H_2O_2 \rightarrow O_2^- \cdot + \cdot OH + H_2O$$

The alkaline conditions can be caused either by potassium hydroxide or by the disproportionation of superoxide. In the presence of superoxide ion, the products are dioxygen and hydroxide ion:

$$O_2^- \cdot + \cdot OH \rightarrow O_2 + OH^-$$

A mechanism for the hydrogen peroxide disproportionation must take into account the fact that it does not occur in protic solvents. The authors therefore suggest a pathway dependent on hydrogen bonding, which would be overpowered by a hydrogen-bonding agent.

$$HOO^- + HOOH \rightarrow \overset{\ominus}{H}\underset{O-O}{\overset{O-O}{(}H} \rightarrow O_2^- \cdot + H_2O + OH\cdot$$

Diacyl peroxides or anhydrides epoxidize olefins in the presence of superoxide. The rate of this reaction, but not the yield, depends on the concentration of crown ether (i.e., solution concentration of superoxide) and on the concentration of "substrate" acetic anhydride. Many details of this reaction indicated it to involve complex free-radical fragmentation.

Danen and Arudi (1978) report the formation of singlet dioxygen on addition of aliphatic diacyl peroxides to a slurry of superoxide in benzene–crown ether. The reaction also gives a quantitative yield of the corresponding carboxylate salts.

$$2O_2^- \cdot + \underset{\displaystyle \overset{\displaystyle \|}{O}}{R-C}-O-O-\underset{\displaystyle \overset{\displaystyle \|}{O}}{C}-R \rightarrow 2O_2 + 2RCO^-$$

The excited state of the dioxygen product was detected by DPBF. Tetramethylethylene was also used, but gave only low yields of 3-hydroxy-2,3-dimethyl-1-butene. 1,2-Dimethylcyclohexene gave better yields of allylic hydroperoxide products (Figure 4-1), but the ratios of the three products varied between benzoyl peroxide and lauroyl peroxide. Quenching of DPBF oxidation by carotene was also observed.

Superoxide ion will react with ferrohemes in aprotic solvents to produce peroxoheme complexes (McCandish et al., 1980). It is not known whether this type of complex occurs anywhere in biological systems or not. The reaction of superoxide ion with ferrihemes merely reduces the heme to the ferroheme and produces dioxygen.

4.4 Detection of ·OH Radicals

The presence of the Haber–Weiss reaction can best be detected by the detection of ·OH radicals. Hydroxyl radical is extremely reactive. It reacts at a high rate with almost any compound containing a hydrogen (Anbar et al., 1966), e.g., alkanes (Paraskevopoulos and Nip, 1980; Darnall et al., 1978) including

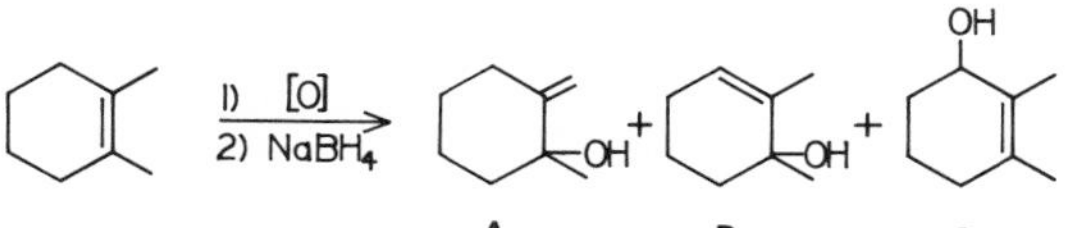

Figure 4-1. Detection of singlet dioxygen by 1,2-dimethylcyclohexene in the reaction of superoxide ion with diacyl peroxides. Compounds A and B are products of singlet dioxygen reactions, and the main product, C, is the result of a triplet-ground-state-dioxygen reaction.

methane (Jeong and Kaufman, 1982), alkenes (Darnall *et al.*, 1976), aldehydes (Audley *et al.*, 1981), alkyl halides (Jeong and Kaufman, 1982), and alcohols (Overend and Paraskevopoulos, 1978). The rate of reaction depends on the C–H bond dissociation enthalpy (Heicklen, 1981), but in general reacts rapidly with any hydrogen.

Hydroxyl radical also reacts with methional, methionine, and ketomercaptobutyric acid to form ethylene. The ethylene can be detected chromatographically. Superoxide does not generally react with these trapping agents, but other radicals have been shown to initiate the same reaction with methional (Pryor and Tang, 1978).

The reaction of hydroxyl radical with methionine has béen studied (Hiller *et al.*, 1981) and found to be similar to, but more complex than, the reaction with simple thioethers. Below pH 3, the oxidation produces the same strong optical absorption in the visible region observed with thioethers. The absorption is centered at 480 nm, has a microsecond-scale lifetime, and is attributed to the $\sigma \rightarrow \sigma^*$ transition of complex A (Figure 4-2), which is formed by the mechanism shown.

Hydroxyl radical can also react, in this pH range, by removing a hydrogen atom from either the methyl group or the methylene adjacent to sulfur, producing an absorption at 290 nm, or a proton may be lost from cation radical to produce the same neutral carbon-centered radical.

At higher pH, ionization of the carboxyl group ($pK = 2.2$) affects the course of the reaction. In this pH range, the intitial product is also $(R_2S)_2OH\cdot$. However, CO_2 becomes one of the main products. Comparison of the pH dependence with that for other sulfur carboxyl compounds indicates that sulfur cation radical does not directly oxidize the carboxylate group. Instead, the decarboxylation proceeds by an intramolecular oxidation–reduction to form a nitrogen-centered radical (Figure 4-3) that undergoes decarboxylation (for which the carboxylate anion is required). The amine oxidation appears to be sterically

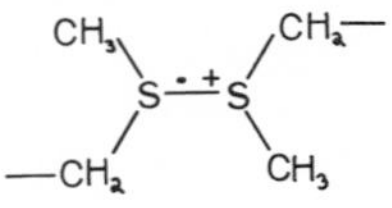

A

$$R_aS + OH\cdot \longrightarrow R_aS(OH)\cdot$$

$$R_aS(OH)\cdot + R_aS \longrightarrow (R_aS)_a(OH)\cdot$$

$$(R_aS)_a(OH)\cdot \longrightarrow A + OH^-$$

Figure 4-2. Mechanism of formation of radical cation A from methionine and hydroxyl radical. Both $R_2S(OH)\cdot$ and $(R_2S)_2(OH)\cdot$ have the radical centered on sulfur.

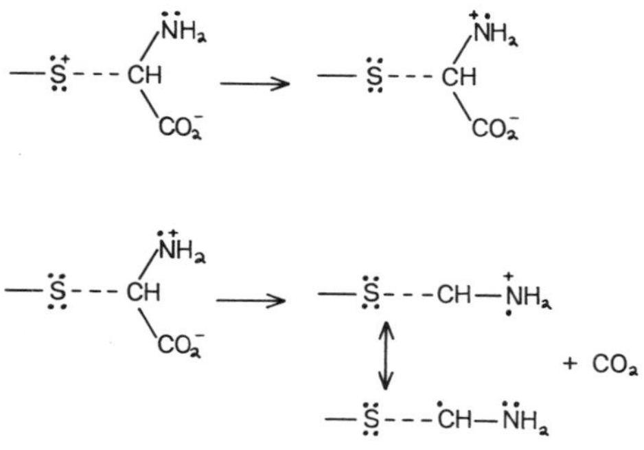

Figure 4-3. Mechanism of decarboxylation of the methionine radical cation.

assisted. The radical resulting from decarboxylation is a reducing species that loses a single electron to form a protonated imine. The imine, of course, hydrolyzes to give methional and ammonia. (In the absence of an electron acceptor, disproportionation of this radical gives the same products plus the amine.)

Direct trapping to form nitroxide radicals has been suggested, but has many disadvantages. Among them is the fact that some nitroxides are good quenchers of singlet dioxygen (Bellus, 1979). The difficulties and methods involved here have been reviewed (Finkelstein *et al.*, 1980).

In general, then, determining the intermediacy of reactive oxygen species is at best a complicated process. Hydroxyl radical is perhaps the easiest of these, since it is so reactive that the agents used to trap it are stable to other oxygen species. Caution is advised on three counts: (1) Since the concentration of reactive (toward OH·) dissolved materials in cells is so high, it is puzzling that millimolar concentrations of trapping agent would have an effect (Czapski, 1978). (2) The trapping of OH· generally results in radicals, some of which can be very reactive. (3) Even though superoxide cannot react with the OH·-trapping agents directly, there is good evidence that a catalyzed form of the "Haber–Weiss process" does take place in biological systems, so that a test for OH· could appear positive even if superoxide were the species involved in catalysis (Fielden *et al.*, 1978).

4.5   Reactions of Superoxide Ion with Organic Compounds

Superoxide ion is not as reactive as might be expected. For example, superoxide ion does not react with acetate ion, EDTA, Tris, imidazole, or tartrate ion (Bielski and Richter, 1977). Despite the strong oxidizing potential of superoxide ion when protons are available, the reactions of superoxide ion with organic substrates are often not as an oxidizing agent. The reaction to add an electron to form the high-energy O_2^{2-} is virtually impossible, and that with one proton and an electron to form HO_2^- is very difficult. Superoxide ion can act as a base or as a nucleophil, can abstract a hydrogen atom, and can donate electrons as a reducing agent.

Superoxide ion acts as a base because protons are consumed in the disproportionation reaction. For example, in the presence of acetonitrile, superoxide ion will produce the acetonitrile carbanion (Sawyer and Gibian, 1979). Even with compounds as easily oxidized as catechols, the predominant reaction is proton removal.

Sawyer et al. (1978) found that 3,5-di-t-butyl catechol anion is inert to superoxide anion in dry solvent and that while monomethyl and dimethyl catechol ethers are inert to superoxide, monomethyl catechol catalyzes the disproportion of superoxide. They propose a mechanism (Figure 4-4) for the oxidation of 3,5-di-t-butyl catechol in the presence of protons.

Superoxide ion often acts as a strong nucleophile. The factors accounting for the potent nucleophilicity of superoxide are not well understood. Clearly, it is in the category of nucleophile that exhibit an "α-effect," like hydroxylamine and hydrazine. An explanation of the "α-effect," itself, however, is not well established. The α-effect is the property of some molecules to exhibit enhanced nucleophilicity, relative to basicity, in comparison to a series of nucleophiles, the enhancement not being explained by polarizability. The molecules that exhibit the effect are those that have an unbonded electron pair localized on the atom

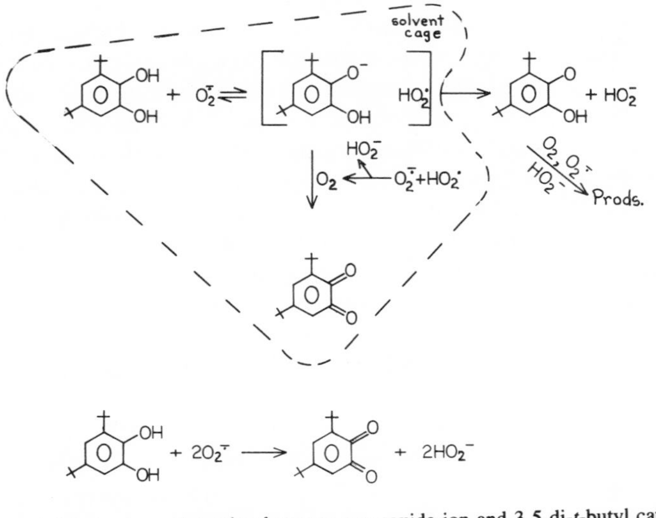

Figure 4-4. Mechanism of reaction between superoxide ion and 3,5-di-t-butyl catechol.

adjacent to the nucleophilic center (including the examples mentioned above). The effect has been rationalized in a number of different ways (McIsaac *et al.*, 1972). The extent to which superoxide behaves as an α-effect nucleophil may be a measure of the difference in orbital hybridization between superoxide and dioxygen. If the electrons were in orbitals with predominantly π character, an α-effect would not be expected to be strong. Since superoxide is a good nucleophile, it may be that the orbitals are more like the lone-pair orbitals of peroxide dianion.

Many of these nucleophilic displacement reactions in aprotic solvents are useful synthetic reactions (Sawyer and Gibian, 1979). Superoxide ion reacts with saturated compounds containing a good leaving group initially as a nucleophile. The displacement has been used to stereoselectively replace halides and sulfonates to form alcohols, in cases in which the normal nucleophiles for this conversion (hydroxide or carboxylate) cause racemization or other side products (Corey *et al.*, 1975). An interesting facet of this reaction is the mechanistic sequence following nucleophilic attack. The displacement product, which is an organic peroxy radical (Merritt and Johnson, 1977), is reduced to the corresponding anion by a second molecule of superoxide with liberation of dioxygen. The peroxide anion (pK_a of H_2O_2 is 10), which is considerably more basic than superoxide (pK_a of HO_2 is 4.69), can then either react with solvent (DMSO to form an alcohol and dimethylsulfone) or react nucleophilically with a second molecule of organic halide to produce a dialkyl peroxide (Johnson and Nidy, 1975). Thus, in benzene, treatment of alkyl halides with potassium superoxide–crown ether gave primarily dialkyl hydroperoxides, while in DMSO, the same reagents led directly to the alcohols (Corey *et al.*, 1975; Gibian and Ungermann, 1976; San Filippo *et al.*, 1976). Furthermore, stoichiometry requiring two equivalents of superoxide per equivalent of alcohol produced has been demonstrated (Danen and Warner, 1977). Rate constants were measured for the displacement reaction with a group of representative bromides in DMSO (Danen and Warner, 1977) and chlorides (Merritt and Sawyer, 1970). The reactivity order of substrates showed the reaction to be definitely nucleophilic. Furthermore, comparison of the rate constants with those of other nucleophiles (albeit in somewhat different solvents) showed superoxide to be more reactive than methoxide or azide by a large margin and at least as reactive as several other nucleophiles, the reactivities of which are much more easily explained. Under identical conditions, 1-bromooctane showed a half-life of 20 hr in the presence of potassium iodide as opposed to 45 sec in the presence of potassium superoxide (San Filippo *et al.*, 1976).

Roberts and Sawyer (1981) have electrochemically studied the reaction of superoxide with chlorinated hydrocarbons such as carbon tetrachloride, chloroform, and DDT in DMF (dimethylformamide) and DMSO:

$$CCl_4 + 6O_2^- \cdot \rightarrow CO_4^{2-} + 4Cl^- + 4O_2$$

$$CHCl_3 + 4O_2^- \cdot \rightarrow HC\overset{\overset{\displaystyle O}{\|}}{-}OO^- + 3Cl^- + 5/2O_2$$

$$DDT + 3O_2^- \cdot \rightarrow prods. + 2Cl^- + 3/2O_2$$

The reaction rates are higher than one might expect (bimolecular rate constant varied from 9 $M^{-1}S^{-1}$ for CH_2Cl_2 to 1300 $M^{-1}S^{-1}$ for CCl_4) compared to thiocynate ion (10 $M^{-1}S^{-1}$). Elimination from $CHCl_3$ and DDT does not occur. The rate of reaction with CCL_4 is very much faster than the rate of reaction with CH_3Cl or CH_2Cl_2, which indicates that the reaction is not a simple S_N2 process (steric hindrance would make the CCl_4 reaction much slower in a true S_N2 mechanism), although the initial bimolecular reaction is rate-limiting. Instead, the initial reaction may be an electron transfer to produce CCl_4^- and dioxygen.

Superoxide also acts as a nucleophile toward carboxylic acid derivatives such as esters and acid chlorides. Like the reaction with aliphatic substrates, these reactions are S_N2 displacements during which dioxygen is evolved and peroxy compounds are implicated (San Filippo *et al.*, 1976).

Superoxide does not react with aliphatic halides to give elimination products. Indeed, the lack of this reaction is one reason for its value in organic synthesis. However, toward *N*-chloro amines, it reacts very easily as a base to give imines (Scully and Davis, 1978). The conditions for this reaction are milder than those required for dehydrohalogenation with KOH. Inorganic reaction products are KCl and KOOH. Neither potassium acetate, which has the same *pK* as superoxide, nor sodium hydrogen peroxide, a probable intermediate in the superoxide reaction, gives the imine.

Toward substrates that have relatively weak second-row-atom–hydrogen bonds, superoxide can act as an oxidant by abstracting hydrogen atoms. Sawyer *et al.* (1978) point out that superoxide anion is an extremely poor single electron acceptor. Oxidation reactions become possible when deprotonation is less favorable, even with the dismutation driving force, than hydrogen-atom abstraction. The oxidation produces hydrogen peroxide anion and the substrate radical. The remaining organic radical can then generally be oxidized by atmospheric oxygen before it can react with superoxide anion.

In many cases, single-electron oxidation reactions resemble autoxidation reactions. Generally, the resulting radicals are oxygenated by molecular oxygen, and in one case, the reaction with electrochemically generated superoxide was accelerated by oxidizable metal salts. For example, the oxidation of 9,10-dihydroanthracene has been studied by Tezuka *et al.* (1975). The voltage change for the generation of the superoxide on inclusion of ferrous or cobalt phthalocyanin indicated the initial formation of metal-superoxide complex, as expected from studies of ferroheme–dioxygen complexes. Measuring the rate of oxygen uptake, the authors were able to show that the reaction is a free-radical chain reaction and that the initiation step is first-order in both superoxide and anthracene, but they did not measure the rate constant for this step. The color and electron spin resonance (ESR) spectrum of anthraquinone anion radical were observed, as could be expected from the work of Frimer and Rosenthal (1978) with chalcones and tetracyclone and the work of Poupko and Rosenthal (1973) with quinones. The latter authors outlined the mechanism shown in Figure 4-5. Potassium superoxide in DMSO caused this same reaction, which led to anthracene in nitrogen and to anthraquinone in air. Several other molecules react in the same way, including 1,3- and 1,4-cyclohexadiene and several phenolic and hydroquinone-like molecules. Allylic and benzylic substrates were not labile enough to be oxidized under these conditions. However, o- or p-nitrobenzyl groups appear to be capable of undergoing hydrogen-atom abstraction and are oxidized to the corresponding acids or ketones via the aldehyde or alcohol, respectively (Sagae et al., 1977, 1980).

Hydrazine derivatives give similar reactions (Chern and San Filippo, 1977). Those derivatives that can form diazene intermediates (monosubstituted aryl or alkyl hydrazines) lose molecular nitrogen quickly, generating alkyl or phenyl radicals. 1,2-Dialkylhydrazines do not react, while 1,2-diaryl derivatives are oxidized to azo compounds. 1,2-Diarylhydrazines form nitrosamines, secondary amines form nitroxides (Poupko and Rosenthal, 1973), and aryl hydrazones generally dimerize with elimination of N_2 to form azines. Also, some diaryl diazo compounds are oxidized to the corresponding ketones and others are not.

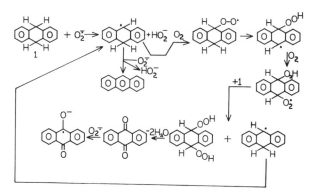

Figure 4-5. Mechanism of the reaction of superoxide ion with anthracene in the presence and absence of dioxygen.

The product distribution and coupling products, both of which vary with solvent, indicate the reaction to be a free-radical chain mechanism, probably initiated by hydrogen-atom abstraction.

It is surprising that as strong an oxidizing agent as superoxide ion can actually act as an electron donor.

Toward highly polarized olefins, superoxide acts as an electron donor. Frimer *et al.* (1977) studied the reaction of superoxide with a variety of highly electron-deficient olefins and aromatics in the presence of molecular oxygen. They found consistently that 18 label in the superoxide salt is not incorporated into the product. Thus, the radicals generated by electron donation from superoxide to nitrobenzenes, chalcones, tetracyclone, or phenyldinitro- or diacyano-olefins are stable enough to diffuse away from the oxygen molecule formed in the reaction (Figure 4-6). This indicates that formation of dioxygen (as opposed to a peroxy radical adduct) is the driving force for the substrate radical anion formation. In the case of 1,1-diphenyl-2,2-dinitroethylene, the radical anion was observed directly by ESR spectroscopy.

The products obtained from the phenyl dinitro olefin were phenyl ketones that further oxidized to benzoic acids, as confirmed by control experiments. The mechanism offered (Figure 4-7) involved a free-radical chain reaction. However, formation of unlabeled superoxide was not shown. Likewise, the sufficiency of a catalytic amount of superoxide or some other radical generating agent was not shown, though molar equivalents of the radical initiators are not required by the mechanisms.

Also, several of these mechanisms require single-electron reductions of undetected intermediate radicals. For instance, for tetracyclone, two equivalents of electrons are required by the mechanism for conversion (Figure 4-8). Some other interesting observations came out of this series of experiments: (1) Poly-phenolic compounds were not observed as products of the reaction. Thus, in contrast to the usual terminology of organic chemistry, hydroxyl group substitution has, in this case, a deactivating effect on the aromatic nucleus. (2) Ni-

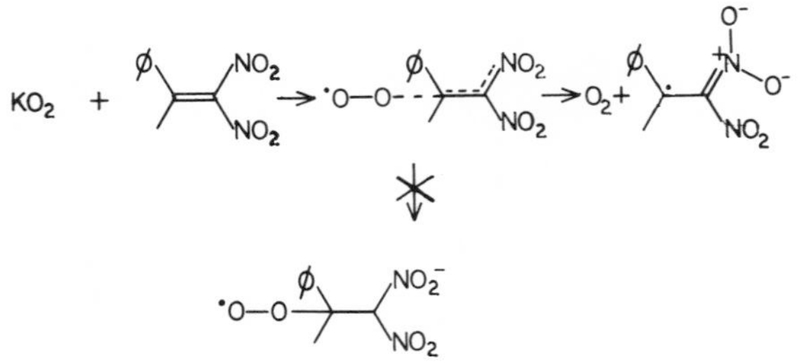

Figure 4-6. Reaction of superoxide ion with a highly electron-deficient olefin.

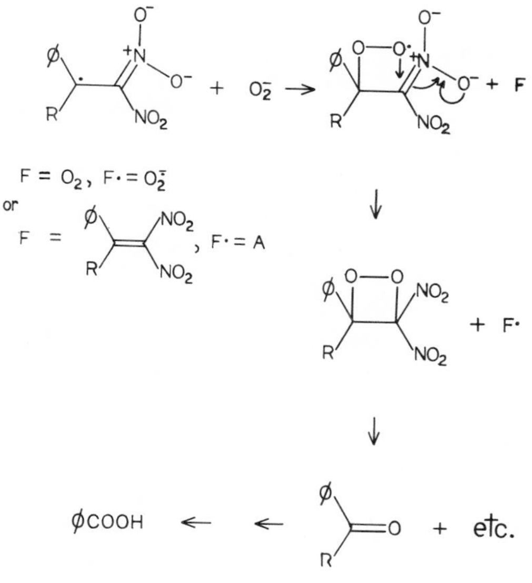

$$F = O_2, \quad F\cdot = O_2^{\overline{\cdot}}$$

or

$$F = \underset{R'}{\overset{\phi}{\diagdown}}C=C\underset{NO_2}{\overset{NO_2}{\diagup}}, \quad F\cdot = A$$

$\phi COOH \leftarrow \leftarrow$ $+$ etc.

Figure 4-7. Mechanism of reaction of superoxide ion with a phenyl dinitro ethylene.

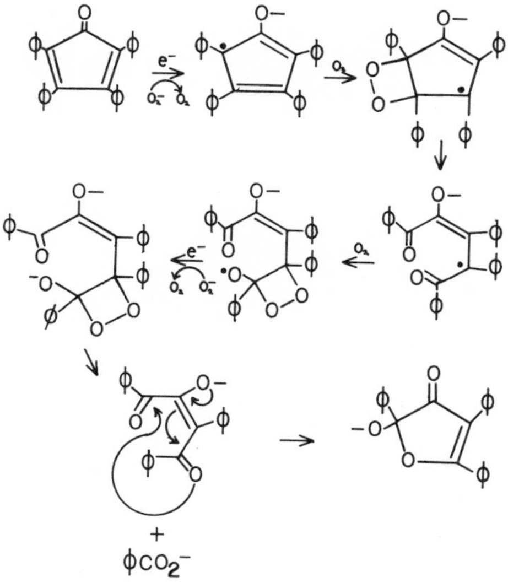

Figure 4-8. Reaction of superoxide ion with tetracyclone.

$+$

ϕCO_2^-

trobenzene (Poupko and Rosenthal, 1973) and *m*-bromonitrobenzene react with superoxide to form a radical detectable by ESR spectroscopy, but the radical does not react further to form phenolic products, while radicals formed from more highly electron-withdrawn aromatic systems do so react. This is exactly contrary to what one would expect considering the attack by molecular oxygen to be electrophilic.

Superoxide anion undergoes several complex reactions involving combinations of these simpler reactions.

Superoxide reacts with the naphthoquinone nucleus of vitamin K to give the epoxide and phthalic acid in modest yields (Figure 4-9) (Saito *et al.*, 1979). The acid is believed to be produced via degradation of the peroxide, as judged from control experiments, though a mechanism for this oxidation is not apparent. The epoxide is thought to be formed by electron donation to the electron-withdrawn olefin and subsequent attack by dioxygen. Reduction of the resulting peroxy radical gives a peroxy anion, which is similar to the addition product of hydrogen peroxide and unsaturated ketone in base, known to yield epoxy ketones (House, 1972). However, this mechanism would seem to be impossible in this case. The solvent was rigorously dry, and there was no other proton source. This being the case, it is difficult to imagine how elimination of OH⁻ from the dianion could occur or how the dianion could remain in solution in benzene.

Two other superoxide reactions are also of direct biochemical interest. It has been reported (San Filippo and Chern, 1976) that 2-halo-, hydroxy-, carbonyl-, or carboxyl-ketones are oxidized with carbon–carbon bond cleavage to the corresponding acids if there is a hydrogen on the same carbon (in the saturated compounds). Thus, 2-hydroxycyclohexanone was oxidized to adipic acid, while 1-hydroxycycloheptanecarboxylic acid did not react. Moreover, α-ketoglutaric acid was converted to succinic acid by this reagent, strengthening the hypothesis

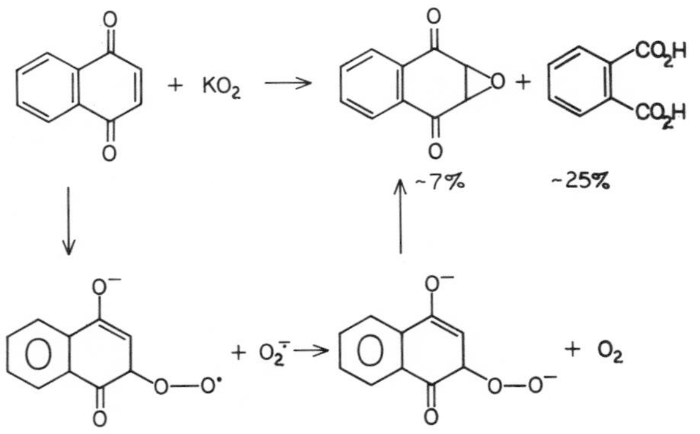

Figure 4-9. Reaction of superoxide ion with 1,4-naphthaquinone.

(see Chapter 19) that the biochemical reaction could occur through an iron–superoxide salt. A mechanism for the uncatalyzed reaction is not established.

Finally, the dangers of superoxide ion in biological systems are emphasized by the reaction of superoxide with an α-tocopherol model compound (Matsumoto and Matsuo, 1977). The reaction observed is shown in Figure 4-10. The reaction was carried out in tetrahydrofuran (THF),* and the yield increased almost 5-fold when performed under an oxygen atmosphere. It seems likely that the initial step in the reaction is hydrogen-atom abstraction from the phenolic function, or proton abstraction, followed by oxidation of the anion, but the remainder of the reaction pathway is unclear. Hydroquinones and catechols are oxidized by superoxide via the semiquinone radical to quinones and more highly oxidized products (Moro-Oka and Foote, 1976). These reactions were done under vigorously dry conditions in both the presence and absence of external oxygen. In view of Sawyer's results (Sawyer and Valentine, 1981), the possibility that the oxidizing agents are actually products of the proton-catalyzed disproportionation of superoxide, namely, dioxygen and hydroperoxide anion, must be examined. Hydroperoxide anion certainly would be expected to be no better an oxidant than superoxide. However, if another proton were available, either from a second hydroquinone molecule, from the hydroquinone anion, or from some impurity, hydrogen peroxide could be the oxidant. Dioxygen could also be the oxidant, though the experimental conditions appear to have minimized this possibility (Moro-Oka and Foote, 1976).

The mechanism offered for the cleavage of the aromatic nucleus in the extensive oxidation of catechols observed on reaction with superoxide is very similar to the mechanism discussed for those analogous biochemical cleavages in which the enzyme rests in the ferrous form before binding dioxygen (Moro-Oka and Foote, 1976).

The enzymatic reaction of superoxide in L-tryptophan oxygenase remains unexplained by these studies. Frimer and Rosenthal (1978) have reported observing a slow reaction with indole, but a mechanistic explanation awaits further studies.

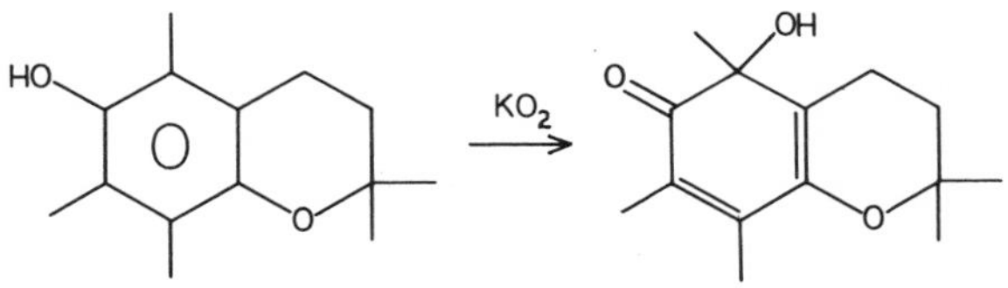

Figure 4-10. Reaction of superoxide ion with a model for α-tocopherol.

* Although superoxide reactions are highly solvent-dependent, membranes and protein clefts may furnish an environment closer to that of THF than that of an aqueous solution.

Superoxide is, in summary, a very complex molecule. It can undergo free-radical or anionic reactions. It can either oxidize or reduce various organic compounds readily. One of the most interesting observations about superoxide is that it does not seem to interact with molecular oxygen. In a control experiment to ensure the validity of their radioactive-label-incorporation data, Frimer and Rosenthal (1978) bubbled air through a solution of potassium superoxide, 18-crown-6, in benzene, in which the superoxide was labeled with ^{18}O. They detected no exchange; i.e., the expected equilibrium did not take place.

$$^{18}O_2^- + {}^{16}O_2 \rightleftharpoons {}^{18}O_2 + {}^{16}O_2^-$$

This is very surprising in that electron donation to other electrophilic substrates is very rapid and that $O_4^- \cdot$ is a well-known species in the gas phase, the thermodynamic data of which are known (Varney et al., 1973).

4.6 Superoxide Dismutase

Biochemical interest in superoxide ion develops primarily from the fact that superoxide ion can be formed as a by-product in certain oxidations. There are enzymes called superoxide dismutases (SOD) that destroy superoxide ion by a disproportionation reaction. The reactions of superoxide are claimed to be cytotoxic, so that the dismutation enzymes would serve primarily as a defensive mechanism (Fridovich, 1978b). In accord with this rationale, it has been shown that (1) the only cells without this enzyme activity are strict obligate anaerobes, (2) increasing an organism's exposure to dioxygen during growth induces an increase in the level of activity of this enzyme, and (3) there is a correlation between a cell's normal exposure to dioxygen and its SOD level (Fridovich, 1978b). However, this neat correlation of oxygen exposure to SOD level has been questioned by Fee (1980). For a further discussion of the role of superoxide ion in oxygen toxicity, see Chapter 20.

Superoxide ion can be formed in the cell by any number of enzymatic and nonenzymatic oxidations by dioxygen. The xanthine oxidase (McCord and Fridovich, 1968) reaction, the autoxidation of hemoglobin (Misra and Fridovich, 1972b; Wever et al., 1973; Wallace et al., 1974), the oxidation of epinephrine (Misra and Fridovich, 1972a; May et al., 1973), NADPH-cytochrome c reductase (Aust et al., 1972), and certain nonheme ion proteins all act as sources of superoxide ion in the cell.

One source of superoxide ion may be from the disproportionation of hydrogen peroxide in basic solution mentioned in Section 4.3 (Wilshire and Sawyer, 1979). The reaction is greatly shifted toward disproportionation by the addition of pyridine, which scavenges the hydroxyl radicals:

$$H^+ + H_2O_2 + HO_2^- \rightleftharpoons HO_2\cdot + \cdot OH + H_2O$$

$$\cdot OH + Pyr\cdot \rightarrow PyrOH$$

There are certainly many compounds that would react with the hydroxyl radical in a biological system.

Both superoxide ion and HO_2 are thermodynamically unstable with respect to dioxygen and hydrogen peroxide. The disproportionation of HO_2 is highly exergonic, with a free energy of -37.5 kcal. However, the disproportionation can be reversed in the presence of tetranitromethane. Hydrogen peroxide and dioxygen combine to give superoxide, which is oxidized by the tetranitromethane. The tetranitromethane is reduced to nitroform (Hodgson and Fridovich, 1973).

As would be expected, the rate is slowest when two negatively charged ions must approach each other. Because the pK of $HO_2\cdot$ is 4.69 and the reaction is fastest between the anion and the acid, the rate of dismutation is fastest at pH of 4.69. In the presence of SOD, the reaction rate, even with O_2^-, is 2×10^9 M^{-1} sec^{-1}. This rate is probably the fastest enzyme-catalyzed reaction known. The decrease in rate on addition of glycerol to the solution probably indicates the reaction to be diffusion-controlled (Rotilio $et\ al.$, 1972a). The rates of the nonenzymatic dismutation reaction to form oxygen and hydrogen peroxide are appreciable, but too slow to prevent damage to the cell (Rabani and Nielsen, 1969; Behar $et\ al.$, 1970; Marklund, 1976):

$$HO_2\cdot + HO_2\cdot \qquad k = 8 \times 10^5\ M^{-1}\ sec^{-1}$$

$$HO_2\cdot + O_2^-\cdot \qquad k = 8 \times 10^7\ M^{-1}\ sec^{-1}$$

$$O_2^-\cdot + O_2^-\cdot \qquad k = 0.2\ M^{-1}\ sec^{-1}$$

The enzymes that catalyze this reaction contain either iron, manganese (of which there are two types), or copper and zinc. These four types of SOD (Fridovich, 1974) all appear to have the same mechanism of action. One of the manganese enzymes is found in mitochondria (Weisiger and Fridovich, 1973; Britton $et\ al.$, 1978) and the other in bacteria (Keele $et\ al.$, 1970; Vance $et\ al.$, 1972). In acidic solution, the Mn(II) can be replaced by Co(II), Ni(II), Zn(II), Fe(II), or Cu(II). Surprisingly, none of these re-formed enzymes has activity. However, activity can be restored by returning the managanese (Ose and Fridovich, 1979). The iron-containing SOD is found in the perisplasmic space of $Escherichia\ coli$ (Yost and Fridovich, 1973; Gregory $et\ al.$, 1973). The manganese- and iron-containing enzymes are built from basic subunits of molecular weight 20,000. Two of these subunits form the bacterial enzyme and four form the mitochondrial enzymes.

The copper–zinc enzyme is found in the cytosol of eukaryotic cells (Steinman and Hill, 1973), but is also found in one bacterium (*Photobacterium leiognathi*) that is symbiotic with a fish (Reichelt *et al.*, 1977). One explanation is that the bacterium obtained the enzyme through a gene transfer from the fish (Fridovich, 1978a).

The copper–zinc enzymes from many eukaryotic sources are blue-green and contain two Cu(II) atoms and two zinc ions per molecule. The absorption peak is at 680 nm (Weser *et al.*, 1971). The molecular weight is 32,000. The complete primary structure is known. There are two identical subunits (Steinman *et al.*, 1974) held together by noncovalent bonds (Abernethy *et al.*, 1974). A crystal-structure determination has shown that the enzyme has a 2-fold axis between the two subunits, which have identical secondary structure. There is extensive β-structure (Thomas *et al.*, 1974). The electron paramagnetic resonance spectrum of the cupric ion (Rotilio *et al.*, 1971, 1972c; Lieberman and Fee, 1973) is best interpreted in terms of distorted tetragonal symmetry around the copper.

Crystal-structure studies (Richardson *et al.*, 1975) of the copper–zinc enzyme have shown that the copper and the zinc are about 6 Å apart. There is one histidine as a common ligand to both metals. There are four histidines around the copper in a distorted planar arrangement and three histidines and an aspartate around the zinc in a tetrahedral arrangement.

The copper–zinc dismutase may be distinguished easily from the manganese and iron enzymes because the former is sensitive to cyanide ion, whereas the latter are not.

Metals can catalyze the dismutation of superoxide by either of two mechanisms. As mentioned for the enzymatic case, metals with the correct oxidation–reduction potentials can alternatively donate and accept one electron to and from the superoxide ion. Another possibility is that the metal (X^{n+}) is reduced by the superoxide ion to an oxidation level ($X^{(n-1)+}$) that will bind dioxygen:

$$X^{n+} + O_2^- \cdot \rightarrow X^{(n-1)} - O_2 \cdot$$

$$X^{(n-1)} - O_2 \cdot - O_2^- \cdot \rightarrow X^{(n-1)} - O_2^- + O_2$$

$$X^{(n-1)} - O_2^- + H^+ \rightarrow X^{n+} + HO_2^-$$

The mechanism of the enzymatic reaction appears to be the first one in which superoxide ions alternately donate and accept one electron from the dismutase (SOD):

$$O_2^- \cdot + SOD \rightarrow SOD^- + O_2$$

$$O_2^- \cdot + 2H^+ + SOD^- \rightarrow H_2O_2$$

The copper in the copper–zinc enzyme is oxidized and reduced between the $1+$ and $2+$ oxidation levels. The enzyme form called SOD would represent the $2+$ copper and SOD$^-$ the $1+$ copper. There are two atoms of copper per molecule of enzyme, and both of these will function catalytically at higher concentrations of substrate. If SOD^{2-} represents doubly reduced enzyme, the following reaction will also occur (Klug-Roth et al., 1973):

$$SOD^- + O_2^- \rightarrow SOD^{2-} + O_2$$

$$SOD^{2-} + O_2^- \rightarrow SOD^- + H_2O_2$$

The potential for the reduction of dioxygen is 0.16 V (when the standard state of dioxygen is taken as 1 M) and the potential for the reduction of superoxide ion to hydrogen peroxide is $+0.89$ at pH 7. Thus, the catalyst should have a potential somewhere between these two values (Sawyer and Valentine, 1981). The potentials for all the SOD enzymes fall in a narrow range between these limits as predicted by Sawyer and Valentine (1981). (The potentials of the various SOD enzymes are $+0.26$ (Cu-Zn), $+0.31$ (Mn), $+0.29$ (Fe) and $+0.23$V (Fe).

The cupric form of the enzyme can be reduced to the doubly reduced form by hydrogen peroxide (Rotilio et al., 1972b, 1973; Bannister et al., 1973):

$$SOD + H_2O_2 \rightarrow SOD^{2-} + 2H^+ + O_2$$

This reaction results in the loss of the visible 680 nm band.

The manganese and iron enzymes (Lavelle et al., 1977) are oxidized and reduced between the $2+$ and $3+$ oxidation states (Pick et al., 1974). Certain iron porphyrin compounds show SOD activity. These compounds have about 1–3% of the activity of the enzymes.

The question arises as to whether the SOD would produce singlet or triplet dioxygen.* One could conceive of a mechanism in which an oxidant with an unpaired electron abstracts an electron from superoxide to form triplet dioxygen without a change in total spin:

$$Cat\ (\uparrow) + O_2^-\ (\uparrow \downarrow \uparrow) \rightarrow Cat\ (\uparrow \downarrow) + {}^3O_2\ (\uparrow \uparrow)$$

This is important because forming highly reactive singlet dioxygen would be harmful to the cell. This step would be followed by a one-electron reduction of superoxide ion by the catalyst:

$$H^+ + O_2^-\ (\uparrow \downarrow \uparrow) + Cat\ (\uparrow \downarrow) \rightarrow Cat\ (\uparrow) + HO_2^-\ (\uparrow \downarrow \uparrow \downarrow)$$

---

* Note the discussion on the product of noncatalyzed superoxide dismutation in Section 4.1.

References

Abernethy, J. L., Steinman, H. M., and Hill, R. L., 1974. Bovine erythrocyte superoxide dismutase: Subunit structure and sequence location of the intrasubunit disulfide bond, *J Biol. Chem.* 249:7339–7347.

Anbar, M., Meyerstein, D., and Neta, P., 1966. Reactivity of aliphatic compounds towards hydroxyl radicals, *J. Am. Chem. Soc. (B)* 1966:742–747.

Andrews, L., 1968. Matrix infrared spectrum and bonding in the lithium superoxide molecule, L_1O_2, *J. Am. Chem. Soc.* 103:4965–4966.

Aubry, J. M., Rigaudy, J., Ferradini, C., and Pucheault, J., 1981. A search for singlet oxygen in disproportionation of superoxide anion, *J. Am. Chem. Soc.* 103:4965–4966.

Audley, G. J., Baulch, D. L., and Campbell, I. M., 1981. Gas-phase reactions of hydroxyl radicals with aldehyde in flowing H_2O_2 + NO_2 + CO mixtures, *J. Chem. Soc. Faraday Trans. I* 1981:2451–2549.

Aust, S. E., Roerig, D. L., and Pederson, T. C. P., 1972. Evidence for superoxide generation by NADPH-cytochrome *c* reductase of rat liver microsomes, *Biochem. Biophys. Res. Commun.* 47:1133–1137.

Azzi, A., Montecrucco, C. and Richiter, C., 1975. The use of acetylated ferricytochrome *c* for the detection of superoxide radicals produced in biological membranes, *Biochem. Biophys. Res. Commun.* 65:597–603.

Bannister, J. V., Bannister, W. H., Bray, R. C., Fielden, E. M., Roberts, P. B., and Rotilio, G., 1973. The superoxide dismutase activity of human erythrocuprein, *FEBS Lett.* 32:303–306.

Beauchamp, C., and Fridovich, L., 1970. A mechanism of production of ethylene from methional: The generation of hydroxyl radical by xanthine oxidase, *J. Biol. Chem.* 245:4641–4646.

Behar, D., Czapski, G., Rabini, J., Dorfman, L. M., and Schwarz, H., 1970. The acid dissociation constant and decay kinetics of the perhydroxyl radical, *J. Phys. Chem.* 74:3209–3213.

Bellus, D., 1979. Physical quenchers of singlet molecular oxygen, in *Advances in Photochemistry, Vol. 11*, J. M. Pitts, Jr., G. S. Hammond, K. Gollnick, and D. Grosjean (eds.), John Wiley, New York, pp. 105–205.

Bielski, B. H. J., 1978. Reevaluation of the spectral and kinetic properties of HO_2 and O_2^- free radicals, *Photochem. Photobiol.* 28:645–649.

Bielski, B. H. J., and Allen, A. D., 1977. Mechanism of the disproportionation of superoxide radicals, *J. Phys. Chem.* 81:1048–1050.

Bielski, B. H. J., and Richter, H. W., 1977. A study of the superoxide radical chemistry by stopped-flow radiology and radiation-induced oxygen consumption, *J. Am. Chem. Soc.* 99:3019–3023.

Bielski, B. H. J., and Saito, E., 1971. Deuterium isotope effect on the decay kinetics of perhydroxyl radical, *J. Phys. Chem.* 75:2263–2266.

Bors, W., Saran, M., Longfelder, E., Michel, C., Fuchs, C., and Frenzel, C., 1978. Detection of oxygen radicals in biological systems, *Photochem. Photobiol.* 28:629–638.

Bors, W., Michel, C., and Saran, M., 1979. Superoxide anions do not react with hydroperoxides, *FEBS Lett.* 107:403–406.

Britton, L., Malinouski, D. P., and Fridovich, I., 1978. Superoxide dismutase and oxygen metabolism in *Streptococcus faecalis* and comparisons with other organisms, *J. Bacteriol.* 134:229–236.

Chern, C.-I., and San Filippo, J., 1977. The reaction of superoxide with hydrazines, hydrazones, and related compounds, *J. Org. Chem.* 42:178–180.

Corey, E. J., Nicolaou, K. C., Shibasaki, M., Machida, Y., and Shiner, C. S., 1975. Superoxide ion as a synthetically useful oxygen nucleophile, *Tetrahedron Lett.* 1975:3183–3186.

Czapski, G., 1978, *Photochem. Photobiol.* 28:926 (in the Discussion).

Czapski, G., and Ilan, Y. A., 1978. On the generation of the hydroxylation agent from superoxide radical: Can the Haber–Weiss reaction be the source of ·OH radicals?, *Photochem. Photobiol.* 28:651–653.

Danen, W. D. and Arudi, R. L., 1978. Generation of singlet oxygen in the reaction of superoxide anion radical with diacyl peroxides, *J. Am. Chem. Soc.* 100:3944–3945.

Danen, W. C., and Warner, R. J., 1977. The remarkable nucleophilicity of superoxide anion radical: Rate constants for reactions of superoxide with aliphatic bromides, *Tetrahedron Lett.* 1977:989–992.

Darnall, K., Winer, A. M., Lloyd, A. C., and Pitts, J. N., Jr., 1976. Relative rate constants for the reaction of OH radicals with selected C_6 and C_7 alkanes and alkenes at 305 ± 2°K, *Chem. Phys. Lett.* 44:415–418.

Darnall, K. R., Atkinson, R., and Pitts, J. N., Jr., 1978. Rate constants for the reaction of the OH radical with selected alkanes at 300°K, *J. Phys. Chem.* 82:1581–1584.

Fee, J. A., 1980. A comment on the hypothesis that oxygen toxicity is mediated by superoxide, in *Oxygen and Life,* Burlington House, London, Royal Society of London, pp. 77–97.

Fee, J. A., and Hildebrand, P. G., 1974. On the development of a well-defined source of superoxide ion for studies with biological studies, *FEBS Lett.* 39:79–82.

Ferradini, C., Foos, J., Houee, C., and Pauchault, J., 1978. The reaction between superoxide anion and hydrogen peroxide, *Photochem. Photobiol.* 28:697–700.

Fielden, E. M., Cohen, G., Bors, W., and Czapski, G., 1978. *Photochem. Photobiol.* 28:674–675 (in the Discussion).

Finkelstein, E., Rosen, G. M., and Rauchman, E. J., 1980. Spin trapping of superoxide and hydroxyl radical: Practical aspects, *Arch. Biochem. Biophys.* 200:1–6.

Foote, C. S., 1978. Untitled paper, *Photochem. Photobiol.* 28:718–740.

Foote, C. S., Shook, F. C., and Abakerli, R. A., 1980a. Chemistry of superoxide ion. 4. Singlet oxygen is not a major product of dismutation, *J. Am. Chem. Soc.* 102:2503–2504.

Foote, C. S., Abakerli, R. B., Clough, R. L., and Shook, F. C., 1980b. On the question of singlet oxygen production in leucocytes, macrophages, and the dismutation of superoxide anion, in *Biological and Clinical Aspects of Superoxide and Superoxide Dismutase,* W. H. Bannister and J. V. Bannister (eds.), Proceedings of the Federation of European Biochemical Societies Symposium No. 62, Elsevier, New York, pp. 222–230.

Fridovich, I., 1974. Superoxide dismutase, *Adv. Enzymol.* 41:35–97.

Fridovich, I., 1978a. The biology of oxygen radicals, *Science* 201:875–880.

Fridovich, I., 1978b. Superoxide radicals, superoxide dismutases, and the aerobic lifestyle, *Photochem. Photobiol.* 28:733–740.

Frimer, A. A., and Rosenthal, I., 1978. Chemical reactions of superoxide anion radical in aprotic solvents, *Photochem. Photobiol.* 28:711–719 (and Discussion following).

Frimer, A. A., Rosenthal, I., and Hoz, S., 1977. The reaction of superoxide anion radical with electron poor olefins, *Tetrahedron Lett.* 1977:4631–4634.

Gibian, M. J., and Ungermann, T., 1976. Reaction of *tert*-butyl hydroperoxide anion with dimethyl sulfoxide: On the pathway of the superoxide–alkyl halide reaction, *J. Org. Chem.* 41:2500–2502.

Gregory, E. M., Yost, F. J., Jr., and Fridovich, I., 1973. Superoxide dismutases of *Escherichia coli:* Intracellular localizations and functions, *J. Bacteriol.* 115:987–991.

Guiraud, H. J., and Foote, C. S., 1976. Chemistry of superoxide ion. III. Quenching of singlet oxygen, *J. Am. Chem. Soc.* 98:1984–1986.

Haber, F., and Weiss, J., 1934. The catalytic decomposition of hydrogen peroxide by iron salts, *Proc. R. Soc. London Ser. A* 147:332–351.

Heicklen, J., 1981. The correlation of rate coefficients for H-atom abstraction by HO radicals with C–H bond dissociation enthalpies, *Int. J. Chem. Kinet.* 13:651–665.

Hiller, K. O., Mastoch, B., Gobt, M., and Asmus, L. D., 1981. Mechanism of OH radical induced oxidation of methionine in aqueous solution, *J. Am. Chem. Soc.* 103:2734–2743.

Hirata, F., and Hayaishi, O., 1975. Studies on indoleamine 2,3-dioxygenase. I. Superoxide anion as substrate, *J. Biol. Chem.* 250:5960–5966.

Hodgson, E. K., and Fridovich, I., 1973. Reversal of the superoxide dismutase reaction, *Biochem. Biophys. Res. Commun.* 54:270–274.

House, H. O., 1972. *Modern Synthetic Reactions,* Benjamin/Cummings, Menlo Park, California, pp. 307–312.

Howard, D. J., 1980. Kinetic study of the equilibrium $HO_2 + NO \rightleftharpoons OH + NO_2$ and the thermochemistry of HO_2, *J. Am. Chem. Soc.* 102:6937–6941.

Ilan, Y. A., Meisel, D., and Czapski, G., 1974. The redox potential of the O_2–O_2 system in aqueous media, *Isr. J. Chem.* 12:891–895.

Jeong, K.-M., and Kaufman, F., 1982. Kinetics of the reaction of hydroxyl radical with methane and with nine Cl- and F-substituted methanes. 2. Calculation of rate parameters as a test of transition state theory, *J. Phys. Chem.* 88:1816–1821.

Johnson, R. A., and Nidy, E. G., 1975. Superoxide chemistry: A convenient synthesis of dialkyl peroxides, *J. Org. Chem.* 40:1680–1681.

Keele, B. B., Jr., McCord, J. M., and Fridovich, I., 1970. Superoxide dismustase from *Escherichia coli* B, *J. Biol. Chem.* 245:6176–6181.

Khan, A. U., 1978. Activated oxygen: Singlet molecular oxygen and superoxide anion, *Photochem. Photobiol.* 28:615–626 (and Dicussion following).

Klug-Roth, D., Fridovich, I., and Rabani, J., 1973. Pulse radiolytic investigations of superoxide catalyzed disproportionation: Mechanism for bovine superoxide dismutase, *J. Am. Chem. Soc.* 95:2786–2790.

Kobayashi, S., and Ando, W., 1979. Co-oxidation of 1,3-diphenylisobenzofuran by the Haber–Weiss reaction: Is singlet oxygen concerned in this reaction?, *Biochem. Biophys. Res. Commun.* 188:676–681.

Koppenol, W. H., and Butler, J., 1977. Mechanism of reactions involving singlet oxygen and the superoxide anion, *FEBS Lett.* 83:1–6.

Koppenol, W. H., Butler, J., and vanLeeuwen, J. W., 1978. The Haber–Weiss cycle, *Biochem. Biophys. Res. Commun.* 188:655–658 (and Discussion following).

Lavelle, F., McAdam, M. E., Fielden, E. M., Roberts, P. B., Puget, K., and Michelson, A. M., 1977. A pulse-radiolysis study of the catalytic mechanism of the iron-containing superoxide dismutase from *Photobacterium leiognathi*, *Biochem. J.* 161:2–11.

Lee-Ruff, E., 1977. The organic chemistry of superoxide, *Chem. Soc. Rev.* 6:195–214.

Lieberman, R. A., and Fee, J. A., 1973. Preliminary report on the electron paramagnetic resonance spectra of singlet crystals of bovine erythrocyte superoxide dismutase, *J. Biol. Chem.* 248:7617–7619.

Liotta, C. L., and Harris, H. P., 1974. The chemistry of "naked" anions. I. Reactions of the 18-crown-6 complex of potassium fluoride with organic substrates in aprotic solvents, *J. Am. Chem. Soc.* 96:2250–2252.

Marklund, S., 1976. Spectrophotometric study of spontaneous disproportionation of superoxide anion radical and sensitive direct assay for superoxide dismutase, *J. Biol. Chem.* 251:7504–7507.

Matsumoto, S., and Matsuo, M., 1977. The reaction of a α-tocopherol model compound with KO_2, a new oxidation product of 6-hydroxy-2,2,5,7,8-penta-methylchroman, *Tetrahedron Lett.* 1977:1999–2000.

May, S. W., Abbott, B. J., and Felix, A., 1973. On the role of superoxide in reactions catalyzed by rubredoxin of *Pseudomonas oleovorans*, *Biochem. Biophys. Res. Commun.* 54:1540–1545.

Mayer, R., Widom, J., and Que, Jr., L., 1980, Involvement of superoxide in the reactions of the catechol dioxygenases, *Biochem. Biophys. Res. Commun.* 92:285–291.

McCandish, E., Miksztal, A. R., Nappa, M., Springer, A. Q., Valentine, J. S., Stong, J. D., and Spiro, T. G., 1980. Reactions of superoxide with iron porphyrins in aprotic solvents: A high spin ferric porphyrin peroxo complex, *J. Am. Chem. Soc.* 102:4268–4271.

McCord, J. M., and Day, E. D., 1978. Superoxide dependent production of hydroxyl radical catalyzed by iron–EDTA complex, *FEBS Lett.* 86:139–142.

McCord, J. M., and Fridovich, I., 1968. The reduction of cytochrome *c* by milk xanthine oxidase, *J. Biol. Chem.* 243:5753–5760.

McElroy, A. D., and Hashman, J. S., 1964. Synthesis of tetramethylammonium superoxide, *Inorg. Chem.* 3:1798–1799.

McIsaac, J. E., Subbaraman, L. R., Subbaraman, J., Mulhausen, H. A., and Behrman, E. J., 1972. The nucleophilic reactivity of peroxy anions, *J. Org. Chem.* 37:1037–1041.

Merritt, M. V., and Johnson, R. A., 1977. Spin trapping, alkylperoxy radicals, and superoxide–alkyl halide reactions, *J. Am. Chem. Soc.* 99:3713–3719.

Merritt, H., and Sawyer, D. T., 1970. Electrochemical studies of the reactivity of superoxide ion with several alkyl halides in dimethyl sulfoxide, *J. Org. Chem.* 35:2157–2159.

Milligan, D. E., and Jacox, M. E., 1963. Infrared spectroscopic evidence for the species HO_2, *J. Chem. Phys.* 38:2627–2631.

Misra, H., and Fridovich, I., 1972a. The role of superoxide anion in the autoxidation of epinephrine and a simple assay for superoxide dismustase, *J. Biol. Chem.* 247:3170–3175.

Misra, H., and Fridovich, I., 1972b. The generation of superoxide radical during the autoxidation of hemoglobin, *J. Biol. Chem.* 247:6960–6962.

Moro-Oka, Y., and Foote, C. S., 1976. Chemistry of superoxide ion. I. Oxidation of 3,5-di-*tert*-butylcatechol with KO_2, *J. Am. Chem. Soc.* 98:1510–1514.

Nilsson, R. and Kearns, D. R., 1974. Some useful heterogeneous systems for photosensitized generation of singlet oxygen, *Photochem and Photobiol.* 19:181–184.

Ose, D. E., and Fridovich, I. 1979. A manganese-containing superoxide dismutase from *Escherichia coli:* A reversible resolution and metal replacements, *Arch. Biochem. Biophys.* 194:360–364.

Overend, R., and Paraskevopoulos, G., 1978. Rates of OH radical reaction. 4. Reactions with methanol, ethanol, 1-propanol, and 2-propanol at 296°K, *J. Phys. Chem.* 82:1329–1333.

Ozawa, T., Hanaki, A., and Yamamoto, H. 1977. On a spectrally well-defined and stable source of superoxide ion, O_2^-, *FEBS Lett.* 74:99–102.

Paraskevopoulos, G., and Nip, W. S., 1980. Rates of OH radical reactions. VII. Reactions of OH and OD radicals with $n\text{-}C_4H_{10}$, $n\text{-}C_4D_{10}$, H_2 and D_2, and of OH with $n\text{-}C_5H_{12}$ at 297°K, *Can. J. Chem.* 58:2146–2149.

Pedersen, C. J., 1967. Cyclic polyethers and their complexes with metal salts, *J. Am. Chem. Soc.* 89:7017–7036.

Peters, J. W., and Foote, C. S., 1976. Chemistry of superoxide ion. II. Reaction with hydroperoxides, *J. Am. Chem. Soc.* 98:873–875.

Pick, M., Rabani, J., Yost, F., and Fridovich, I., 1974. The catalytic mechanism of the manganese-containing superoxide dismutase of *Escherichia coli* studied by pulse radiolysis, *J. Am. Chem. Soc.* 96:7329–7333.

Poupko, R., and Rosenthal, I., 1973. Electron transfer interactions between superoxide ion and organic compounds, *J. Chem. Phys.* 77:1722–1724.

Pryor, W. A., and Tang, R. H., 1978. Ethylene formation from methional, *Biochem. Biophys. Res. Commun.* 81:498–503.

Rabani, J., and Nielsen, S. O., 1969. Absorption spectrum and decay kinetics of O_2^- and HO_2 in aqueous solutions by pulse radiolysis, *J. Phys. Chem.* 73:3736–3744.

Reichelt, J. L., Nealson, K., and Hastings, J. W., 1977. The specificity of symbiosis: Pony fish and luminescent bacteria, *Arch. Microbiol.* 112:157–161.

Richardson, J. S., Thomas, K. A., Rubin, B. H., and Richards, D. C., 1975. Crystal structure of bovine Cu, Zn in superoxide dismutase at 3 Å resolution: Chain tracing and metal ligands, *Proc. Natl. Acad. Sci. U.S.A.* 72:1349–1353.

Roberts, J. L., Jr., and Sawyer, D. T., 1981. Facile degradation by superoxide ion of carbon tetrachloride, chloroform, methylene chloride and *p,p'*-DDT in aprotic media, *J. Am. Chem. Soc.* 103:712–714.

Rotilio, G., Finazzi-Agro, A., Calabrese, L., Bossa, F., Guerrieri, P., and Mondovi, B., 1971. Studies of the metal sites of copper proteins, ligands of copper in hemocuprein, *Biochemistry* 10:616–621.

Rotilio, G., Bray, R. C., and Fielden, E. M., 1972a. A pulse radiolysis study of superoxide dismutase, *Biochim. Biophys. Acta* 286:605–609.

Rotilio, G., Calabrese, L., Bossa, F., Barra, D., Finazzi-Agro, A., and Mondovi, B., 1972b. Properties of the apo-protein and role of copper and zinc in protein conformation and enzyme activity of bovine superoxide dismutase, *Biochemistry,* 11:2182–2187.

Rotilio, G., Morpurgo, L., Gioviagnoli, C., Calabrese, L., and Mondovi, B., 1972c. Studies of the metal sites of copper proteins: Symmetry of copper in bovine superoxide dismutase and its functional significance, *Biochemistry* 11:2187–2192.

Rotilio, G., Morpurgo, L., Calabresse, L., and Mondovi, B., 1973. On the mechanism of superoxide dismutase reaction of the bovine enzyme with hydrogen peroxide and ferrocyanide, *Biochim. Biophys. Acta* 302:229–235.

Sagae, H., Fujihira, M., Osa, T., and Lund, H., 1977. Oxidation of nitroalkylbenzene with electro-generated superoxide ion, *Chem. Soc. Jpn. Chem. Lett.* 1977:793–796.

Sagae, H., Fujihira, M., Lund, H., and Osa, T., 1980. Oxidation of nitrotoluenes with electro-generated superoxide ion, *Bull. Chem. Soc. Jpn.* 53:1537–1541.

Saito, I., Otsuki, T., and Matsuura, T., 1979. The reaction of superoxide ion with vitamin K_1 and its related compounds, *Tetrahedron Lett.* 1979:1693–1696.

San Filippo, J., and Chern, C.-I., 1976. Oxidative cleavage of α-keto, α-hydroxy, and α-halo ketones, esters, and carboxylic acids by superoxide, *J. Org. Chem.* 41:1077–1078.

San Filippo, J., Romano, L. J., Chern, C.-I., and Valentine, J. S., 1976. Cleavage of esters by superoxide, *J. Org. Chem.* 41:586–588.

Sanka, J., and Martinsons, V., 1968. Chemical properties of potassium peroxide, *Chem. Abstr.* 68:45846A.

Sawyer, D. T., and Gibian, M. J., 1979. The chemistry of superoxide ion, *Tetrahedron* 35:1471–1481.

Sawyer, D. T., and Valentine, J. S., 1981. How super is superoxide?, *Acc. Chem. Res.* 14:393–399.

Sawyer, D. T., Gibian, M. J., Morrison, M. M., and Seo, E. T., 1978. On the chemical reactivity of superoxide ion, *J. Am. Chem. Soc.* 100:627–628.

Scully, F. E., Jr., and Davis, R. C., 1978. Superoxide in organic synthesis: A new mild method for the oxidation of amines to carbonyls via *N*-chloroamines, *J. Org. Chem.* 43:1467–1468.

Seyb, E., and Kleinberg, J., 1951. Determination of superoxide oxygen, *Anal. Chem.* 23:115–117.

Stanley, J. P., 1980. Reactions of superoxide with peroxides, *J. Org. Chem.* 45:1413–1418.

Steinman, H. M., and Hill, R. L., 1973. Sequence homologies among bacterial and mitochondrial superoxide dismutases, *Proc. Natl. Acad. Sci. U.S.A.* 70:3725–3729.

Steinman, H. M., Vishweshwar, R. N., Abernethy, J. L., and Hill, R. L., 1974. Bovine erythrocyte superoxide dismutase: Complete amino acid sequence, *J. Biol. Chem.* 249:7326–7338.

Tezuka, M., Ohkatsu, Y., and Osa, T., 1975. Reactivity of electrogenerated superoxide ion. I. Autoxidation of 9,10-dihydroanthracene, *Bull. Chem. Soc. Jpn.* 48:1471–1474.

Thomas, K. A., Rubin, B. H., Bier, J. C., Richardson, J. S., and Richardson, D. C., 1974. The crystal structure of bovine Cu^{2+}, Zn^{2+} superoxide dismutase at 5.5 Å resolution, *J. Biol. Chem.* 249:5677–5683.

Thomas, M. J., Mehl, K. S., and Pryor, W. A., 1978. The role of superoxide anion in the xanthine oxidase-induced autoxidation of linoleic acid, *Biochem. Biophys. Res. Commun.* 83:927–932.

Valentine, J. S., and Curtis, A. B., 1975. A convenient preparation of solutions of superoxide anion and the reaction of superoxide anion with a copper(II) complex, *J. Am. Chem. Soc.* 97:224–226.

Vance, D. G., Keele, B. B., Jr., and Rajagopalan, K. V., 1972. Superoxide dismutase from *Streptococcus mutans, J. Biol. Chem.* 247:4782–4786.

Varney, R. N., Pahl, M., and Mark, T. D., 1973. Properties of the ionic system $N_4^+\cdot$, $O_4^+\cdot$, and $O_4^-\cdot$, *Acta Phys. Austriaca* 38:287–294.

Wallace, M. J., Maxwell, J. C., and Caughey, W. S., 1974. The mechanisms of hemoglobin autoxidation: Evidence for proton-assisted nucleophilic displacement of superoxide by anions, *Biochem. Biophys. Res. Commun.* 57:1104–1110.

Weinstein, J., and Bielski, B. H. J., 1979. Kinetics of the interaction of HO_2 and O_2^- radicals with hydrogen peroxide: The Haber–Weiss reaction, *J. Am. Chem. Soc.* 101:58–62.

Weisiger, R. A., and Fridovich, I., 1973. Superoxide dismutase, *J. Biol. Chem.* 248:3582–3592.

Weser, U., Bunnenberg, E., Cammack, R., Djerassi, C., Flohe, L., Thomas, G., and Voelter, W., 1971. A study on purified bovine erythrocuprein, *Biochim. Biophys. Acta* 243:203–213.

Wever, R., Oudega, B., and Van Gelder, B. F., 1973. Generation of superoxide radicals during the autoxidation of mammalian oxyhemoglobin, *Biochim. Biophys. Acta* 302:475–478.

Wilshire, J., and Sawyer, D. T., 1979. Redox chemistry of dioxygen species, *Acc. Chem. Res.* 12:105–110.

Yost, F. J. and Fridovich, I., 1973. An iron containing superoxide dismutase from *Escherichia coli*, *J. Biol. Chem.* 248:4905–4908.

Dialkyl Peroxides

5

5.1 Introduction (Table 5-1)

Hydroperoxides are common intermediates in the autoxidation of hydrocarbons, lipids, and other organic compounds. The chemistry of these compounds is discussed in Chapter 2 in connection with the autoxidation of hydrocarbons. The dialkyl peroxides are discussed in this chapter.

Dialkyl peroxides are prepared by the alkylation reactions of hydrogen peroxide or of peroxide salts. Hydrogen peroxide can be alkylated with an alkyl bromide in the presence of silver ion (Porter *et al.*, 1979, 1980). Dialkyl peroxides may also be prepared by the reaction of sodium superoxide with an alkyl bromide (Johnson *et al.*, 1977) or by the reaction of an alkyl hydroperoxide (M. Salomon *et al.*, 1976) or of bis(tri-*n*-butyl-tin) peroxide with a secondary alkyl trifluoromethane sulfonate (R. G. Salomon and M. F. Salomon, 1977). Dialkyl peroxides may also be prepared by the peroxymercuration of monosubstituted ethylenes (Ballard and Bloodworth, 1971):

$$R—CH{=}CH_2 + t\text{-BuOOH} + Hg(OAc)_2 \rightarrow$$
$$t\text{-BuOOCHR—CH}_2\text{—HgOAc} + OAc^-$$

The acetoxy mercury group may be removed with sodium borohydride.

5.2 Naturally Occurring Dialkyl Peroxides Other Than Prostaglandin Intermediates

Dialkyl peroxides are found primarily in seaweed and sponges. The great

75

Table 5-1. Properties of Peroxide

Property		Reference
HOOH		Blair and Goddard (1982)
O–H bond distance	0.967 Å	
O–O bond distance	1.464 Å	
O–O–H angle	99.9°	
Dihedral angle	119.1°	
CH$_3$OOH		Blair and Goddard (1982)
O–H bond distance	0.967 Å	
O–O bond distance	1.452 Å	
O–C bond distance	1.446 Å	
O–O–H angle	99.5°	
Dihedral angle	126°	
O$_2^{2-}$ frequency ≈ 770 cm^{-1}		Sawyer and Gibian (1979)
H$_2$O$_2$ frequency (O–O) 880 cm^{-1}		Milligan and Jacox (1963)

majority of these compounds have no known biological function and are of interest only as chemical curiosities. An exception is the endoperoxides that are intermediates in prostaglandin biosynthesis. Because of their importance, the prostaglandins are discussed separately.

Many types of peroxides are found in natural sources. These have been reviewed by W. Adam and Bloodworth (1978), but we shall mention a few here. The structures of some typical natural dialkyl peroxides and one hydroperoxide are shown in Figures 5-1 to 5-9. Some of these compounds may be the products of singlet-dioxygen trapping reactions. Many of the molds and seaweeds that contain cyclic peroxides are highly colored with dyes that are good photosensitizers to convert ground-state dioxygen to singlet dioxygen. The dye in *Penicillium rubrum,* called mitorbrium, is about 20% as effective as methylene blue

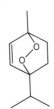

Figure 5-1. Ascaridol, a component of *Chenopodium* oil.

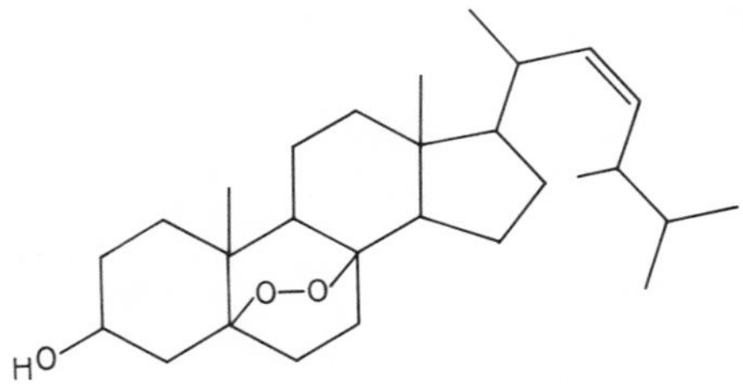

Figure 5-2. Ergosterol peroxide, a peroxide found in *Aspergillus fumigatus* (Wieland and Prelog, 1947) and also in *Penicillium rubrum* and *Gibberella fujikoroi* (Bates and Reid, 1976).

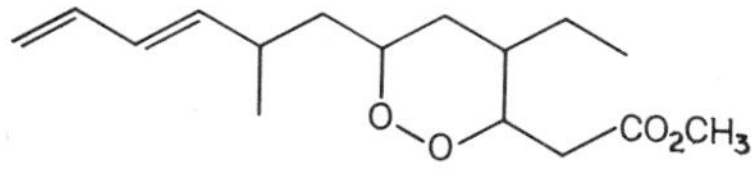

Figure 5-3. Plakortin, an antibiotic found in *Pla-korita* (Higgs and Faulkner, 1978).

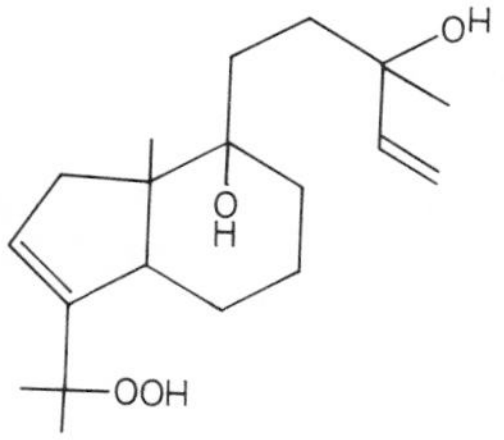

Figure 5-4. Neococinndiol hydroperoxide, a peroxide found in red seaweed [*Laurencia synderiae* (Howard *et al.*, 1977)]. The precursor is probably cocinndiol (Figure 5-5). Neococinndiol hydroperoxide is believed to be formed enzymatically with oxidation of the resulting olefin to the allyl hydroperoxide.

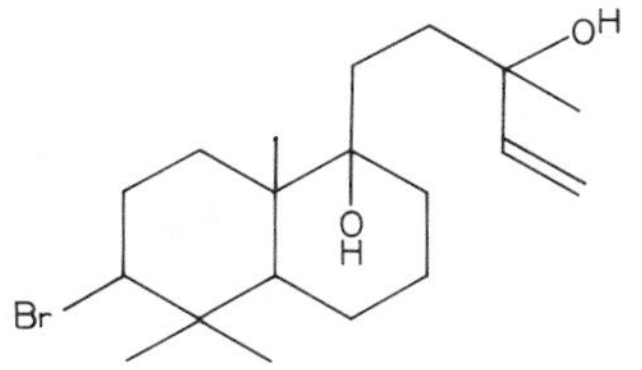

Figure 5-5. Cocinndiol, the precursor of neococinndiol hydroperoxide.

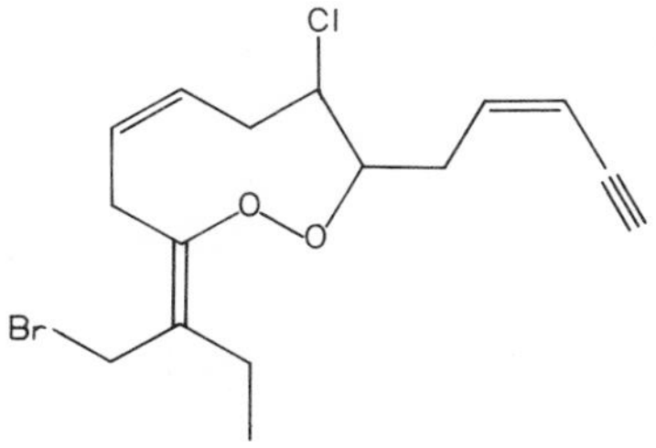

Figure 5-6. Rhodophytin, a vinyl peroxide found in red seaweed [*Laurencia* sp. (Fenical, 1974)].

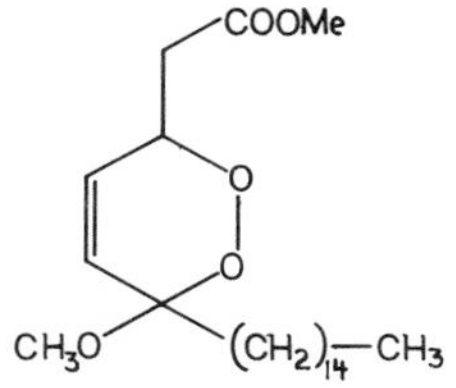

Figure 5-7. A peroxyketal found in a marine sponge [*Chondrilla* (Wells, 1976)].

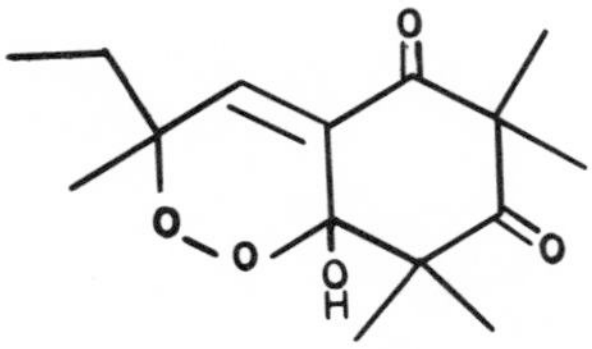

Figure 5-8. Peroxyketals found in *Eucalyptis grandis*. There are two stereoisomers.

in forming singlet dioxygen. Bikarverin is a dye in *Gibberella fujikoroi* that may also be a singlet-dioxygen sensitizer. The singlet dioxygen would be trapped by dienes to form the cyclic peroxides that are found in the organisms.

Peroxides may also be formed enzymatically or may even be on the biosynthetic pathway to other products. It is interesting to note that a chemical synthesis of substituted furans (Figure 5-10) (Kondo and Matsumoto, 1976) utilizes cyclic endoperoxides.

Because olefins can be photochemically oxidized to peroxides, it is often difficult to determine whether these compounds are produced in the organism or formed during isolation procedures. The first peroxide to be isolated was ascaridol (Figure 5-1) in 1908 from *Chenopodium oil*. It is probably formed photochemically in the leaves from the dioxygen oxidation of α-terpinene. Ergosterol peroxide (Figure 5-2) is formed in the dark in molds, but the production is greatly increased by light. Hence, it is probably formed by both enzymatic and photochemical oxidation (H. K. Adam *et al.*, 1967).

A few functions have been proposed for peroxides in nature. Peroxyketals (Figure 5-8) found in eucalyptus act as inhibitors for root formation on cuttings

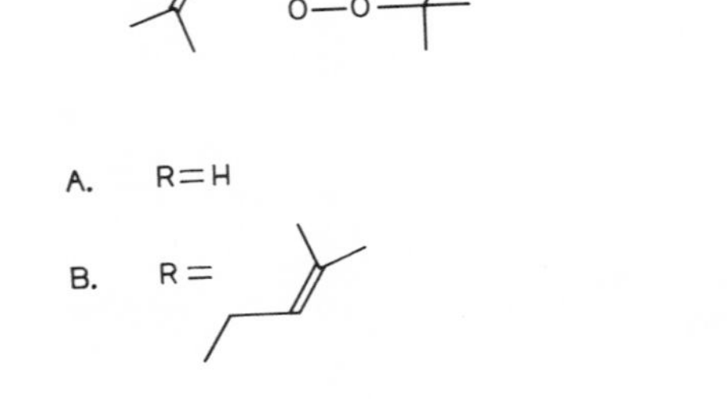

A.   R=H

B.   R=

Figure 5-9. Peroxides found in *Penicillium* and *Aspergillus* species. (A) Verruculogen (Fayos *et al.*, 1974); (B) fumitremorgan A (Eickman *et al.*, 1975).

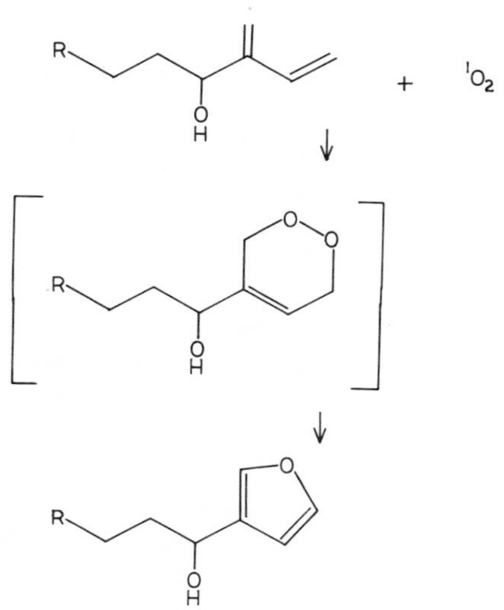

Figure 5-10. A chemical synthesis of a substituted furan that proceeds through a cyclic peroxide.

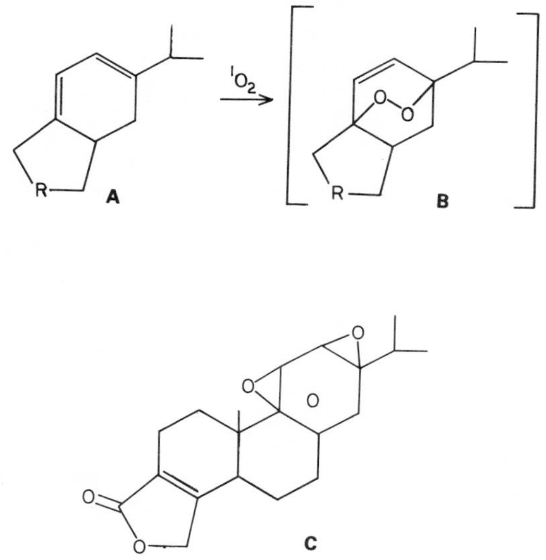

Figure 5-11. Mechanism of biosynthesis of stemolide (C). (A) Cyclic peroxide; (B) parent diterpene.

(Crow *et al.*, 1971). Another peroxide has been proposed as an intermediate in the biosynthesis of stemolide (Figure 5-11), a bisepoxide found in *Stemodia maritima*. The mechanism of formation has been proposed to proceed through the cyclic peroxide formed by singlet dioxygen and the parent diterpene (Manchand and Blount, 1976).

5.3. Prostaglandin

The endoperoxide of the most biochemical interest is the prostaglandin endoperoxide (actually many similar structures), which manifests striking chemical, biochemical, and biological properties (Samuelsson *et al.*, 1978). Members of this class of compounds, themselves with potent biological activity, are enzymatically transformed into several different compounds, some stable and some not, that have markedly diverse biological activities.

Many more stable analogues of the prostaglandin endoperoxide have been synthesized and found to be biologically active, as eventually has the biological compound itself (Porter *et al.*, 1979). Also, thermal rearrangement reactions of the bicycloendoperoxide nucleus have been studied (Coughlin and Salomon, 1979; R. G. Salomon *et al.*, 1978). (This nucleus is conveniently synthesized by diimide reduction of the product of reaction of singlet dioxygen with cyclopentadiene.) These rearrangements are striking for several reasons. First, the thermal rearrangement products are not analogous to the enzymatic products. *In vivo* (Figure 5-12), the endoperoxide is converted to prostaglandin D (PGD) and PGE by isomerization, PGF by reduction, and PGI, thromboxane A (TXA), and 12-hydroxy-5,8,10-heptadecatrienoic acid (Hamberg and Samuelsson, 1966) by rearrangement. Thermolysis (Figure 5-13) of the bis-phenyl-bridgehead-substituted endoperoxide analogue (A in Figure 5-13) does result in rearrangement to ethylene and 1,3-diphenyl-1,3-propanedione (Coughlin and Salomon, 1977), analogous to reaction 6 (Fig. 5-12), but the unsubstituted derivative (B) does not give this reaction. Furthermore, kinetic analysis of the fragmentation of the diphenyl endoperoxide indicated this reaction to be a radical bond scission rather than a concerted rearrangement.

The 3-hydroxycyclopentanone isomerization product (E) analogous to PGD and PGE was detected only when the thermolysis was carried out in water and was formed in about one third the yield of the other aqueous product, 4-oxopentanal (C). The nonenzymatic aqueous rearrangement of the prostaglandin endoperoxide to form PGD and PGE has also been observed to give the same product mixture. Formation of this product (E) only in water is rationalized by a requirement, in the transition state, for a basic molecule to remove the bridgehead proton. Acetic acid cannot assist proton removal and hence does not give the product. Diazabicyclo[2.2.2]-octane (DABCO) (Figure 5-14), on the other hand, catalyzes exclusively formation of this product. These observations suggest

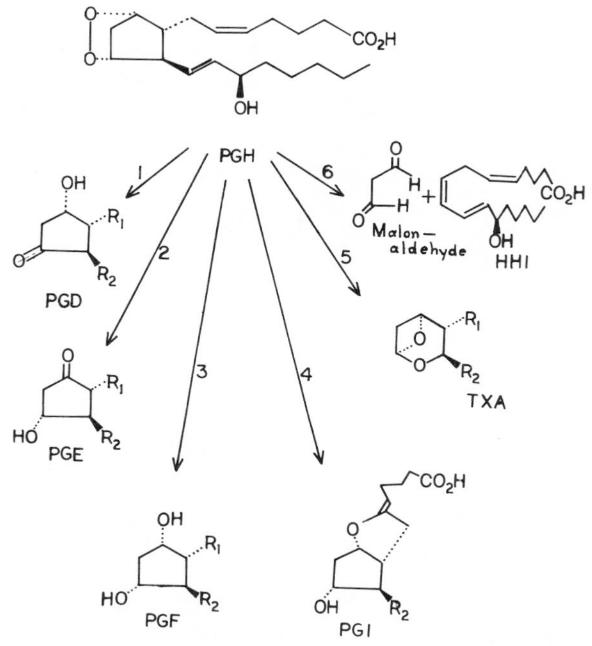

Figure 5-12. *In vivo* conversion of prostaglandin endoperoxide to prostaglandins.

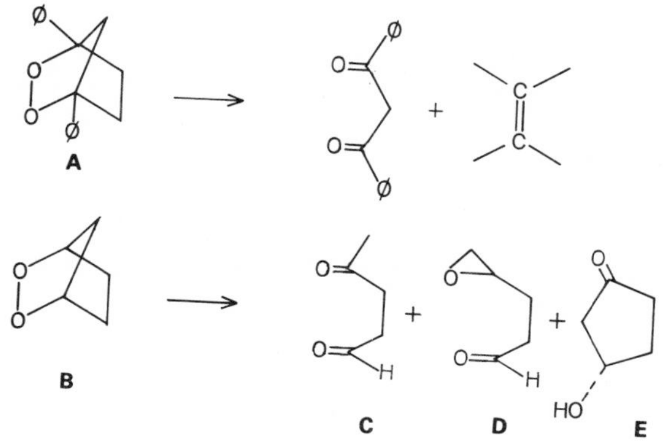

Figure 5-13. Thermolysis of endoperoxides analogous to the prostaglandin endoperoxide.

that the PGD and PGE isomerases could operate primarily by placing acidic and basic groups in strategic positions and that a radical mechanism is not involved.

Indeed, homolysis (Fig. 5-15) of the peroxide bond appears to lead to the epoxy aldehyde product (Figure 5-13D), which is not analogous to any known degradation products of prostaglandin endoperoxides. This conclusion is based on the fact that the rate of formation of the epoxy aldehyde is largely independent of solvent polarity, whereas formation of 4-oxopentanal (C in Figure 5-13) is markedly dependent on solvent polarity. The process is entirely intramolecular, however.

Acidic or polar conditions enhance the rate of formation of the keto aldehyde (Figure 5-16). Differences in (aprotic) solvent polarity from cyclohexane to acetonitrile are sufficient to change product composition from 97% radical product in cyclohexane to 40% in acetonitrile. Glacial acetic acid gives entirely 4-oxopentanal (C in Figure 5-13), and water yields a mixture, 72% of which is this product. This suggests a heterolytic peroxide bond cleavage in which removal of any proton would follow the rate-limiting step (Figure 5-17). Imidazole catalyzes formation of both 3-hydroxycylopentanone and 4-oxopentanal, as expected from its amphoteric nature, and leads to a 1 : 1 mixture of the two products.

Enhanced acid lability for the [2.2.1]bicycloendoperoxide system over other peroxides is expected due to the geometric requirement that the oxygen lone pairs maintain an eclipsed conformation. This is also a factor in the instability of dioxetanes. In fact, the decomposition rate constant for the endoperoxide is an order of magnitude larger in the protic solvents studied (i.e., acetic acid) than in aprotic solvents of similar polarity. Also, the decomposition of the much less rigid [2.2.2]bicyclooctane endoperoxide is enhanced by a much smaller factor in acetic acid over dichloroethylene (Coughlin and Salomon, 1979).

Comparison with [2.2.2]bicyclooctane endoperoxide (Figure 5-18) provides an excellent perspective for the anomalous decomposition of the prostaglandin endoperoxide. Decomposition of the former compound in all solvents gives ethylene and succinaldehyde in a manner not strongly dependent on solvent polarity. This is apparently analogous to reaction 6 (Figure 5-12) of the prostaglandin endoperoxide, which is not observed under uncatalyzed conditions. Hence, reaction 6 probably occurs via a catalyzed free-radical process that somehow suppresses the pathway leading to 4,5-epoxypentanal.

The activation parameters for the nonpolar decompositions are also inter-

Figure 5-14. DABCO, a catalyst in thermolysis of endoperoxides.

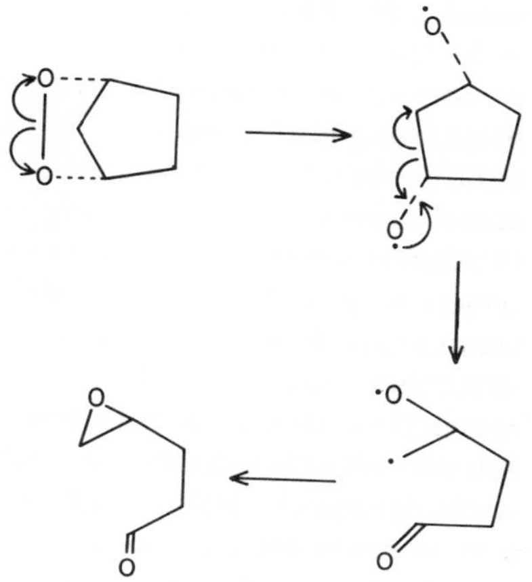

Figure 5-15. Homolytic cleavage of an endoperoxide.

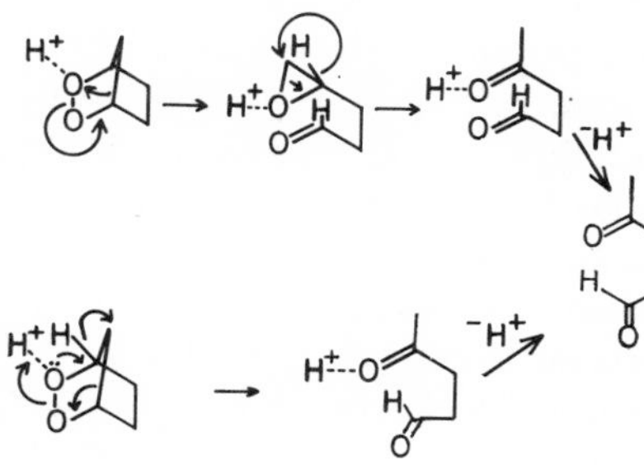

Figure 5-16. Mechanisms of heterolytic cleavage of an endoperoxide.

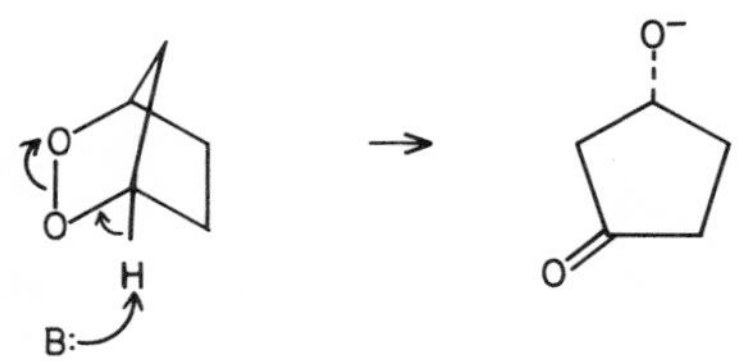

Figure 5-17. A mechanism for endoperoxide cleavage.

esting. The activation entropy of decompositon of di-*t*-butylperoxide is about + 20 e.u., reflecting the fact that rotation and movement of the bulky alkyl groups are easier after the peroxide bond has been stretched to the breaking point. The activation entropy for [2.2.2]bicyclooctane endoperoxide (Figure 5-18) is about + 3 e.u., indicative of some increased freedom of movement in the activated complex, but not nearly so much as if the alkyl groups were not bound together at two points. The activation entropies (Figure 5-19) of the analogous prostaglandin nucleus endoperoxide and endoperoxide (A in Figure 5-19) decomposition, however, are both negative by about 20 e.u. Considering the constraint present in the [2.2.1]bicycloheptane endoperoxide, the activated complex must have very specific geometry requirements. This poorly understood factor actually enhances the thermal stability of the prostaglandin endoperoxides over what one would expect on the basis of bond and strain energies. It also provides a means by which enzymes could catalyze the formation of different products. An enzyme with a nonpolar active site would promote peroxide bond homolysis. If the active-site conformation also prohibited the rigid required activated complex geometry, perhaps by interference at a point remote from the radical center, formation of different products would be facilitated.

The decomposition of the prostaglandin endoperoxide nucleus by amines has been further explored (Zagorski and Salomon, 1980). The reaction in benzene with a catalytic amount of DABCO (Figure 5-14) resulted in the formation of 4-oxopentanal (C in Figure 5-13) and 3-hydroxycyclopentanone (E) in a ratio of about 3 : 1. The catalyst made the activation entropy more negative ($\Delta S\ddagger = -30$ e.u.) than for the uncatalyzed decomposition and decreased the activation enthalpy by one half ($\Delta H\ddagger = 10$ kcal/mole). A large deuterium isotope effect ($k_H/k_D = 8$) suggests that the catalysis involves abstraction of a bridgehead proton, resulting in ketoalkoxide (A in Figure 5-20), which can protonate to form the hydroxycyclopentanone (B) (E in Figure 5-13) or, apparently, undergo a retroaldol cleavage to give the other observed product. The fact that under the reaction conditions, DABCO does not catalyze the isomerization of E in Figure 5-13 to C, while E is more acidic than A, suggests that intermediate A in Figure 5-19 is vibrationally excited. Such vibrational excitation could well result from release of the large steric strain of the endoperoxide.

Another possible explanation for the formation of compound C in Figure 5-13 under these conditions is a solvent-cage effect in which association of the

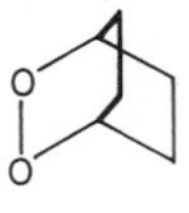

Figure 5-18. An endoperoxide studied because of its similarity to prostaglandin endoperoxide.

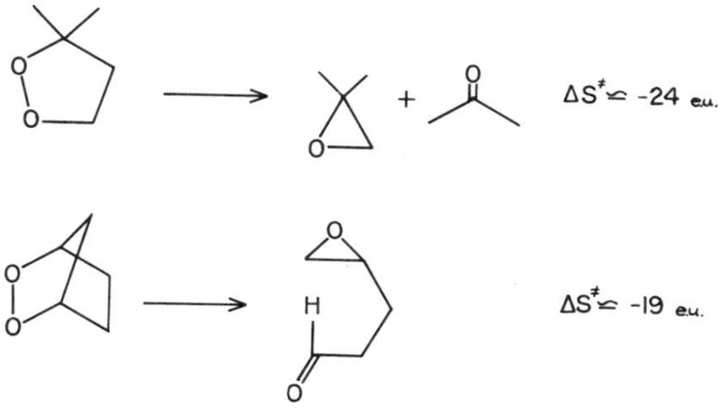

Figure 5-19. Comparative entropies of activation of endoperoxide thermolysis.

protonated base with the newly formed carbonyl promotes enol formation. This effect, if valid, should be strongly dependent on solvent polarity.

From these considerations, it is clear that the enzyme that catalyzes isomerization of the prostaglandin endoperoxide to PGD and PGE must efficiently protonate the alkoxide anion and must absorb some of the intermediate's vibrational energy to avoid the fragmentation reaction if the reaction proceeds via bridgehead-proton removal. Indeed, any pathway leading to breakdown of the bicycloendoperoxide nucleus will result in vibrational excitation unless the product is also strained and is formed in a concerted process.

Though little about the enzymatic conversion of prostaglandin endoperoxides is known with certainty, a few studies have given interesting results. Uncatalyzed aqueous decomposition of the endoperoxides results in isomerization

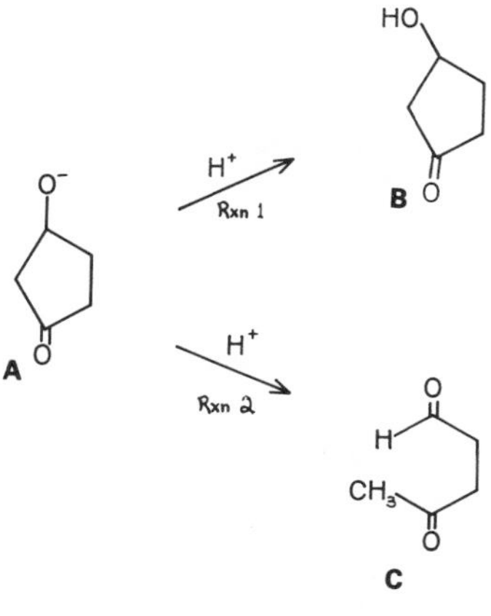

Figure 5-20. Competing reactions of the ketoalkoxide intermediate in endoperoxide decomposition. Reaction 2 could result either from retroaldol rearrangement of the keto anion or form a solvent-cage effect in which the removed bridgehead proton effectively protonates the carbonyl oxygen formed in the rearrangement.

predominantly to PGE. Several types of proteins either accelerate endoperoxide decomposition, change the isomer ratio (PGD/PGE), or catalyze reduction to PGF (Christ-Hazelhof *et al.*, 1976). PGD isomerases catalyze decomposition exclusively into PGD. These enzymes require glutathione, but the reactions do not involve reduction. Serum albumins, which are generally known to bind fatty acids, lead to the formation of a higher percentage of PGD than aqueous solutions. The rate of endoperoxide decomposition, however, varies markedly with the species from which the albumin is isolated, being close to the uncatalyzed rate for rabbit serum albumin. Also, the isomerization is not affected by sulfhydryl blocking groups, which are often required for other isomerases. This suggests that the effect is due to a stereospecific binding of the fatty acid derivative that "directs" the aqueous decomposition toward the PGD isomer and, in some cases, enhances the isomerization rate.

A glutathione enzyme that behaves as an *S*-transferase toward 1-chloro-2,4-dinitrobenzene was isolated from sheep lung and shown to catalyze the formation of PGD and PGF from PGH. A glutathione–prostaglandin adduct was not detected, though this result is largely inconclusive. The purified enzyme shares many superficial properties with other glutathione-*S*-transferases, but the others generally catalyze the formation of a mixture of PGF and PGE, rather than PGD.

Preliminary studies have also been performed on an enzyme that catalyzes the isomerization of the endoperoxide to PGE (Ogino *et al.*, 1977). Glutathione is required catalytically and aids in enzyme protection. However, the requirement is not stoichiometric, and the function is not clear.

Two essentially identical mechanisms have been hypothesized by two groups (Diczfalusy *et al.*, 1977; Fried and Barton, 1977) for the conversion of PGH into TXA, as have mechanisms for formation of malonaldehyde and 12-hydroxyheptadecatrienoic acid, and PGI, all involving electrophilic attack on the peroxide function and electron-deficient rearrangements (Figure 5-21).

Studies of inhibition by several unrelated compounds indicated that fragmentation to malonaldehyde and rearrangement to TXA involve the same enzyme. Advantage was taken of the nucleophilic lability of TXA to show that TXA itself could not be converted to the other observed products.

PGH_1 is converted to the corresponding C-17 hydroxydienoic acid and malonaldehyde, but in only small amounts to a corresponding TXA_1. Reaction of a synthesized *cis*-$\Delta4$-PGH_2 (Diczfalusy and Hammarstrom, 1979) (instead of $\Delta5$) also gave the fragmentation product, but not the corresponding thromboxane product. This was taken as evidence of a stereochemical requirement for a *cis* double bond in the 5-position.

It should be pointed out that these results could be explained by a radical mechanism as well as by a heterolytic mechanism. Indeed, nonenzymatic results with the endoperoxide nucleus (diphenyl bridgehead substituents, I) implied that fragmentation to malonaldehyde and olefin is a nonpolar process. Formation of

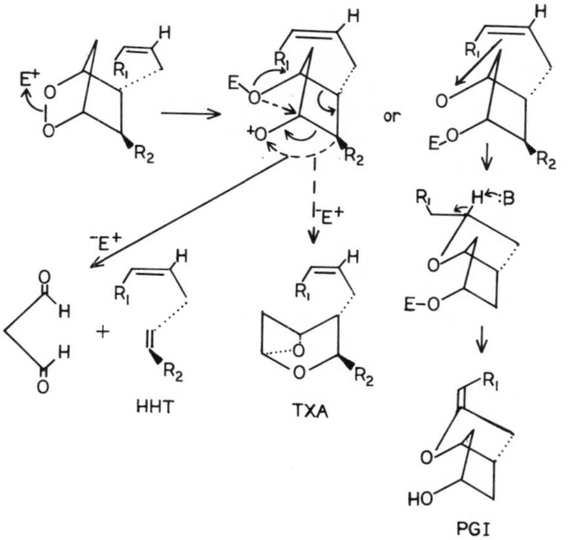

Figure 5-21. Mechanisms of the enzymatic conversions of prostaglandin endoperoxide. (HHT) 12-Hydroxyheptadecatrienoic acid.

a TXA-type nucleus did not occur in the model system, so that this could also be either a radical or a heterolytic process, which is sterically influenced by the enzyme. The prospect that one heterolytic and one free-radical reaction, both catalyzed by the same enzyme and using the same substrate, lead to different products is quite fascinating. Metalloenzymes could, of course, behave in this fashion. It is easier to imagine how the enzyme could contort the substrate side chains to lead to heterolytic PGI formation than to TXA formation. Also, it is geometrically more difficult to imagine the PGI rearrangement as a homolytic process, since the final step would require migration of H· over a long distance. Unfortunately, very little information is currently available on the enzymatic PGI formation reaction.

References

Adam, W., and Bloodworth, A. J., 1978. Organic peroxides, biological and synthetic aspects, *Annu. Rep. Prog. Chem. Sect. B* 75:342–369.

Adam, H. K., Campbell, I. M., and McCorkinlate, N. J., 1967. Ergosterol peroxide: A fungal artefact, *Nature (London)* 216:397.

Ballard, D. H., and Bloodworth, A. J., 1971. Oxymetalation. Part 1. The peroxymercuration of monosubstituted ethylenes: A synthesis of secondary alkyl peroxides, *J. Am. Chem. Soc.* 1971:945–949.

Bates, M. L., and Reid, W. W., 1976. Duality of pathways in the oxidation of ergosterol to its peroxide *in vivo, J. Chem. Soc. Chem. Commun.* 1976:44–45.

Blair, R. A., and Goddard, W. A., II, 1982. *Ab initio* studies of the structures of peroxides and peroxy radicals, *J. Am. Chem. Soc.* 104:2719–2724.

Christ-Hazelhof, E., Nugteren, O. H., and Van Corp, D. A., 1976. Conversions of prostaglandin endoperoxides by glutathione-*S*-transferases and serum albumins, *Biochim. Biophys. Acta* 450:450–461.

Coughlin, D. J., and Salomon, R. G., 1977. Synthesis and thermal reactivity of some 2,3-dioxabicyclo-(2.2.1) heptane models of prostaglandin endoperoxides, *J. Am. Chem. Soc.* 99:655–657.

Coughlin, D. J., and Salomon, R. G., 1979. Extraordinary reactivity of the prostaglandin endoperoxide nucleus: Nonpolar rearrangement of 2,3-dioxabicyclo(2.2.1) heptane and (2.2.2) octane, *J. Am. Chem. Soc.* 101:2761–2763.

Crow, D. W., Nichols, W., and Sterns, M., 1971. Root inhibitors in *Eucalyptus grandis:* Naturally occurring derivatives of the 2,3-dioxabicyclo [4.4.0]-decane system, *Tetrahedron Lett.* 1971:1353–1356.

Diczfalusy, U., and Hammarstrom, S., 1979. A structural requirement for the conversion of prostaglandin endoperoxides to thromboxanes, *FEBS Lett.* 105:291–295.

Diczfalusy, U., Falardeu, P., and Hammarstrom, S., 1977. Conversion of prostaglandin endoperoxides to C_{17}-hydroxy acids catalyzed by human platelet thromboxane synthetase, *FEBS Lett.* 84:271–274.

Eickman, N., Clardy, J., Cole, R. J., and Kirksey, J. W., 1975. The structure of fumetremorgen A′, *Tetrahedron Lett.* 1975:1051–1054.

Fayos, J., Lokensgard, D., Chardy, J., Cole, R. J., and Kirksey, J. W., 1974. Structure of verruculogen, a tremor producing peroxide from *Penicillium verruculosum, J. Am. Chem. Soc.* 86:6785–6787.

Fenical, W., 1974. Rhodophytin, a halogenated vinyl peroxide of marine origin, *J. Am. Chem. Soc.* 96:5580–5581.

Fried, J., and Barton, J., 1977. Synthesis of 13,14-dehydroprostacycline methyl ester: A potent inhibitor of platelet aggregation, *Proc. Natl. Acad. Sci. U.S.A.* 74:2199–2203.

Hamberg, M., and Samuelsson, B., 1966. Novel biological transformations of 8,11,14-eicosatrienoic acid, *J. Am. Chem. Soc.* 88:2349–2350.

Higgs, M. D., and Faulkner, D. J., 1978. Plakortin, an antibiotic from *Plakortis holichondrioides, J. Org. Chem.* 43:3454–3457.

Howard, B. M., Fenical, W., Finer, J., Hirotsu, K., and Chardy, J., 1977. Neoconcinndiol hydroperoxide, a novel marine diterpenoid from the red alga, *Laurencia, J. Am. Chem. Soc.* 99:6440–6441.

Johnson, R. A., Nidy, E. G., Baczynskyj, L., and Gorman, R. R., 1977. Synthesis of prostaglandin H_2 methyl ester, *J. Am. Chem. Soc.* 99:7738–7740.

Kondo, K., and Matsumoto, M., 1976. Synthesis of furanoterpenes: Perillaketone, α-clausenane, ($\pm$)-ipomeamarone, and ($\pm$)-epiiponeamarone, *Tetrahedron Lett.* 1976:4363–4366.

Manchand, P. S., and Blount, J. F., 1976. X-ray structure and absolute stereochemistry of stemolide, a novel diterpene bisepoxide, *Tetrahedron Lett.* 1976:2489–2492.

Milligan, D. E., and Jacox, M. E., 1963. Infrared spectroscopic evidence for the species HO_2, *J. Chem. Phys.* 38:2627–2631.

Ogino, N., Miyamoya, T., Yamamoto, S., and Hayaishi, O., 1977. Prostaglandin endoperoxide E isomerase from bovine vesicular gland microsomes, a glutathione-requiring enzyme, *J. Biol. Chem.* 252:890–895.

Porter, N. A., Byers, J., Holden, K. M., and Menzel, D. B., 1979. Synthesis of prostaglandin H_2, *J. Am. Chem. Soc.* 101:4319–4322.

Porter, N. A., Byers, J. D., Ali, A. E., and Eling, T. E., 1980. Prostaglandin C_2, *J. Am. Chem. Soc.* 102:1183–1184.

Salomon, M., Salomon, R. G., and Glein, R. D., 1976. A synthesis of mixed dialkyl peroxides via reaction of an alkyl hydroperoxide with alkyl trifluoromethane sulfonates, *J. Org. Chem.* 41:3983–3987.

Salomon, R. G., and Salomon, M. F., 1977. 2,3-Dioxabicyclo [2.2.1] heptane: The strained bicyclic peroxide nucleus of prostaglandin endoperoxides, *J. Am. Chem. Soc.* 99:3501–3503.

Salomon, R. G., Salomon, M. F., and Coughlin, D. J., 1978. Prostaglandin endoperoxides. 6. A polar transition state in the thermal rearrangement of 2,3-dioxabicyclo-[2.2.1] heptane, *J. Am. Chem. Soc.* 100:660–662.

Samuelsson, B., Goldyne, M., Granstrom, E., Hamberg, E., Hammarstrom, S., and Malsten, C., 1978. Prostaglandins and thromboxanes, *Annu. Rev. Biochem.* 47:997–1029.

Samuelsson, B., Hammarstrom, S., and Borgeat, P., 1979. Pathways of arachidonic acid metabolism, *Adv. Inflamm. Res.* 1:405–412.

Sawyer, D. T., and Gibian, M. J., 1979. The chemistry of superoxide ion, *Tetrahedron* 35:1471–1481.

Wells, R. J., 1976. A novel peroxyketal from a sponge, *Tetrahedron Lett.* 1976:2637–2638.

Wieland, P., and Prelog, V., 1947. Uber die Isolierung von Ergosterin, Ergosterin-palmitat and Ergosterin-peroxyd aus dem Mycel von *Aspergillus fumigatus*, mut. *helvola, Helv. Chim. Acta* 30:1028–1030.

Zagorski, M. G., and Salomon, R. G., 1980. Prostaglandin endoperoxides. 11. Mechanism of amine catalyzed fragmentation of 2,3-dioxabicyclo-[2.2.0] heptane, *J. Am. Chem. Soc.* 102:2501–2503.

Catalases and Peroxidases

6.1 Catalase

Catalase is the enzyme that catalyzes the decomposition of hydrogen peroxide to water and dioxygen. The usual sources of catalase are bovine liver and bovine erythrocytes. The enzyme exists as a 250,000-dalton tetramer with one heme per monomer. The complete amino acid sequence of the bovine liver enzyme is known (Schroeder *et al.*, 1982) as well as much of the sequence of the bovine erythrocyte catalase (Schroeder *et al.*, 1982). The crystal structure of the beef liver enzyme has been determined (Reid *et al.*, 1981; Murthy *et al.*, 1981). The heme is buried in the enzyme accessible only by a 20-Å hydrophobic channel. There is a tyrosine group on one side of the heme and an asparagine and a histidine on the other. The negative charge of the tyrosine probably aids in the stability of the 4 + iron (Reid *et al.*, 1981) intermediate to be discussed below.

The resting form of the enzyme is high-spin ferric iron that will combine with the substrate hydrogen peroxide, but not with dioxygen. The kinetic mechanism of the primary reaction is rather simple. The ferric form of the enzyme reacts with hydrogen peroxide to form compound I (Figure 6-1), which reacts with another molecule of hydrogen peroxide to form dioxygen and the ferric form of the enzyme again.

There are further reactions (Schonbaum and Chance, 1976) that are relatively slow. Compound I can be reduced by one electron to compound II, which can be again reduced by one electron to the ferric form of the enzyme or oxidized by hydrogen peroxide to compound III. Compound III can be slowly reduced to the ferric enzyme.

Compound I has five oxidizing equivalents, compound II has four, and compound III has six. These are total oxidizing equivalents of iron, oxygen, and the porphyrin ring, as discussed later in Section 6.3.

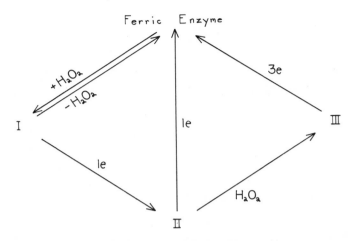

Figure 6-1. Reactions of catalase and peroxidase.

Both oxygens of dioxygen produced by catalase arise from the same hy-drogen peroxide molecule. This fact was shown by Jarnagin and Wang (1958a), who studied the catalase-catalyzed decomposition of singly and doubly ^{18}O-labeled hydrogen peroxide. An equilibrium mixture of singly and doubly labeled hdyrogen peroxide was diluted with unlabeled hydrogen peroxide. The dioxygen product had the same amount of doubly labeled product as the diluted hydrogen peroxide, rather than a smaller amount of doubly labeled product, which would result from combination of oxygens from labeled and unlabeled hydrogen per-oxide. This could be explained only if both oxygens in the dioxygen arise from the same hydrogen peroxide.

There is a small deuterium isotope rate effect. The decomposition of H_2O_2 (in H_2O) is about twice that of D_2O_2 (in D_2O) (Jarnagine and Wang, 1958b). Thus, the hydrogen is transferred in (or before) the rate-determining step.

6.2 Peroxidase

Peroxidase is a rather nonspecific enzyme that catalyzes the one-electron hydrogen peroxide oxidations of many substrates (Dunford and Stillman, 1976). Peroxidase has the same intermediate compounds I, II, and III as catalase. In catalase mechanisms, only compound I seems to be physiologically important, but in peroxidase mechanisms, both compounds I and II are physiologically functional. The kinetic mechanism proceeds through compound I, which has properties very similar to those of compound I of catalase. Compound I then oxidizes the substrate in a one-electron step to produce compound II, and compound II oxidizes the substrate in another one-electron step:

$$Fe^{3+} \; Per \; + \; H_2O_2 \longrightarrow I$$

$$I \; + \; e^- \longrightarrow II$$

$$II \; + \; e^- \longrightarrow Fe^{3+} \; Per$$

At low pH values, the kinetics of the reation through compounds I and II blend to a one-step, two-electron process. For one isozyme of peroxidase (A), this change in mechanism occurs at pH 4.5; for another isozyme, the change is at pH 7.7 (Araiso *et al.*, 1976).

The reaction of dioxygen with reduced peroxidase forms compound III or oxyperoxidase. Compound III can also be formed from compound II and hydrogen peroxide. Compound III probably does not have a physiological function. Horseradish peroxidase catalyzes the hydrogen peroxide oxidation of 1,3-diphenylisobenzofuran in a reaction (Figure 6-2) identical to that of singlet dioxygen (Chan, 1971).

Peroxidase will catalyze several reactions that produce light. These reactions include the oxidation of aldehydes to the next lower aldehyde or ketone and formic acid, the oxidation of aromatic pyruvates to oxalic acid and the next lower aldehyde, the oxidation of indole-3-acetic acid, and the oxidation of unsaturated fatty acids.

Peroxidase oxidizes aldehydes in a mechanism (Figure 6-3) that is believed to proceed through a dioxetane that cleaves by a symmetry-forbidden process to

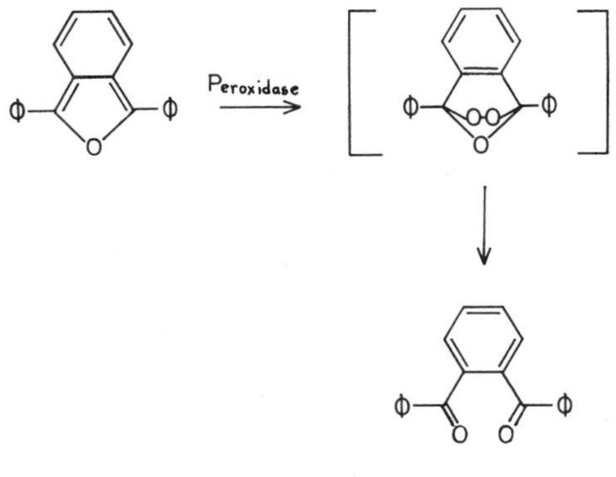

Figure 6-2. Reaction of hydrogen peroxide and 1,3-diphenylisobenzofuran catalyzed by horseradish peroxidase.

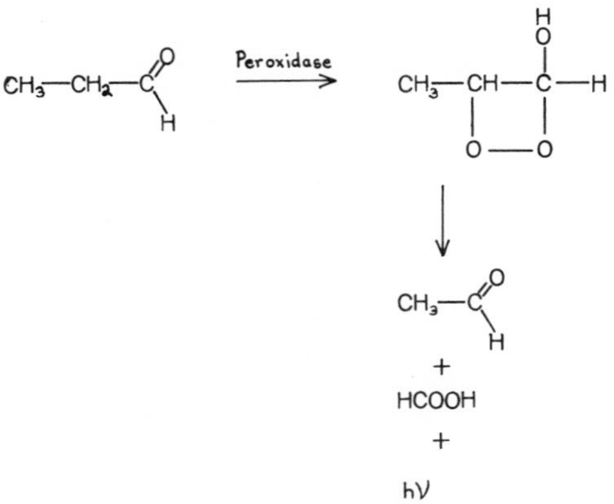

Figure 6-3. Mechanism by which peroxidase, propionaldehyde, and hydrogen peroxide produce light.

formic acid and a smaller aldehyde (Duran *et al.*, 1977; Oliviera *et al.*, 1978). There is a small activity toward long-chain fatty acids (Haun *et al.*, 1980). The products are carbon dioxide and the next lower aldehyde. This reaction also produces excited-state products. Similarly, the peroxidation of lipids to produce malonaldehyde is accompanied by chemiluminescence (Wright *et al.*, 1979). Aromatic pyruvates also produce light with horseradish peroxidase and dioxygen (Cilento, 1975) (Figure 6-4). The products include both carbon monoxide and carbon dioxide. The carbon dioxide must come from a pathway competing for the dioxetane light-producing pathway (Zinner *et al.*, 1980). A similar nonenzymatic reaction, the aerobic oxidation of the thiol-phenyl ester of indole-3-acetic acid, produces light (Duran *et al.*, 1976). Excited-state products produced by the peroxidase-catalyzed reactions may be detrimental to the cell. DNA is damaged during the peroxidase-catalyzed oxidation of isobutanol (Meneghini *et al.*, 1978).

Catalases and peroxidases have a fascinating use in the defense of the bombardier beetle, which protects itself by discharging a hot (100°C) solution of *p*-benzoquinone onto its enemies. This is accomplished by catalases and peroxidases (Hochachka, 1974; Eisner and Meinwold, 1966) rapidly decomposing a 25% hydrogen peroxide–10% hydroquinone mixture in a special chamber with a control valve (Aneshansky *et al.*, 1969).

6.3 Mechanisms of Catalase and Peroxidase

The resting ferric enzymes are probably pentacoordinated (Sievers *et al.*, 1979). The pentacoordinated ferric iron adds hydrogen peroxide to form com-

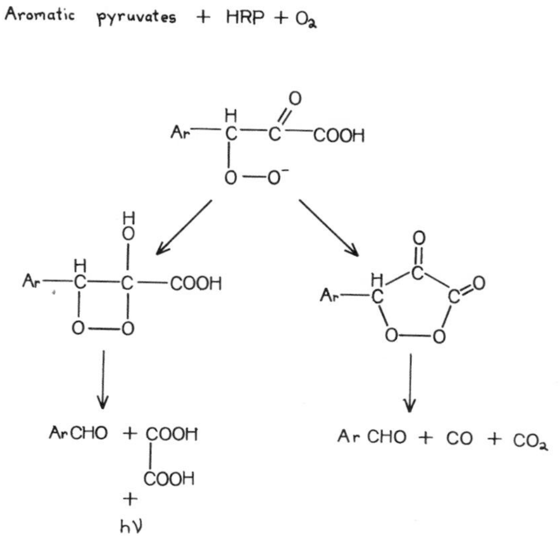

Figure 6-4. Mechanism by which aromatic pyruvates, horseradish peroxidase (HRP), and ground-state dioxygen produce light.

pound I. There has been considerable speculation as to the structure of compound I. This compound contains only one oxygen atom of the peroxide as shown by electron–nuclear double resonance (ENDOR) experiments (Roberts *et al.*, 1981b) and by experiments with *m*-nitrobenzoyl peroxide (Schonbaum and Lo, 1972). Various peroxides, ROOH, will react with the ferric form of peroxidase to form compound I. If R is H or Et, there is no net uptake or release of hydrogen ion during this reaction. However, if R is *m*-nitrobenzoyl, there is 1 mole of hydrogen ion released per mole of complex I formed. This result is consistent with the following reaction, in which only one oxygen atom of peroxide if transferred to the enzyme (E):

$$E + ROOH \longrightarrow EO + ROH$$

If ROH is water or alcohol, no hydrogen ions will appear, but if ROH is *m*-nitrobenzoic acid, one hydrogen ion will appear from the ionization of the acid.

The Mossbauer spectrum has shown that compound I contains iron in the 4+ state (moss *et al.*, 1969; Maeda and Morita, 1967). Yonetani *et al.* (1966) found that compound I has an electron spin resonance spectral line with a value of $g = 2.004$, which corresponds to a free electron. Iizuka *et al.* (1968) have analyzed the magnetic susceptibility data for complex I and found that it fits a 4+ iron and a a free radical. Compounds of 4+ iron are uncommon, but there are examples in organic chemistry in which the iron is chelated by highly negatively charged ligands. Quadrivalent iron compounds are stabilized by arsine (Hazeldean *et al.*, 1966) and phosphine complexes (Warren and Bennett, 1976). Tris-(*N*,*N*-disubstituted dithiocarbamato) Fe(IV) tetrafluoroborates are also known. Interestingly, these are formed by air oxidation of Fe-(III) compounds (Pasek and Straub, 1972).

Dolphin *et al.* (1971) were able to prepare π-cation radicals by the two-electron oxidation of cobaltous octaethyl porphyrin. The similarity of the spectra of these π-cation radicals and the spectra of catalase and peroxidase compound I caused them (Dolphin *et al.*, 1971, 1973; Dolphin and Felton, 1974) to propose that compound I contains iron in the $4+$ oxidation state and a porphyrin cation radical. The $4+$ iron is probably oxygenated. A likely structure would be a ferryl ion, FeO^{2+}. The porphyrin cation radical structure for compound I is supported by nuclear magnetic resonance (NMR) (La Mar and de Ropp, 1980; La Mar *et al.*, 1981), electron paramagnetic resonance (Schultz *et al.*, 1979), ENDOR (Roberts *et al.*, 1981a,b), and theoretical studies (Loew and Herman, 1980). Similar cation radicals derived from chlorophyll have been proposed as intermediates in photosynthesis (Davis *et al.*, 1979a,b) and, as we shall see in Chapter 15, as the hydroxylating agent in P-450 enzyme. Electrochemical oxidation of ferric porphyrins will produce π-cation radicals of iron porphyrins (Shimomura *et al.*, 1981; Phillipps and Goff, 1982).

A rather intriguing consequence of the Dolphin and Felton proposal is that it rationalizes why compound I reacts differently in catalase than it does in peroxidase. The highest filled orbitals of phorphyrin are almost degenerate. One of these has a symmetry of a_{1u} and the other a symmetry of a_{2u}. If an electron is removed from porphyrin, it is quite conceivable that the electron could be removed from different orbitals in the two enzymes. The a_{1u} cation radical has a high unpaired electron density at the 2- and 5-positions of the pyrrole ring, and the a_{2u} orbital has a high unpaired electron density on the nitrogens and on the methylene bridges. Calculations (Hanson *et al.*, 1981) have been able to predict quantitatively the electron densities that are expected at all positions in the porphyrin molecule. The relative stabilities of the a_{1u} and a_{2u} radical cations depend on minor changes in the porphyrin ring. For example, magnesium and zinc tetraphenyl porphyrin cation radicals are a_{2u} symmetry, and magnesium and zinc octaethyl porphyrin radical cations are a_{1u} symmetry (Fajer *et al.*, 1973). By the analysis of spectral data, the π-cation radical of catalase appears to be of a_{1u} symmetry and that of peroxidase to be of a_{2u} symmetry (Fajer *et al.*, 1974; Hanson *et al.*, 1981; Roberts *et al.*, 1981a).

It is not hard to imagine that these two radicals could react very differently. However, there is evidence to indicate that the protein is also important in controlling reactivity. The peroxidase cation radical appears to be changed from a_{1u} by the substitution of deuteroporphyrin for proto-porphyrin in the enzyme. However, the reactivity of peroxidase does not change when this substitution is made, and the modified peroxidase has no catalase activity (Di Nello and Dolphin, 1979).

The emission band (1280 nm) of singlet dioxygen has been detected in the peroxidase reaction (Kanofsky, 1983; Khan, 1983). It is difficult to rationalize this result in terms of the mechanism proposed here for peroxidase unless the singlet dioxygen is a by-product of the reaction.

In the catalase reaction, compound I reacts with another hydrogen peroxide molecule to form water, dioxygen, and the resting enzyme, but in the peroxidase reaction, compound I adds one electron (from the substrate) to form compound II. Compound II of peroxidase again appears to be low-spin $4+$ iron, but without the extra radical. There is both Mossbauer (Simmoneaux et al., 1982; Maeda and Morita, 1967) and NMR spectral (Morishima and Ogawa, 1978) evidence to support the low-spin $4+$ iron state.

Compound II is also believed to be a "ferryl" compound (FeO^{2+}), like compound I but without the porphyrin cation radical. Ferryl porphyrins have been prepared in toluene at $-80°C$ by adding N-methyl-imidazole to peroxo-bridged Fe(III) porphyrins (Figure 6-5) (Chin et al., 1980). The NMR spectrum of compound II from horseradish peroxidase is the same as the NMR spectrum of these known ferryl compounds (La Mar et al., 1982). Iterative extended Hückel calculations on ferryl ions are able to explain the electronic properties of compound II of horseradish peroxidase (Hanson et al., 1981). It is interesting to note that carbon analogues of ferryl ions may be prepared. These consist of a carbene (e.g., dichlorocarbene) bound to the iron of a ferroporphyrin (Mansuy et al., 1977, 1978).

The mechanism of action of catalase must take into account the structure of compound I, the deuterium rate effect, and the observation that all the dioxygen comes from one hydrogen peroxide molecule. The mechanism shown in Figure 6-6 is consistent with these observations.

This mechanism is a two-electron oxidation of hydrogen peroxide by removal of a hydride ion. As we have seen in Chapter 3 two-electron oxidations of hydrogen peroxide are a common chemical method to prepare singlet dioxygen. One could wonder whether catalase also produces singlet oxygen. Catalase is normally thought of as eliminating a potentially hazardous chemical in the body; it would not be of any advantage to produce a more hazardous one.

Experiments with catalase in the presence of the singlet-oxygen-detecting

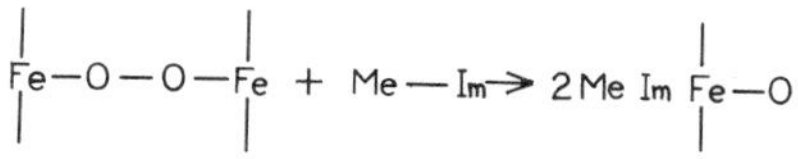

Figure 6-5. A method for the preparation of ferryl heme compounds.

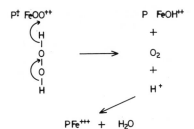

Figure 6-6. A proposed mechanism for catalase action. (P) Porphyrin ring.

reagent 2,5-diphenylfuran have shown that no detectable singlet dioxygen is produced from catalase. The sensitivity is good enough to state that at least 98% of the dioxygen from catalase is produced in the triplet ground state (Porter and Ingraham, 1974). However, catalase shows a broad emission at 1640 nm (Khan, 1983). This could be a solvated singlet-dioxygen emission or an emission from an excited iron–porphyrin complex.

It appears that catalase produces triplet dioxygen directly or catalyzes the rapid conversion of singlet dioxygen to triplet dioxygen. If the reactant is iron in the 4+ state and the product is low-spin ferric iron that would convert to high-spin ferric iron in a subsequent step, then one can write a reaction producing triplet dioxygen from hydrogen peroxide with spin conservation (Figure 6-7). Three unpaired spins in the reactants produce three unpaired spins in the product. Note that in an octahedral field, 4+ iron would have two unpaired electrons. This reaction mechanism requires that the enzyme have a relatively stable low-spin ferric state.

The peroxidase reactions that give products containing two oxygen atoms,

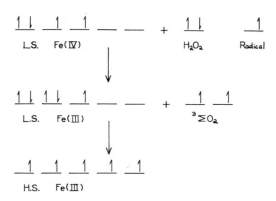

Figure 6-7. Spin conservation in the catalase reaction.

like the furan reactions and dioxetane-mediated luminescent reactions, must arise from compound III, which contains two oxygen atoms.

6.4   Chloroperoxidase

Chloroperoxidase is a heme protein (Champion *et al.*, 1973) that uses peroxide to chlorinate substrates with chloride ion. A typical reaction is the chlorination of 1,3-cyclopentadione (Figure 6-8). Another typical reaction, commonly used for assay, is the chlorination of monochlorodimedone to dichlorodimedone (Figure 6-9) (Hager *et al.*, 1975). Perbenzoic acid will serve as a source of oxidizing agent instead of hydrogen peroxide (Hager *et al.*, 1975).

Chloroperoxidase and similar enzymes iodinate aromatic rings. Lacto-peroxidase of milk will iodinate tyrosine and histidine in proteins that are supplied with H_2O_2 and iodide ion (Mueller and Morrison, 1974). Iodination of the benzene ring to produce triiodothyronine and thyroxine is accomplished by a peroxidase enzyme that utilizes hydrogen peroxide and iodide ion (Tata, 1976).

There are strong similarities between chloroperoxidase and the hydroxylating enzyme P-450 in the magnetic circular dichroism (Dawson *et al.*, 1976) and UV-visible spectra of the CO complexes and other complexes (Hollenberg and Hager, 1973), in the Mossbauer spectra (Champion *et al.*, 1975a), and in the resonance Raman spectra (Champion *et al.*, 1978). The spectral changes in chloroperoxidase are similar to those found for cytochrome P-450. The fifth and sixth ligands on the heme are cysteine and imidazole, identical to those found on cytochrome P-450 (Cramer *et al.*, 1978).

The resonance Raman spectra (Remba *et al.*, 1979) show a low symmetry for the iron and low-frequency vibrations in the porphyrin, indicating a strong donation of electrons from an axial ligand into antibonding π-orbitals of the porphyrin. Chloroperoxidase in the resting state at room temperature is in the high-spin form, which converts to low-spin at lower temperatures. This is one

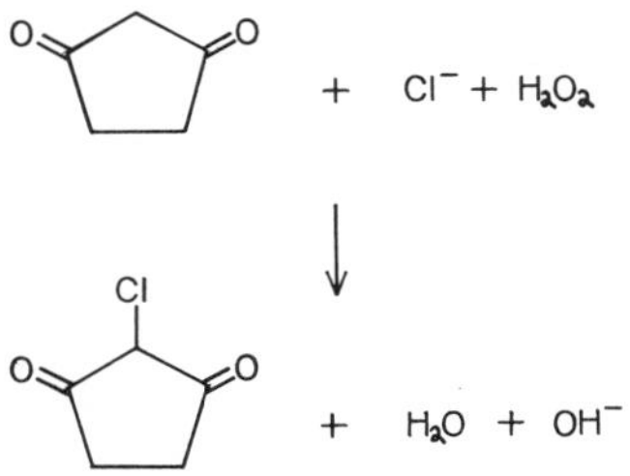

Figure 6-8. A typical reaction catalyzed by chloroperoxidase.

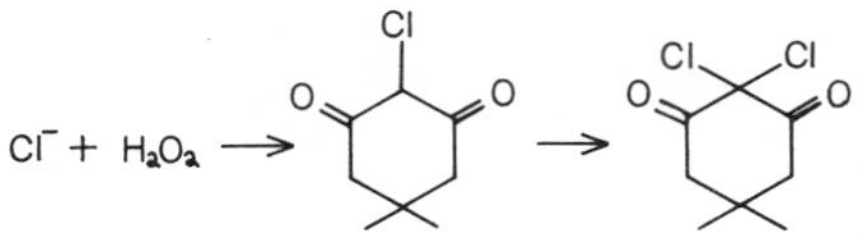

Figure 6-9. Usual assay reaction for chloro-peroxidase.

difference between chloroperoxidase and cytochrome P-450. Cytochrome P-450 is predominantly low-spin in the resting state at room temperature.

The Mossbauer spectra of chloroperoxidase (Champion *et al.*, 1973) show that the resting enzyme is in the high-spin ferric form at room temperature and converts to the low-spin ferric form below about 200°K. The chloroperoxidase–chloride ion complexes are identical to those of the resting enzyme, indicating that the chloride ion is not bound to the iron. The enzyme intermediate has a spectrum identical to that of complex I (Champion *et al.*, 1975b) of peroxidase. A plausible mechanism for chloroperoxidase proceeds through the same complex I as in catalase and peroxidase, followed by reaction with chloride ion to give the chlorinating species (Hager *et al.*, 1975) for the substrate, SH:

$$\text{Porphyrin } Fe^{3+} + H_2O_2 \longrightarrow \text{porphyrin} \cdot{}^+ FeO^{2+} + H_2O$$

$$\text{Porphyrin} \cdot{}^+ FeO^{2+} + Cl^- \longrightarrow \text{porphyrin} \cdot{}^+ FeOCl^+$$

$$H^+ + \text{porphyrin} \cdot{}^+ FeOCl^+ + SH \longrightarrow \text{porphyrin } Fe^{3+} + SCl + H_2O$$

Chloroperoxidase also has an emission band at 280 nm (Khan, 1983; Khan *et al.*, 1983) typical of singlet dioxygen similar to that observed for peroxidase.

6.5   Glutathione Peroxidase

Glutathione peroxidase is an enzyme that catalyzes the nonspecific reduction of organic hydroperoxides or hydrogen peroxide to the alcohol and water, with concomitant oxidation of two molecules of glutathione to form the disulfide adduct:

$$ROOH + 2GSH \longrightarrow ROH + H_2O + GSSG$$

It contains neither heme nor flavin. The enzyme has received substantial attention in recent years (Flohe *et al.*, 1979), partly because of the finding that it contains selenium (Rotruck *et al.*, 1972) as an essential cofactor. The enzyme has been isolated from a large number of sources and appears to be quite homogeneous throughout. X-ray crystallography has been performed on a crystalline form of the enyzme, showing it to be tetrameric with a total molecular weight of 84,000. The subunits are apparently identical, and each contains a selenium atom positioned in a selenocysteine residue at the enzyme's outer surface.

The kinetics exhibited by this enzyme are quite unusual. Like other types of peroxidases, there is no evidence of binding of peroxide to enzyme. The first step is strictly bimolecular. Assuming that selenium participates in this step, its placement is compatible with this observation. The subsequent kinetics are of the Ping-Pong type, suggestive of a cyclic redox-type mechanism. Glutathione does bind very specifically with the oxidized enzyme, and the substituents crucial to binding have been determined. Remarkably, the unoxidized enzyme shows no affinity for glutathione. However, the reduced enzyme is labile to oxidation by peroxides or dioxygen in the absence of glutathione. This oxidation, as would be expected from selenium chemistry, goes beyond the oxidation state involved in catalysis (reaction 4 below) (Splittgerber and Tappel, 1979), so that if the enzyme is not preincubated with glutathione, the reaction exhibits a lag phase. The second sulfhydryl entity does not bind specifically to the enzyme, so that any of the large number of disulfide adducts may be formed (Forstrom and Tappel, 1979).

On the basis of this and other information (Flohe *et al.*, 1979; Forstrom and Tappel, 1979), a mechanism has been proposed for the various steps of the reaction:

Reaction

$$1.\ \text{E-SeH} + \text{ROOH} \longrightarrow \text{ESeOH} + \text{ROH}$$

$$2.\ \text{E-SeOH} + \text{GSH} \longrightarrow \text{E-SeSG} + \text{H}_2\text{O}$$

$$3.\ \text{E-SeSG} + \text{RSH} \longrightarrow \text{E-SeH} + \text{GSSR}$$

Inactivation

$$4.\ \text{E-SeOH} + \text{ROOH} \longrightarrow \text{E-SeO}_2\text{H (inactive)} + \text{ROH}$$

It is known that selenol (as shown in reaction 1) is the active oxidation state of the prosthetic group. Selenium nucleophilically displaces ROH from the peroxide to form the seleninic acid. Water is then nucleophilically displaced from the seleninic acid by glutathione to form a selenosulfide adduct. Finally, a second sulfhydryl group displaces the selonol from this adduct to form a disulfide.

While the mechanism of the glutathione peroxidase reaction seems fairly well established, the physiological role of the enzyme is not clear. Lack of dietary selenium (and therefore of glutathione peroxidase activity) has been associated with lipid peroxide damage leading to hemolysis. However, demonstration of glutathione peroxidase activity in membranes has not been accomplished. Furthermore, its function is closely connected with those of other peroxidases and vitamine E. The physiological distinctions among these peroxide-reactive chemicals are not clear.

McCay *et al.* (1981) have obtained evidence that contrary to the usual assumptions, glutathione neither prevents membrane lipid peroxidation nor reduces membrane lipid peroxides alone. In combination with an uncharacterized heat-labile factor, however, glutathione prevents lipid peroxidation. Experiments with liver microsomes showed that the essential heat-labile factor is not glutathione peroxidase.

References

Aneshansky, D. J., Eisner, T., Widom, J. M., and Widom, B., 1969. Biochemistry at 100°C: Explosive secretory discharge of bombardier beetles (*Brachinus*), *Science* 165:61–63.

Araiso, T., Miyoshi, K., and Yanozaki, I., 1976. Mechanisms of electron transfer from sulfite to horseradish peroxidase–hydroperoxide compounds, *Biochemistry* 15:3059–3063.

Champion, P. M., Munck, E., Debrunner, P. G., Hollenberg, P. F., and Hager, L. P., 1973. Mossbauer investigations of chloroperoxidase and its halide complexes, *Biochemistry* 12:426–435.

Champion, P. M., Lipscomb, T. D., Munck, E., Debrunner, P. and Gunsalus, I. C. 1975a. Mossbauer investigations of high-spin ferrous heme proteins I cytochrome P-450. *Biochem* 14:4151–4158.

Champion, P. M., Chiang, R., Munck, E., Debrunner, P. and Hager, L. P. 1975b. Mossbauer investigations of high-spin ferrous heme proteins II. Chloroperoxidase, horseradish peroxidase, and hemoglobin. *Biochem* 14:4159–4166.

Champion, P. M., Gunsalus, I. C., Wagner, C. C. 1978. Resonance Raman investigations of cytochrome P-450 CAM from *Pseudomonas putida*, *J. Am. Chem. Soc.* 100:3743–3751.

Champion, P. M., Gunsalus, I. C., and Wagner, G. C., 1978. Resonance Raman investigations of cytochrome P450$_{CAM}$ from *Pseudomonas putida*, *J. Am. Chem. Soc.* 100:3743–3751.

Chan, H. W.-S., 1971. Singlet oxygen analogs in biological systems: Peroxidase-catalyzed oxygenation of 1,3-dienes, *J. Am. Chem. Soc.* 93:4632–4633.

Chin, D.-H., Balch, A. L., and LaMar, G. N., 1980. Formation of porphyrin ferryl (FeO$_2^{++}$) complexes through the addition of nitrogen bases to peroxo-bridged iron (III) porphyrins, *J. Am. Chem. Soc.* 102:1446–1448.

Cilento, G., 1975. Dioxetanes as intermediates in biological processes, *J. Theor. Biol.* 55:471–479.

Cramer, S. P., Dawson, J. H., Hodgson, K. O., and Hager, L. P., 1978. Studies on the ferric forms of cytochrome P450 and chloroperoxidase by extended X-ray absorption fine structure: Characterization of the Fe–N and Fe–S distances, *J. Am. Chem. Soc.* 100:7282–7290.

Davis, M. S., Forman, A., and Fajer, J., 1979a. Ligated chlorophyll cation radicals: Their function in photosystem II of plant photosynthesis, *Proc. Natl. Acad. Sci. U.S.A.* 76:4170–4174.

Davis, M. S., Forman, A., Hanson, L. K., Thornber, J. P., and Fajer, J., 1979b. Anion and cation radicals of bacteriochlorophyll and bacteriopheo-phytin *b:* Their role in the primary charge separation of *Rhodopseudomonas viridis*, *J. Phys. Chem.* 83:3325–3332.

Dawson, S. H., Trudell, I. R., Barth, D., Linder, R. D., Bunnenberg, E., Djerassi, C., Chang, R., and Hager, L. P., 1976. Chloroperoxidase evidence for a P450 type heme environment from magnetic circular dichroism spectroscopy, *J. Am. Chem. Soc.* 98:3709–3710.

Di Nello, R. K., and Dolphin, D., 1979. The role of protein and porphyrin in the reactivity of horseradish peroxidase toward hydrogen donors, *Biochem. Biophys. Res. Commun.* 86:190–198.

Dolphin, D., and Felton, R. H., 1974. The biochemical significance of porphyrin cation radicals, *Acc. Chem. Res.* 7:26–32.

Dolphin, D., Forman, A., Borg, D. C., Fajer, J., and Felton, H. H., 1971. Compounds I of catalase and peroxidase: π-Cation radicals, *Proc. Natl. Acad. Sci.* 68:614–618.

Dolphin, D., Muljiani, A., Rousseau, K., Borg, D. C., Fajer, J., and Felton, R. H., 1973. The chemistry of porphyrin π-cations, *Ann. N. Y. Acad. Sci.* 206:107–200.

Dunford, H. B., and Stillman, J. S., 1976. On the function and mechanism of action of peroxidases, *Coord. Chem. Rev.* 19:187–251.

Duran, N., Zinner, K., De Baptista, C., Vidigal, C. C. C., and Cilento, G., 1976. Chemiluminescence from the oxidation of auxin derivatives, *Photochem. Photobiol.* 24:383–388.

Duran, N., Oliviera, O. M. M. F., Haun, M., and Cilento, G., 1977. Enzyme generated triplet acetone, *J. Chem. Soc. Chem. Commun.* 1977:442–443.

Eisner, T., and Meinwald, J., 1966. Defensive secretions of arthropods, *Science* 153:1341–1350.

Fajer, J., Borg, D. C., Forman, A., Felton, R. H. Vegh, L., and Dolphin, D., 1973. ESR studies of porphyrin π-cations: The $^2a_{1u}$ and $^2a_{2u}$ states, *Ann. N. Y. Acad. Sci.* 206:349–364.

Fajer, J., Borg, D. C., Forman, A., Alder, A. D., and Varad, V., 1974. Cation radicals of tetraalkyl porphyrins, *J. Am. Chem. Soc.* 96:1238–1239.

Flohe, L., Günzler, W. A., and Loschen, G., 1979. The flutathione reaction: A key to understand the selenium requirement of mammals, in *Trace Metals in Health and Disease*, N. Kharasch (ed.), Raven Press, New York, pp. 263–286.

Forstrom, J. W., and Tappel, A. L., 1979. Donor substrate specificity and thiol reduction of glutathione disulfide peroxidase, *J. Biol. Chem.* 254:2888–2891.

Hager, L. P., Hollenberg, P. F., Rand-Meir, T., Chiang, R., and Daubek, D., 1975. Chemistry of peroxidase intermediates, *Ann. N. Y. Acad. Sci.* 244:80–92.

Hanson, L. K., Chang, C. K., Davis, M. S., and Fajer, J., 1981. Electron pathways in catalase and peroxidase enzymic catalysis: Metal and macrocycle oxidations of iron porphyrins and chlorins, *J. Am. Chem. Soc.* 103:663–670.

Haun, M., Duran, N., Augusto, O., and Cilento, G., 1980. Model studies of the α-peroxidase system: Formation of an electronically excited product, *Arch. Biochem. Biophys.* 200:245–252.

Hazeldean, G. S. F., Nyholm, R. S., and Parish, R. V., 1966. Octahedral ditertiary arsine complexes of quadrivalent iron, *J. Chem. Soc. A* 1966:162–165.

Hochachka, P. W., 1974. Regulation of heat production at the cellular level, *Fed. Proc. Fed. Am. Soc. Exp. Biol.* 33:2162–2169.

Hollenberg, P. F., and Hager, L. P., 1973. The P450 nature of the carbon monoxide complex of ferrous chloroperoxidase, *J. Biol. Chem.* 148:2630–2633.

Iizuka, T., Kotani, M., and Yonetani, T., 1968. A thermal equilibrium between high and low-spin states in ferric cytochrome *c* peroxidase and some discussion on the enzyme–substrate complex, *Biochim. Biophys. Acta* 167:257–267.

Jarnagin, R. C., and Wang, J. H., 1958a. Investigation of the catalytic mechanisms of catalase and other ferric compounds with doubly labeled O^{18} labeled hydrogen peroxide, *J. Am. Chem. Soc.* 80:786–787.

Jarnagin, R. C., and Wang, J. H., 1958b. Further studies on the catalytic decomposition of hydrogen peroxide by triethylenetetramine–Fe(III) complex and related substances, *J. Am. Chem. Soc.* 80:6477–6481.

Kanofsky, J. R., 1983. Singlet oxygen production by lactoperoxidase, *J. Biol. Chem.* 258:5991–5993.

Khan, A. U., 1983. Enzyme systems generation of singlet ($^1\Delta g$) molecular oxygen observed directly by 1.0–1.8 μM luminescence spectroscopy, *J. Am. Chem. Soc.* 105:7195–7197.

Khan, A. U., Gebauer, P., and Hager, L. P., 1983. Chloroperoxidase generation of singlet Δ molecular oxygen observed directly by spectroscopy in the 1- to 1.6-μm region, *Proc. Natl. Acad. Sci. U.S.A.* 80:5195–5197.

La Mar, G. N., and de Ropp, J. S., 1980. Proton nuclear magnetic resonance characterization of the electronic structure of horseradish peroxidase compound I, *J. Am. Chem. Soc.* 103:395–397.

La Mar, G. N., de Ropp. J. S., Smith, K. M., and Langry, K. C., 1981. Proton nuclear magentic resonance investigation of the electronic structure of compound I of horseradish peroxidase, *J. Biol. Chem.* 256:237–243.

La Mar, G. N., de Ropp, J. S., Latos-Gratzynski, L., Balch, A. L., Johnson, R. B., Smith, K. M., Parish, D. W., and Cheng, R.-J., 1982. Proton NMR characterization of the ferryl group in model heme complexes and hemoproteins: Evidence for the $Fe^{IV} = O$ group in ferryl myoglobin and compound II of horseradish peroxidase, *J. Am. Chem. Soc.* 105:782–787.

Loew, G. H., and Herman, Z. S., 1980. Calculated spin densities and quadrupole splittings for model horseradish peroxidase compound I: Evidence for iron(IV) porphyrin (S = 1) π-cation radical electronic structure, *J. Am. Chem. Soc.* 102:6173–6174.

Maeda, Y., and Morita, Y., 1967. Mossbauer effect in peroxidase–hydrogen peroxide compounds, *Biochem. Biophys. Res. Commun.* 29:680–685.

Mansuy, D., Lange, M., Chottard, J. C., Guerin, P., Morlierre, P., Brault, D., and Rougee, M., 1977. Reaction of carbon tetrachloride with 5,10,15,20-tetraphenylporphinato–iron (II) [(TPP)Fe^{II}]: Evidence for the formation of the carbene complex [(TPP) Fe^{II} (CCl$_2$)], *J. Chem. Soc. Chem. Commun.* 1977:648–649.

Mansuy, D., Lange, M., Chottard, J. C., Bartoli, J. F., Cherrier, B., and Weiss, R., 1978. Dichlorocarbene complexes of iron (II)–porphyrins—crystal and molecular structure of Fe(TPP)(CCl$_2$)(H$_2$O), *Ang. Chem. Int. Ed. Engl.* 17:781–782.

McCay, P. G., Gibson, D. D., and Hornbrook, K. R., 1981. Glutathione-dependent inhibition of lipid peroxidation by a soluble, heat-labile factor not glutathione peroxidase, *Fed. Proc. Fed. Am. Soc. Exp. Biol.* 40:199–205.

Meneghini, R., Hoffman, M. E., Duran, N., Faljoni, A., and Cilento, G., 1978, DNA damage during the peroxidase-catalyzed aerobic oxidation of isobutanol, *Biochim. Biophys. Acta* 518:177–180.

Morishima, I., and Ogawa, S., 1978. Nuclear magnetic resonance characterization of compounds I and II of horseradish peroxidase, *J. Am. Chem. Soc.* 100:7125–7127.

Moss, T. H., Ehrenberg, A., and Bearden, A. J., 1969. Mossbauer spectroscopic evidence for the electronic configuration of iron in horseradish peroxidase and its peroxide derivatives, *Biochemistry* 8:4159–4162.

Mueller, T. J., and Morrison, M., 1974. The transmembrane proteins in the plasma membrane of normal human erythrocytes, *J. Biol. Chem.* 249:7568–7573.

Murthy, M. R. N., Reid, T. J., III, Sicignano, A., Tanaka, N., and Rossman, M. G., 1981. Structure of beef liver catalase, *J. Mol. Biol.* 152:465–499.

Oliviera, O. M. M. F., Haun, M., Duran, N., O'Brien, P. J., O'Brien, C. R., Bechara, E. J. H., and Cilento, G., 1978. Enzyme-generated electronically excited carboxyl compounds, *J. Biol. Chem.* 253:4707–4712.

Pasek, E. A., and Straub, D. K., 1972. Tris (*N,N*-Disubstituted dithiocarbamato) iron(IV) tetrafluoroborates, *Inorg. Chem.* 11:259–263.

Phillipps, M. A., and Goff, H. M., 1982. Electrochemical synthesis and characterization of the single electron oxidation products of ferric porphyrins, *J. Am. Chem. Soc.* 104:6026–6034.

Porter, D. J. T., and Ingraham, L. L., 1974. Concerning the formation of singlet O$_2$ during the decomposition of H$_2$O$_2$ by catalase, *Biochim. Biophys. Acta* 334:97–102.

Reid, T. J., III, Murthy, M. R. N., Sicignano, A., Tanaka, N. Musick, W. D. L., and Rossman, B. G., 1981. Structure and heme environment of beef liver catalase at 3.5 Å resolution, *Proc. Natl. Acad. Sci. U.S.A.* 78:4767–4771.

Remba, R. D., Champion, P. M., Fitchen, D. B., Chiang, R., and Hager, L., 1979. Resonance Raman investigations of chloroperoxidase, horseradish peroxidase, and cytochrome *c* using Soret band laser excitation, *Biochemistry* 18:2280–2290.

Roberts, J. E., Hoffman, B. M., Rutter, R., and Hager, L. P., 1981a. Electron-nuclear double resonance of horseradish peroxidase compound I, *J. Biol. Chem.* 256:2118–2121.

Roberts, J. E., Hoffman, B. M., Rutter, R., and Hager, L. P., 1981b. ^{17}O ENDOR of horseradish peroxidase I, *J. Am. Chem. Soc.* 103:7654–7656.

Rotruck,. J. T., Hoekstra, W. G., Pope, A. L., Ganther, A., Swanson, A., and Hafeman, D., 1972. Relationship of selenium to GSH peroxidase, *Fed. Proc. Fed. Am. Soc. Exp. Biol.* 31:691.

Schonbaum, G. R., and Chance, B., 1976. Catalase, in *The Enzymes*, P. D. Boyer (ed.), Academic Press, New York, pp. 363–408.

Schonbaum, G. R., and Lo, S., 1972. Interaction of peroxidases with aromatic peracids and alkyl peroxides, *J. Biol. Chem.* 247:3353–3360.

Schroeder, W. A., Schelton, J. R., Shelton, J. B., Robberson, B., Apell, G., Fang, R. S., and Bonaventura, J., 1982. The complete amino acid sequence of bovine liver catalase and the partial sequence of bovine erythrocyte catalase, *Arch. Biochem. Biophys.* 214:397–421.

Schultz, D. E., DeVaney, P. W., Winkler, H., Debrunner, P. G. Doan, N., Chiang, R., Rutter, R., and Hager, L. P., 1979. Horseradish peroxidase compound I: Evidence for spin coupling between the heme iron and a free radical, *Eur. J. Biochem.* 103:102–105.

Shimomura, E. T., Phillipps, M. A., and Goff, H. M., 1981. Infrared spectroscopy of oxidized metalloporphyrins: Detection of a band diagnostic of porphyrin-centered oxidation, *J. Am. Chem. Soc.* 81:6778–6780.

Sievers, G., Asterlund, K., and Ellfolk, N., 1979. Resonance Raman study on yeast cytochrome *c* peroxidase: Effect of coordination and axial ligands, *Biochim. Biophys. Acta* 581:1–14.

Simmoneaux, G., Scholtz, W. F., Reed, C. A., and Lang, G., 1982. Mossbauer spectra of unstable iron porphyrins: Models for compound II of peroxidase, *Biochim. Biophys. Acta* 716:1–7.

Splittgerber, A. G., and Tappel, A. L., 1979. Steady state and pre-steady state kinetic properties of rat liver selenium–glutathione peroxidase, *J. Biol. Chem.* 254:9807–9813.

Tata, J. R., 1976. Thyroglobulin mystery solved?, *Nature (London)* 259:527–528.

Warren, L. F., and Bennett, M. A., 1976. Comparative study of tertiary phosphine and arsine coordination to the transition metals: Stabilization of high formal oxidation state by *o*-phenyline-based chelate ligands, *Inorg. Chem.* 15:3126–3140.

Wright, J. R., Rumbaugh, R. C., Colby, H. D., and Miles, P. R., 1979. The relationship between chemiluminescence and lipid peroxidation in rat hepatic microsomes, *Arch. Biochem. Biophys.* 192:344–351.

Yonetani, T., Schleyer, H., and Ehrenberg, A., 1966. Studies of cytochrome peroxidase. VII. Electron paramagnetic resonance absorptions of the enzyme and its complex ES in dissolved and crystalline forms, *J. Biol. Chem.* 241:3240–3243.

Zinner, K., Vidigal-Martinelli, C., Duran, N., Marsaioli, A. J., and Cilento, G., 1980. A new source of carbon oxide in biochemical systems: Implications regarding dioxetane intermediates, *Biochem. Biophys. Res. Commun.* 92:32–37.

Dioxygen as a Terminal Oxidant and the Formation of Dioxygen

7

7.1 Fitness of Dioxygen as a Terminal Oxidant

From the previous discussion of the chemistry of dioxygen, it is clear that dioxygen must be activated before it can be utilized by organisms as a terminal oxidant. In this chapter, we shall discuss how dioxygen can be activated, with special reference to the methods used in biology. However, we first wish to point out that the requirement for activation and certain other properties make dioxygen an ideal terminal oxidant for life processes.

Any prospective terminal oxidant for a biological system would necessarily be fairly powerful so that the organism could extract as much energy as possible from its food supply. Thus, the first requirement of a terminal oxidant is that it be a strong oxidizing agent.

The strong oxidizing agents are elements in groups VI and VII of the periodic table, particularly the lighter elements. George (1965) has pointed out that possible strong oxidants for use as a terminal oxidant would include fluorine, chlorine, bromine, iodine, oxygen, and sulfur. The oxidizing potentials of these elements are listed in Table 7-1. Compared with other strong oxidants, dioxygen is rather good. The clear exception is fluorine, which is far more powerful than all the other agents. As we shall see in the following discussion, fluorine has some obvious drawbacks.

Iodine and sulfur are both solids and would therefore place severe restrictions on the organism. These oxidants could not be ubiquitous so would require oxidant storage in the organism. The other elements, fluorine, chlorine, bromine, and oxygen, are all gases, so storage would not be necessary if they occurred in the atmosphere. There are immediate problems that arise with fluorine, chlorine, and bromine. It is hard to visualize a possible life chemistry without water. Chlorine and bromine both react with water, producing, respectively, hypochlorous acid and hypobromous acid:

$$Cl_2 + H_2O \longrightarrow Cl^- + H^+ + HOCl$$

$$Br_2 + H_2O \longrightarrow Br^- + H^+ + HOBr$$

Fluorine will react with water to produce oxygen:

$$2 F_2 (g) + 2 H_2O (liters) \longrightarrow O_2 (g) + 4HF \text{ Aq. } \Delta G = -16.83 \text{ kcal}$$

Fluorine is also extremely reactive toward any organic compound.

Dioxygen is also an ideal terminal oxidant because the products, carbon dioxide and water, are innocuous. Imagine using chlorine or bromine as a terminal oxidant and producing large quantities of hydrochloric or hydrobromic acid in the cell. Another important factor is the solubility in water. Even though dioxygen

Table 7-1. Oxidizing Potentials of the Halogens, Oxygen, and Sulfur

	$E°$
Acidic solution $(H^+) = 1$	
$\frac{1}{2}F_2 + H^+ + e^- \longrightarrow HF$	$+3.06$
$\frac{1}{2}Cl_2 + e^- \longrightarrow Cl^-$	$+1.36$
$\frac{1}{4}O_2 + H^+ + e^- \longrightarrow \frac{1}{2}H_2O$	$+1.23$
$\frac{1}{2}Br_2 + e^- \longrightarrow Br^-$	$+1.07$
$\frac{1}{2}I_2 + e^- \longrightarrow I^-$	$+0.54$
$\frac{1}{2}S + H^+ + e^- \longrightarrow \frac{1}{2}H_2S$	$+0.14$
Neutral solution	
$\frac{1}{2}F_2 + e^- \longrightarrow F^-$	$+2.87$
$\frac{1}{2}Cl_2 + e^- \longrightarrow Cl^-$	$+1.36$
$\frac{1}{2}Br_2 + e^- \longrightarrow Br^-$	$+1.07$
$\frac{1}{4}O_2 + H^+ + e^- \longrightarrow \frac{1}{2}H_2O$	$+0.82$
$\frac{1}{2}I_2 + e^- \longrightarrow I^-$	$+0.54$
$\frac{1}{2}S + 0.74H^+ + e^- \longrightarrow 0.24H_2S + 0.26HS^-$	-0.27

is not as soluble in water as chlorine and bromine, it has a solubility (1.4×10^{-3} M per 1 atm dioxygen) sufficient for reaction.

However, probably the most important factor is that dioxygen is relatively unreactive and must be activated before reaction. Thus, an activator (such as an enzyme) has control over which reactions can and cannot occur. Because there are barriers to oxidation by dioxygen, life does not burn up. The dioxygen must be activated before it can react in biological reactions, and this is one of the most important subjects in oxygen metabolism. The high negative free energy of the oxidation, coupled with its low reactivity, is an ideal combination of characteristics for a terminal oxidant.

The reactive but low-energy excited states of dioxygen could be a problem, but these are not produced by simple light irradiation. Since both ground and excited states are symmetrical with respect to charge distribution, there is no transition dipole for the excitation. Hence, the energetically accessible excited states, which are dangerously reactive, cannot be formed directly by sunlight but only via the intermediacy of dyes that are generally not available in biological systems. A major exception to this is chlorophyll in plants. Chlorophyll is an extremely good singlet-dioxygen sensitizer. Carotenoids, also present in large quantities in plant cells, are extremely efficient singlet-dioxygen quenchers and hence effectively protect the plants from this reactive species.

The catalysis of dioxygen reactions in biological systems seems to be accomplished in at least three different ways. The simplest method is merely to use reactions that are allowed, so that activation of the dioxygen is not necessary. Essentially, the substrate is activated instead of the dioxygen. Chapter 2 discusses the allowed reaction of triplet dioxygen with a radical, a reaction that is utilized in biological systems. An example of this reaction is provided by D-amino acid oxidase, the enzyme that catalyzes the reaction between dioxygen and an amino acid to form an imino acid:

$$O_2 + 2FADH \longrightarrow 2FAD + H_2O_2$$

$$2FAD + \text{amino acid} \longrightarrow \text{imino acid} + 2FADH$$

The cofactor for this reaction is flavin adenine dinucleotide. It is interesting to note that dioxygen reacts with the highly reactive FADH· radical, so that the dioxygen need not be activated (Wellner and Meister, 1961): The details of the second reaction are discussed in Chapter 12.

Another reaction of triplet dioxygen that is allowed is the reaction with a substrate to form a low-lying triplet state. This also occurs with flavin enzymes. As mentioned in Chapter 2, dioxygen can combine with a reduced flavin to form a low-energy triplet complex that can decay to a singlet product. The details of this type of reaction are discussed in Chapter 12 (see also Chapter 17 for the reaction of triplet dioxygen with a carbanion).

The barriers to triplet-dioxygen reaction discussed in Chapter 2 are the difficulty of adding the first electron to the dioxygen and also the restrictions resulting from its triplet character. Both these barriers are eliminated when dioxygen is bound to a metal. The details of this type of activation of dioxygen, which probably comprises the most common biological dioxygen reactions, are dicussed in Chapter 8.

7.2 Formation of Dioxygen

The characteristics that make oxygen an ideal oxidant for respiration processes also make water an ideal reductant for carbon dioxide reduction. This occurs via photosynthesis. Water is, of course, readily available in solution; the oxidation product, O_2, is relatively innocuous unless specifically activated, is easily removed, and causes a very large energy difference between reactants and products.

The photosynthetic process of water oxidation is known as the Hill reaction and has received a great deal of attention, especially since photolytic water oxidation has emerged as a possible future energy source. Water oxidation to dioxygen involves removal of four electrons at an energy of 1.23 eV for each. Four protons are also lost to the solvent in this process.

Chloroplasts (Cheniae, 1970) that have been stored in darkness and then pulsed periodically with light show a periodicity to the quantity of dioxygen evolved. That is, little or no dioxygen is evolved on pulses 1, 5, 9, . . . and maxima are obtained on pulses 3, 7, 11, ... Hence, the periodicity is 4, and the series is damped out within a few cycles.

A relatively successful model has been developed on the basis of this periodicity. According to the model, there are five "S-states" that represent molecular internal states in the overall oxidation. Important conclusions from the light-pulsed experiments are that the electrons are removed one at a time from water and that the intermediates involved are relatively stable. Their lifetimes can be judged from the maximum duration between pulses that does not result in interconversion among states. The observed stabilities are much greater than one would expect for the simple water oxidation intermediates: $\cdot OH$, H_2O_2, and $HO_2\cdot$. Molecular identification of the S-state has received much study.

Both Mn^{2+} and Cl^- are known to be essential to the Hill reaction (Cheniae, 1970). Some Mn^{2+} can be removed from chloroplasts by rupturing cell walls and washing with EDTA. Removal of the Mn^{2+} critical to water oxidation requires further treatment: pH 8 tris wash, NH_2OH extraction, or mild heat treatment. This membrane-bound MN^{2+} pool is electron spin resonance (ESR)-

silent, but gives the characteristic six-line ESR signal when released to solution by these treatments. Approximately one third of the membrane-bound Mn^{2+} is not released by these treatments, but oxygen evolution is reduced by 90% or more. The effects of Mn^{2+} depletion can be reversed by addition of more Mn^{2+}.

Chloride ion is also required for the Hill reaction. This requirement can be met by other anions, but chloride is the only one present in sufficient physiological concentration to fulfill the requirement.

Proton losses to solvent during photosynthesis can be measured with difficulty. It has been shown (Sauer, 1980) that the protons are not all lost at the final step, but rather proton loss is associated with transitions among the S-states. Exactly which transitions give proton loss is a matter of dispute. However, it is agreed that two protons are lost in the final oxidation step and that there is one step in which no protons are lost.

Sauer (1980) observed that preceding heat treatment with illumination gives less Mn^{2+} ESR signal than does dark heat treatment (2 min at 55°C). It was shown that this is because some of the manganese is released as MnO_2, which is ESR-silent. MnO_2 results from high Mn oxidation states associated with the different S-states—quantification of the Mn^{2+} released as a function of S-state led to the conclusion that each S-state contains two Mn ions. Reactions that allowed conversion to detectable Mn^{2+} ions showed which Mn oxidation states were involved in each S-state.

The model that best accommodates these results has both Mn ions in the II state in the resting form with at least one water ligand each. The first electron transfer oxidizes the complex to II–III. The increased positive charge of the complex is neutralized by addition of an OH^- ligand, which results in release of H^+ to the solution. Next, the complex is oxidized to III–III by a second two-photon reaction. Lack of proton release to solution could be explained by complex charge neutralization by a negative ion ligand such as Cl^-. Removal of the third electron results in Mn II–III states, indicating transfer to two electrons from an oxygen-containing species, such as complexed water to Mn. At this state, the binuclear complex may be bridged by a bond. The final two protons are released after removal of the fourth electron. Dioxygen is also released from this fourth state, which then returns to the resting II–II state.

Preliminary spectroscopic results indicate that each Mn has another metal (similar to, and possibly another, Mn) as a nearest neighbor at a distance that corresponds to distances for known binuclear Mn dioxygen model compounds with two O atom bridges.

The species that directly removes electrons from the binuclear Mn complex is not known. Spector and Winget (1980) have isolated an Mn-containing protein from chloroplast membranes. This protein exhibits the property of giving up its Mn^{2+} to pH 8 tris buffer. It has a molecular weight of 65,000. The Mn content was measured at 1.61 and 2.65 moles per mole of protein, from which the authors concluded that each molecule contains two Mn^{2+} ions. When depleted thylakoid membranes (photosomes) were treated with this protein, oxygen evolution resumed. This led to the conclusion that the isolated protein is the enzyme in which the water oxidation occurs. Unfortunately, isolation of this protein has not been reported by other researchers.

References

Cheniae, G. M., 1970. Photosystem II and O_2 evolution, *Annu. Rev. Plant Physiol.* 21:467–498.

George, P., 1965. The fitness of oxygen, in *Oxidases and Related Redox Systems*, T. E. King, H. S. Mason, and M. Morrison (eds.), John Wiley, New York, pp. 3–32.

Sauer, K., 1980. A role for manganese in oxygen evolution in photosynthesis, *Acc. Chem. Res.* 13:249–256.

Spector, H., and Winget, G. D., 1980. Purification of a manganese-containing protein involved in photosynthetic oxygen evolution and its use in reconstituting an active membrane, *Proc. Natl. Acad. Sci. U.S.A.* 77:957–959.

Wellner, D., and Meister, A., 1961. Studies on the mechanism of action of L-amino acid oxidase, *J. Biol. Chem.* 236:2357–2364.

Metal–Dioxygen Complexes

8.1 Metal–Dioxygen Bonding

There has been considerable interest in metal chelates that will complex with dioxygen. These chelates are of interest to biochemists because they provide a comparison to the biological dioxygen carriers, hemoglobin and hemocyanin, and show what characteristics are crucial to enzymes that combine with dioxygen.

The structure of biochemical oxygen-carrying complexes has long been a matter of conjecture and hypothesis. Decades ago, Pauling proposed a structure for oxyhemoglobin that is now believed to be correct. The Pauling structure for oxyhemoglobin is one in which the dioxygen is bound to the metal through only one oxygen and the line connecting the iron–oxygen–oxygen atoms is bent away from linearity. A lone pair on one oxygen of the dioxygen is donated to an empty orbital of the metal (Figure 8-1). The orbital on dioxygen containing the lone pair is mixed with p-character (approximately sp^2) and consequently is not the same direction as the oxygen–oxygen bond. This oxygen-to-metal donation forms a σ-bond. Experimental evidence for the bent structure is found in the fact that oxyhemoglobin is diamagnetic. This indicates that the π_x^* and π_y^* orbitals are not degenerate in the complex as they are in free dioxygen. A linear metal–oxygen–oxygen structure would result in degenerate π_x^* and π_y^* orbitals and a triplet state.

In addition to the σ-bond, there are two types of interactions between the d-orbitals of the metal and the π^* orbitals of the dioxygen. One is through an overlap of the $3d_{xz}$ of the metal with the π_x^* orbitals perpendicular to the Fe–O–O plane (Figure 8-2). The other is through the overlap of the $3d_{z^2}$ orbital of the metal with the π_y^* orbital in the Fe–O–O plane (Figure 8-3). (Note that the directions x and y are arbitrary.) Overlap of the $3d_{z^2}$ orbital can be demonstrated in a BF_3 derivative (Figure 8-4) of the diphenyl glyoxime complex of cobalt II. This compound will reversibly bind oxygen. Cobalt has seven d-electrons so

Figure 8-1. Donation of a lone pair of electrons from dioxygen to a metal to form the metal–dioxygen σ bond in metal–dioxygen complexes.

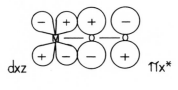

dxz   ∏x*

Figure 8-2. Double-bond character in a metal–dioxygen complex.

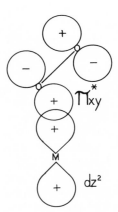

∏*xy

dz²

Figure 8-3. Further bonding in a metal–dioxygen complex.

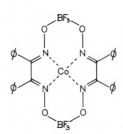

Figure 8-4. A cobalt complex that will reversibly bind dioxygen.

that the resulting complex is paramagnetic, unlike oxyhemoglobin. In a planar field around the cobalt, the unpaired electron would be in the $3d_z^2$ orbital of cobalt. The observation from electron paramagnetic resonance studies that the unpaired electron is in a π^* orbital shows that there must be strong overlap of the $3d_z^2$ orbital with the π_y oxygen orbital that is in the M–O–O plane (Torrog et al., 1976). The scheme is summarized (Summerville et al., 1979) in Figure 8-5. In ferrous complexes of dioxygen, there are only six valence electrons, so that the antibonding orbital formed between the d_{xz} and π_x^* orbitals is empty. In summary, lone-pair electrons (not shown) are donated from dioxygen to the metal, $3d_{xz}$ electrons are donated to the π_x^* orbital of dioxygen, and π_y^* electrons are donated back to the d_z^2 orbital of the metal. Calculations have shown (Olafson and Goddard, 1977; Huynh et al., 1977; Summerville et al., 1979) that this donation and acceptance leaves the dioxygen molecule essentially neutral. However, studies of model heme compounds have shown that the effect of ring substituents on the binding of oxygen is best explained by a slight dipolar character to the iron–oxygen bond, in which the iron is positive and the dioxygen is slightly negative (Traylor et al., 1981).

Griffith proposed another structure for oxyhemoglobin that was subsequently found not to occur in oxyhemoglobin, but that does occur in certain model

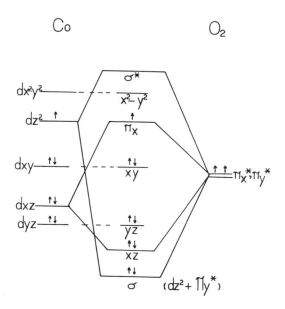

Figure 8-5. Bonding in cobalt–dioxygen complexes.

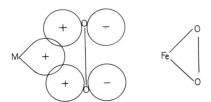

Figure 8-6. Back-bonding in a Griffith-type metal–dioxygen complex.

dioxygen complexes of metals. Griffith (1956) pointed out that the π-electrons of dioxygen are more easily donated to an empty orbital of iron than the non-bonding pair on one oxygen atom. This results in a structure in which the two oxygen atoms are equivalent.

The Griffith structure also allows for a strong back-donation of electrons (Figure 8-6) from the metal into the antibonding oxygen orbitals, as does the Pauling structure. Clearly, there are some large changes in the electronic structure of dioxygen when it binds to a metal.

8.2 Activation of Dioxygen

Chapter 2 discusses the two barriers to oxidation by ground-state dioxygen. One barrier results from the difficulty of adding one electron to the antibonding π-orbital of dioxygen. After the first electron is added, subsequent electrons may be added provided there are protons available to neutralize the negative charges developed. Essentially, the metal accomplishes the activation of dioxygen by adding this first electron. There is strong donation of π-electrons into π^* orbitals of dioxygen in all these metal–dioxygen complexes.

The other barrier to oxidations by ground-state dioxygen is the result of its triplet character. We note that the metal–dioxygen complexes are diamagnetic so that this barrier is also eliminated in the metal–dioxygen complex. The electronic structure resembles that of singlet dioxygen. We have noted earlier that many biological reactions of dioxygen more closely resemble singlet dioxygen than ground-state dioxygen.

Thus, metals may not only function as carriers for dioxygen but also act as activators for dioxygen.

8.3 Classification of Metal–Dioxygen Complexes

The classification of metal–dioxygen complexes with respect to electron donation from metal to dioxygen is useful because it forms a basis for the relative chemistry of these compounds.

The structure of the dioxygen varies from almost unperturbed dioxygen (except for a change in spin state) to a structure close to peroxide. However, like many chemical classifications, the lines between classes are arbitrary. Unfortunately, it is not possible to classify metal–dioxygen complexes with respect to the valence of the metal because the electrons are shared between the metal and oxygen, and without a molecular orbital calculation, one does not know where to assign the electron. However, the bond order between the two oxygens of the dioxygen varies among complexes, depending on how many electrons are donated to the π^* orbitals of the dioxygen. Also, the number of metal atoms required to bond each dioxygen varies from one complex to another and may also be used to classify complexes. The number of electrons donated to the π^* orbitals of dioxygen, or the bond order, may be measured by comparison of two experimental parameters with those of the known oxidation states of dioxygen. The two experimental parameters are the O–O distance, which increases with π^* occupation, and the O–O bond force constant, which decreases with π^* occupation. There are obvious minor problems with such a comparison because the dioxygen in metal complexes tends to have electrons paired, is affected by the metal charge, and has a different charge than the oxidation state of dioxygen (Drago and Corden, 1980). The O–O distances and force constants (as observed frequencies) are listed for four oxidation states of dioxygen in Table 8-1.

The bond order and stoichiometry may be used to classify metal–oxygen complexes (Summerville *et al.*, 1979; Vaska, 1976; Collman, 1977) as superoxide ion and one metal (1A), superoxide ion and two metals (1B), peroxide and one metal (2A), and peroxide and two metals (2B). These classifications are convenient even though it is clear that the metal–oxygen bond is covalent (Drago and Corden, 1980). They do not imply that the dioxygen is negatively charged in the superoxide complexes, or doubly negatively charged in the peroxide complexes, because there is a compensating donation of σ electrons from the dioxygen to the metal. The superoxide description means that the dioxygen

Table 8-1. Bond Orders, Bond Distances, and Vibrational Frequencies of Certain Oxidation States of Dioxygen

Oxidation state	Bond order	Distance (Å)	Frequency (cm^{-1})
O_2^+	2.5	1.12	1905
O_2	2.0	1.21	1580
O_2^-	1.5	1.33	1138[a]
$O_2^{}$	1.0	1.49	802

[a] This value is given by Drago and Corden (1980) for the frequency in HO_2 after correction of the observed frequency coupling to the OH frequency. The frequency for a covalent HO_2 molecule is more appropriate for discussion of transition metal–dioxygen complexes than the usual value of 1097 cm^{-1} from ionic KO_2 or NaO_2, since molecular orbital calculations show that dioxygen bound to the transitional metal is neutral as in HO_2.

Table 8-2. Vibrational Frequencies for Certain Metal–Dioxygen Structures

Structure	Frequency (cm^{-1})
1A	1103–1195
1B	1075–1122
2A	800– 932
2B	790– 884

contains about one extra electron, and the peroxide structures two extra electrons, in the π^* orbitals. The O–O frequencies for these classifications are shown in Table 8-2.

There are many nonbiological complexes that will bind dioxygen. One example is an iridium compound discovered by Vaska (1963), $Ir[(PtC_6H_5)_3]_2ClCO$. This compound reversibly reacts with oxygen at room temperature. The crystal structure of the dioxygen adduct was determined by Ibers and La Placa (1964), who found that the structure was that suggested by Griffith for oxyhemoglobin. Another example is manganese(II) phthalocyanine, which will bond dioxygen reversibly in N,N-dimethylacetamide (Lever et al., 1979). The frequency at 1154 cm^{-1} corresponds to a complex of type 1A.

8.4 Bonding vs. Oxidation

Biological oxygen carriers use cuprous copper or ferrous iron as the metals that bind the dioxygen. This appears to be a strange choice because both ferrous and cuprous ions are readily oxidized to cupric ion and ferric ion by dioxygen. The question is, why do the ferrous ions and cuprous ions in biological carriers bind dioxygen instead of being oxidized? To answer this question, we must look at the aerobic oxidation of ferrous and cuprous ions.

The mechanism of oxidation of ferrous ion to ferric ion by dioxygen is not a one-electron process because a one-electron oxidation is thermodynamically unfavorable. Oxidations of both cuprous and ferrous ions by dioxygen in one-electron processes have positive free energies of reaction:

$$Cu^+ + O_2 \longrightarrow Cu^{2+} + O_2^- \qquad +16.5 \text{ kcal}$$

$$Fe^{2+} + O_2 \longrightarrow Fe^{3+} + O_2^- \qquad +30.8 \text{ kcal}$$

As discussed previously, dioxygen is a poor oxidant for one-electron processes. The rate of oxidation of ferrous ion by $^3\Sigma_g$ dioxygen is proportional to

$[Fe^{2+}]^2 \times [O_2]$, indicating an intermediate that must contain two irons and one dioxygen (George, 1954). The structure of the intermediate must be that of two iron atoms bridged by a dioxygen molecule. Similar results are found for the autoxidation of ferrohemes, although the kinetics are complicated by the presence of pyridine as a fifth and sixth ligand (Cohen and Caughey, 1968) and the pyridines must leave before the dioxygen can form a bridge between the two ions. Direct nuclear magnetic resonance evidence for such an intermediate is found in the oxidation of ferroporphyrins (La Mar and Balch, 1977; Chin *et al.*, 1980). Similar Co(III)–peroxy-Co(III) complexes are known (Stadtherr *et al.*, 1973).

The addition of dioxygen to ferroheme and to many other ferrous chelates results in oxidation of the iron, not an oxygen–iron complex. To prevent the oxidation of iron, it is necessary to prevent the formation of $FeOOFe^{4+}$. This can be accomplished by adding enough bulky groups around the iron to prevent the dimer from forming.

Another factor that affects the oxidation of ferrous iron is the low dielectric constant. Electrons must be transferred from iron to the oxygen in the oxidation, followed by the separation of a negatively charged dioxygen species and a more positively charged iron. Such a process should be inhibited by a medium with a low dielectric constant.

A third factor that influences the dioxygen binding is the stability of a 5-coordinated iron that has one position free to bind the dioxygen. When most amines such as imidazole are added to heme, both the 5- and 6-positions are filled. Binding to the 6-position is more favorable than binding to the 5-position. An exception to this is 2-methyl imidazole, which does form a 5-coordinated complex of heme (Brault and Rougee, 1974). A dioxygen carrier, then, must have one position open for the binding of dioxygen.

References

Brault, D., and Rougee, M., 1974. Binding of imidazole and 2-methyl-imidazole by hemes in organic solvents: Evidence for five coordination, *Biochem. Biophys. Res. Commun.* 57:654–659.

Chin, D.-H., La Mar, G. N., and Balch, A. L., 1980. On the mechanism of autooxidation of iron(II) porphyrins: Detection of a peroxo-bridged iron(II) porphyrin dimer and the mechanism of its thermal decomposition to the oxo-bridged iron(II) porphyrin dimer, *J. Am. Chem. Soc.* 102:4344–4350.

Cohen, I. A., and Caughey, W. S., 1968. Substituted deuteroporphyrins. IV. On the kinetics and mechanism of reaction of iron(II) porphyrins with oxygen, *Biochemistry* 7:636–641.

Collman, J. P., 1977. Synthetic models for the oxygen-binding hemoproteins, *Acc. Chem. Res.* 10:265–272.

Drago, R. S., and Corden, B. B., 1980. Spin pairing model of dioxygen binding and its application to various transition-metal systems as well as hemoglobin cooperativity, *Acc. Chem. Res.* 13:353–359.

George, P., 1954. The oxidation of ferrous perchlorate by molecular oxygen, *J. Chem. Soc.* 1954:4349–4359.

Griffith, J. S., 1956. On the magnetic properties of some hemoglobin complexes, *Proc. R. Soc. London Ser. A* 235:23.

Huynh, B. H., Case, D. A., and Karplus, M., 1977. Nature of the iron oxygen bond in hemoglobin, *J. Am. Chem. Soc.* 99:6103–6105.

Ibers, J. A., and La Placa, S. J., 1964. Structure of a synthetic molecular oxygen carrier in *Proc. 8th Int. Conf. Coord. Chem.* Vienna, Sept. 7, Ed. by V. Gutmann, Springer-Verlag, New York, pp. 95–97.

La Mar, G. N., and Balch, A. L., 1977. Detection and characterization of the long-postulated Fe–O–O–Fe intermediate in the autooxidation of ferrous porphyrin, *J. Am. Chem. Soc.* 99:5486–5488.

Lever, A. B. P., Wilshire, J. P., and Quan, S. K., 1979. A manganese phthalocyanine–dioxygen molecular adduct, *J. Am. Chem. Soc.* 101:3668–3669.

Olafson, B. D., and Goddard, W. A., 1977. Molecular descriptions of dioxygen bonding in hemoglobin, *Proc. Natl. Acad. Sci. U.S.A.* 74:1315–1319.

Stadtherr, L. G., Prados, R., and Martin, R. B., 1973. Mono- and dibridged peroxo complexes of cobalt(III), *Inorg. Chem.* 12:1814–1818.

Summerville, D. A., Jones, R. D., Hoffman, B. M., and Bassolo, F., 1979. Assigning oxidation states to some metal dioxygen complexes of biological interest, *J. Chem. Ed.* 56:157–162.

Torrog, B. S., Kitko, D. J., and Drago, R. S., 1976. Nature of bound O_2 in a series of cobalt dioxygen adducts, *J. Am. Chem. Soc.* 98:5144–5153.

Traylor, T. G. White, D. K., Campbell, D. H., and Berzinis, A. D., 1981. Electronic effects of the binding of dioxygen and carbon monoxide to hemes, *J. Am. Chem. Soc.* 103:4932–4936.

Vaska, L., 1963. Oxygen-carrying properties of a simple synthetic system, *Science* 140:809–810.

Vaska, L., 1976. Dioxygen–metal complexes: Toward a unified view, *Acc. Chem. Res.* 9:175–182.

Biological Iron Dioxygen Carriers

9.1 Hemoglobin and Myoglobin

In biological systems, dioxygen is carried from the point of entry in the organism to the point of consumption by very specialized molecules. These molecules use either iron or copper to carry the dioxygen. The iron dioxygen carriers are hemoglobin (for a review, see Buchler, 1978; Brunori *et al.*, 1982), which carries dioxygen in the blood of higher animals, and myoglobin, which carries dioxygen in the muscles of higher animals. Hemerythrin carries dioxygen in certain nonvertebrates. Hemoglobin and myoglobin contain the iron chelated as ferrous ion in a protoporphyrin IX bound to the protein. Hemerythrin has no porphyrin; the iron is chelated by amino acids in the proteins.

The protein structure of hemoglobin is known in detail from X-ray studies (Perutz, 1976). The structures from quite varied sources are remarkably similar (Ten Eyck, 1979). Hemoglobin consists of four polypeptide chains: two chains designated α and two similar chains designated β. Each of these polypeptide chains contains a heme group, giving a total of four dioxygen-binding sites in hemoglobin. The binding strength at each of these four sites depends on how many other sites are occupied. The heme group consists of a protoporphyrin containing a ferrous ion bound to the imidazole of a histine. When the ferrous iron is oxidized to ferric, the hemoglobin is called methemoglobin.

Hemoglobin binds dioxygen in a cooperative manner; i.e., the binding strength increases as dioxygen is bound. The binding can be expressed mathematically in terms of the Hill equation:

$$Y/(1 - Y) = K\,(pO_2)^n$$

The fractional saturation of binding sites, Y, depends on a constant, K, the partial

121

pressure of dioxygen, pO$_2$, and a Hill coefficient, n. A value of $n = 1$ shows no cooperativity. Cooperativity occurs whenever n is greater than 1. The Hill coefficient for human hemoglobin is 2.8.

This is the result of two types of protein conformation: a resting conformation called an R-structure and a tension conformation called a T-structure. These two forms vary in crystal structure, solubility, and ligand affinity. The R-form has a high affinity and the T-form a low affinity for dioxygen.

Dissociation of the hemoglobin into subunits produces heme compounds that are always in the R-structure, so that the T-structure must be the result of interactions between the polypeptide chains. Oxygenation triggers changes among the binding sites that transform T-structures to R-structures. One theory is that the triggering mechanism for T-to-R conformational changes is the result of a change in iron position during oxygenation (Perutz, 1979; Perutz *et al.*, 1976).

Drago and Corden (1980) have suggested another cause of the interaction among sites triggered by dioxygenation. The dioxygen complex of heme causes the iron to become more acidic because of electron donation from iron to dioxygen. In turn, this causes the imidazole to become more strongly attached to the iron and to move toward the iron. This causes tension that finally triggers a T-to-R conversion when a sufficient number of hemes have been dioxygenated.

The kinetics of myoglobin binding dioxygen have been thoroughly studied (Austin *et al.*, 1975). From free dioxygen in solution to dioxygen bound to the iron, there are four energy barriers. If we define the dioxygen state in solvent as E and the bound dioxygen to iron as A, the barriers shown in Table 9-1 are found. Note that the last two barriers are given reference to stale C.

These barriers have been rationalized to terms of molecular structure. The E–D barrier is due to the loss of hydration shell to enter the hydrophobic cavity. The D–C barrier is ascribed to the narrow entrance to the cavity leading to weak bonding to the cavity (state C). The third barrier (C–B) is ascribed to breaking this weak bond to the cavity, and the fourth barrier (B–A) is ascribed to the actual chemical reaction between the heme and dioxygen.

In oxymyoglobin, the dioxygen is bound end-on to the iron with an

Table 9-1. Barriers between Intermediate States as Dioxygen Is Bound to Hemoglobin

States	Energy (kcal)
A–B	7.1
B–C	8.1
C–D	5.8
C–E	22.3

iron–oxygen–oxygen angle of 121° (Phillips, 1978). This is the Pauling structure for metal–dioxygen complexes. A similar iron–dioxygen structure is found in insect hemoglobin. In insect oxyhemoglobin, the dioxygen is bound in an end-on manner with an Fe–O–O angle of 170° (Weber *et al.*, 1978). The Fe–O–O angle in human oxyhemoglobin is 156° (Shaanan, 1982). Cobalt oxyhemoglobin has one unpaired electron. The electron paramagnetic resonance spectrum of the O^{17}–O^{16} adduct (Gupta *et al.*, 1975) shows two peaks corresponding to two cobalt O^{17} distances. This shows that the oxygens are nonequivalent and supports the Pauling structure for this complex.

There is a histidine group on the distal side of the porphyrin ring that probably aids in the binding of oxygen by donating an electron pair to the inner positive oxygen (Figure 9-1) of the bound dioxygen.

Steric effects of the distal histidine cause hemoglobin to bind carbon monoxide in a nonlinear manner, unlike the structure in model compounds (Peng and Ibers, 1976) (see also Section 9.2). This decreases the binding affinity by orders of magnitude, thus preventing our hemoglobin from being hopelessly tied up by metabolically generated CO. As with the dioxygen complexes of hemoglobin and myoglobin, the iron in carbonmonoxy complexes is in the plane of the prophyrin ring (Norvell *et al.*, 1975). However, the carbon monoxide is not on the heme axis. X-ray analysis of the carbon monoxide adduct of horse hemoglobin shows that the oxygen of the carbon monoxide is pushed off the heme axis (Heidner *et al.*, 1976). The carbon atom of the carbon monoxide cannot be detected, so it is not known whether the Fe–C–O is linear or bent. In insect hemoglobin (Huber *et al.*, 1970), the angle between the CO and the porphyrin ring is 145°.

The Fe^{2+}–Fe^{3+} potential for hemoglobin shows a large dependence on pH. Above about pH 8, the potential begins to decrease. This effect has been thoroughly studied (Brunori *et al.*, 1971) and has been found to be primarily the result of the ionization of heme-linked water. This ionization occurs at about pH 8.8.

Deoxyhemoglobin is a 5-coordinated high-spin complex, whereas oxyhemoglobin is a 6-coordinated low-spin complex. In addition, the electrons on the dioxygen are paired so that oxyhemoglobin is essentially diamagnetic (Pauling and Coryell, 1936; Taylor and Coryell, 1938; Coryell *et al.*, 1939). However,

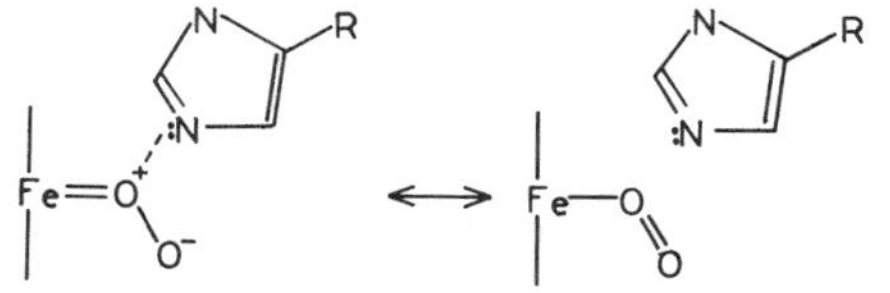

Figure 9-1. Stabilization of the hemoglobin–dioxygen complex by the distal imidazole.

there is a low-lying triplet state that contributes a small amount of magnetic moment through a terminal equilibrium between the singlet ground state and a triplet state at 146 cm^{-1} (Cerdonio et al., 1977, 1978). Theoretical calculations [with intermediate neglect of differential overlap (INDO) with considerable configuration interaction] predict a low-lying triplet state in oxyhemoglobin at 129 cm^{-1} (Herman and Loew, 1980) in excellent agreement with the experimental position for the triplet state.

The change in spin from high-spin to low-spin iron during oxygenation causes changes in the structure. In high-spin iron–porphyrin chelates, the iron is out of the plane of the porphyrin ring, and in low-spin iron–porphyrin chelates, the iron is in the plane of the porphyrin ring (Hoard et al., 1965; Countryman et al., 1969; Brown, 1970; Collins et al., 1972). This is the result of electron repulsion between the partially filled dx^2, dy^2, and dz^2 orbitals and the filled p-orbitals on nitrogen. The conversion of hemoglobin from high spin to low spin during dioxygenation correspondingly moves the iron from a nonplanar to a planar position (Ten Eyck, 1979). In both ferro and ferri hemes, the high-spin complexes always have a longer iron–nitrogen bond than do the low-spin hemes (Perutz, 1979).

The state of oxygen bound to hemoglobin has been studied by means of the infrared spectrum. The oxygen–oxygen vibrational frequency of the dioxygen complex of hemoglobin is at 1107 cm^{-1} (Barlow et al., 1973), which is very close to the value of 1097 cm^{-1} for superoxide ion. The corresponding frequency is 1103 cm^{-1} in oxymyoglobin (Maxwell et al., 1974), showing that the oxy-myoglobin and oxyhemoglobin structures are very similar. The relatively low O–O vibrational frequency shows that about one electron has been donated from the iron into the π^* orbital of the dioxygen.

One can roughly understand that back-donation from iron to dioxygen is essential for dioxygen binding by the following arguments (Jones et al., 1979): Singlet dioxygen is about 22.5 kcal above ground-state dioxygen, and the enthalpy for binding dioxygen to myoglobin is about -15 kcal/mole. Since dioxygen is bound in a diamagnetic state, we can roughly add these values to give a bond strength between iron and dioxygen of 37.5 kcal. This value is higher than one would expect for an iron–dioxygen single bond because there is partial double-bond character between the iron and dioxygen. This double-bond character between the iron and dioxygen prevents rotation of the iron–oxygen–oxygen plane. Calculations (INDO) have shown that there is a surprisingly high barrier of 7 kcal for rotation of the Fe–O–O plane with respect to the heme (Herman and Loew, 1980).

The dioxygen-binding side of the heme is exposed to solution. It is also reactive toward other ligands such as cyanide ion, nitric oxide, and carbon monoxide. This position will allow oxidation of the iron to the ferric state by alkyl halides. Ferrohemoglobin and ferromyoglobin can be oxidized to the ferri form by bromomalonitrile, by bromodiethylmalonate (Wade and Castro, 1973a), or by other alkyl halides (Wade and Castro, 1973b). The aklyl halides are reduced to malononitrile and diethylmalonate, respectively. The oxidation does not occur for cytochrome c, in which the porphyrin ring is exposed instead of the iron. The reaction with hemoglobin and myoglobin occurs in two steps, the first step being production of the bromo complex of the ferri form, together with an alkyl radical, followed by reduction of the alkyl radical to a carbon–hydrogen bond by another ferroheme. These reactions at least partially account for the toxicity of alkyl halides.

9.2 Hemoglobin Models

Baldwin and Huff (1973) have prepared an iron chelate that carries dioxygen by building enough steric hindrance around the iron that the dioxygenated dimer cannot form but the dioxygenated monomer can still form. The compound shown in Figure 9-2 was found to combine with 1 mole of dioxygen at $-78°C$ in a tetrahydrofuran–dimethoxyethane solution containing 4% pyridine.

The bridgehead structure prevents the benzene rings from lying flat, thus producing enough steric interference to prevent formation of the dioxygen–dimetal complex. Note that the bulky hydrocarbon groups and the low

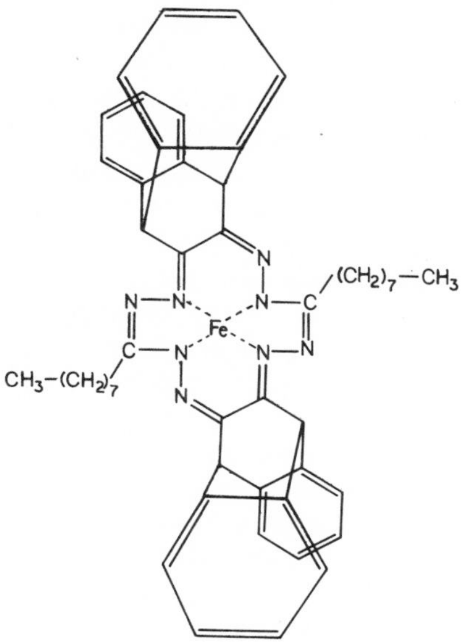

Figure 9-2. A metal chelate that will reversibly bond dioxygen.

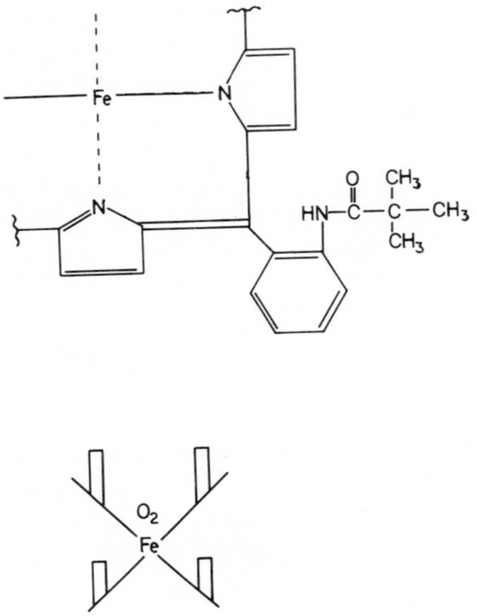

Figure 9-3. Structure of "picket-fence" porphyrin.

dielectric constant of the solvent place the dioxygen in a very hydrophobic environment.

Collman *et al.* (1973) have similarly prepared ferrous oxygen carriers with steric hindrance toward dimer formation by placing *t*-butyl amido groups on tetraphenyl porphyrins (Figure 9-3) (Collman, 1977). Because of steric repulsion between the hydrogens at the *ortho* positions of the phenyl groups and the hydrogens at the 3- and 4-positions of the pyrrole, the plane of the phenyl groups is perpendicular to the plane of the porphyrin ring, so that the *t*-butyl amide groups all stick up like posts on a picket fence above the plane of the porphyrin ring. Again, this complex has steric hindrance to prevent dimer formation and also places the dioxygen in a very hydrophobic environment. The other side of the porphyrin, away from the posts, can be filled with an imidazole chelated to the iron. Steric hindrance prevents the diimidazole complex from forming.

The dioxygen complex of the picket-fence porphyrin was found to have the Pauling structure (Collman *et al.*, 1975a). Two types of dioxygen complexes with the Pauling structure are formed: one in which the Fe–O–O plane is parallel to the imidazole plane and another in which the two planes are perpendicular. The parallel complex has an Fe–O–O angle of 135° and an O–O distance of 1.23 Å. The perpendicular complex has a corresponding angle of 137° and a corresponding distance of 1.26 Å. The bond lengths indicate that the metal-electron donation into the π^* orbitals of oxygen differs (Collman *et al.*, 1974;

Jameson *et al.*, 1978). The oxygen–oxygen stretch frequency of 1159 cm^{-1} (Collman *et al.*, 1976) is considerably below the value of 1556 cm^{-1} for ground-state dioxygen, indicating that there is considerable back-bonding into the antibonding π-orbitals of the bound dioxygen. There is also a donation of σ electrons from the oxygen to the iron.

The free energy, enthalpy, and entropy of dioxygen binding with these model compounds are very similar to those for myoglobin (Collman *et al.*, 1975b; Collman, 1977). This means that the protein in hemoglobin must not contribute significantly to the binding of dioxygen. The primary role of the heme protein is to protect the heme from oxidation.

When 2-methylimidazole instead of imidazole is used as the fifth ligand in oxyheme complexes, the iron is displaced toward the 2-methylimidazole by 0.399 Å from the plane of the ring. The methyl group does not allow the nitrogen to approach the porphyrin as closely as the nitrogen of imidazole. As a result, the iron–oxygen bond is increased to 1.898 Å (Jameson *et al.*, 1980). The stretched iron–oxygen bond in the 2-methylimidazole model is believed to be a model for the T-form.

The dioxygen binding to the picket-fence porphyrin is thus quite comparable to that of myoglobin and hemoglobin with respect to thermodynamics and structure. However, the picket-fence porphyrins bind carbon monoxide much more strongly than myoglobin or hemoglobin. The reason is that the picket-fence porphyrins can bind the carbon monoxide linearly, perpendicular to the porphyrin ring (Collman *et al.*, 1979), but hemoglobin and myoglobin bind carbon monoxide at an angle to the porphyrin ring. This phenomenon is discussed more thoroughly in Section 9.1, but essentially there is a terminal histidine on the hemoglobin that hinders carbon monoxide binding, forcing it to bind in a less stable manner. The CO frequency appears to be related to the binding constant (Collman *et al.*, 1976). The CO frequency for the picket-fence adduct (1969 cm^{-1}) is comparable to that for the carbon monoxide adduct of *N*-methylimidazole heme (1979 cm^{-1}). However, the myoglobin adduct is at 1945 cm^{-1} and the hemoglobin adduct is at 1951 cm^{-1}.

A group of model dioxygen carriers similar to the picket-fence porphyrins have been prepared in which all four chains (pickets) terminate in a benzene ring cap (Figure 9-4) (Almog *et al.*, 1975a). These compounds bind dioxygen to form crystalline complexes in the presence of a base such as imidazole. The dioxygen is bound underneath the cap (Almog *et al.*, 1975b) (Figure 9-5). These capped compounds have a lower affinity for dioxygen than do other synthetic dioxygen carriers because of steric interaction between dioxygen and the cap. The noncapped side of the heme binds the imidazole. When dioxygen is added before imidazole, the ferrous ion is oxidized via an iron–dioxygen–iron complex. When 1,2-dimethylimidazole is the ligand on the open side of the heme, instead of 1-methylimidazole, the affinity for dioxygen is reduced by a factor of about

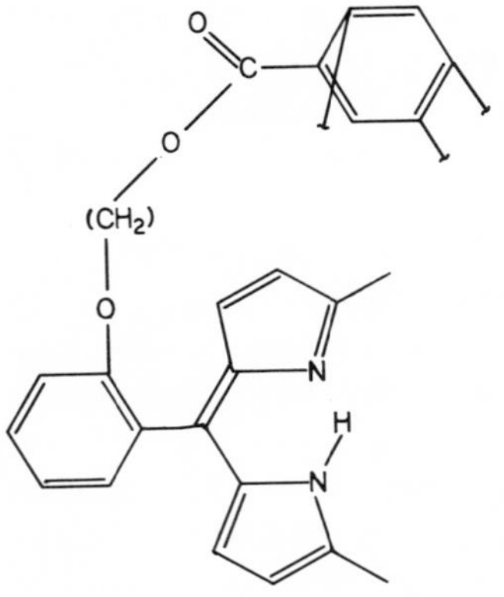

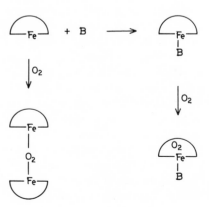

Figure 9-4. Structure of "capped" porphyrin.

180 (Linard *et al.*, 1980). This is quite similar again to the difference between the dioxygen binding of the R- and T-forms of hemoglobin.

Chang and Traylor (1973a,b) prepared porphyrin derivatives in which an imidazole is attached by a long chain to the porphyrin (Figure 9-6) (for a review of these and other types of model hemoglobins, see Traylor, 1981). The imidazole is able to coordinate with the iron. In this manner, it is possible to prepare monoimidazole complexes to porphyrin. Even though there is no steric hindrance to dimer formation, the complex will carry dioxygen in solvents of low dielectric constant.

A derivative in which the terminal amine is a pyridine group instead of imidazole was compared with the imidazole derivative. The partial pressures at which these model compounds are half-saturated with CO (23–25°C) and O_2 (45°C) in methylene chloride are shown in Table 9-2. Despite the relatively small

Figure 9-5. Binding of a base and dioxygen by a "capped" porphyrin.

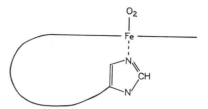

Figure 9-6. Structure of a model hemoglobin.

difference in affinities for CO, there is a large preference of the imidazole derivative for dioxygen over the pyridine derivative. This effect has been ascribed to a high π-basicity of imidazole relative to pyridine (cf. Figure 9-7). When a compound A-B is bound to an iron porphyrin, the π-back-bonding to the bound compound A-B becomes more crucial to binding as the σ-basicity of A-B decreases and the π-electronegativity of A-B increases. One would expect the dioxygen adduct to require more back-bonding than the carbon monoxide adduct because the π^* orbitals of dioxygen are more electropositive than the π^* orbitals of carbon monoxide. Hence, the observation that dioxygen is much more sensitive to the π-basicity of the ligand than is carbon monoxide is in accord with this model of the complex.

9.3 Hemerythrin

Despite its name, hemerythrin is a nonheme iron dioxygen carrier (for review, see Klotz *et al.*, 1976; Klotz and Kurtz, 1984). It serves as a respiratory protein in certain invertebrate marine organisms. The protein is an octamer of subunits, each with two iron atoms. The complete primary structure of the hemerythrin found in muscle tissue has been determined (Klippenstein *et al.*, 1976). The protein is capable of binding one dioxygen per subunit (Kurtz *et al.*,

Table 9-2. Partial Pressures of CO and O_2 Binding to Hemoglobin Models as a Function of the Base

Model	$p\frac{1}{2}$ mm O_2	$p\frac{1}{2}$ mm CO
Imidazole derivative	0.2	0.088
Pyridine derivative	760	1.20

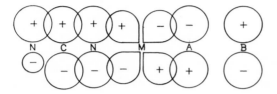

Figure 9-7. Effect of π-basicity of a *trans* ligand on the binding of an unsaturated ligand (A-B) to a metal (M).

1977; Stenkamp and Jensen, 1979). Several X-ray analyses have been made of hemerythrin, with slightly differing results (Hendrickson *et al*, 1975, 1982; Stenkamp *et al*., 1977, 1982; Stenkamp and Jensen, 1979). The iron atoms are between 3.30 and 3.34 Å apart (Klotz and Kurtz, 1984). The two irons are bridged (Stenkamp *et al*., 1981; Elam *et al*., 1982) by a glutamic carboxyl group and an aspartic carboxyl group. In addition, there is an oxo bridge. One iron is bound to three histidines and the other iron to two histidines. The azido and dioxygen ligands bind to the iron with only two histidines.

The oxo bridge is also consistent with both magnetic susceptibility measurements (Dawson *et al*., 1972) and electronic spectral data (Garbett *et al*., 1969). Oxo bridges between two ferrihemes are well known in porphyrin chemistry (Sadasivan *et al*., 1969; Fleischer and Srivastava, 1969). They can be formed by autoxidation of ferrohemes (Alben *et al*., 1968) and by dimerization of hydroxyferrihemes (Cohen, 1969).

Mossbauer spectral studies (York and Bearden, 1970; Garbett *et al*., 1971) show that the two irons in methemerythrin are both high-spin ferric, but the spins are coupled to produce diamagnetism. Both irons may be reduced to the ferrous state. In the ferrous state, both irons are still high-spin ($S = 2$), but there is no coupling between them.

The dioxygen is bound by one oxygen atom to one iron atom with the Fe–O bond perpendicular to the Fe–Fe line as shown in Figure 9-8. Addition of singly ^{18}O-labeled dioxygen to hemerythrin produces a resonance Raman spectrum with two O–O stretch frequencies in agreement with binding by only one oxygen atom (Kurtz *et al*., 1976). binding to only one iron is shown by extended X-ray absorption fine-structure data (Elam *et al*., 1982). Spectral dichroism studies show that the Fe–O bond is perpendicular to the Fe–Fe line (Gay and Solomon, 1978).

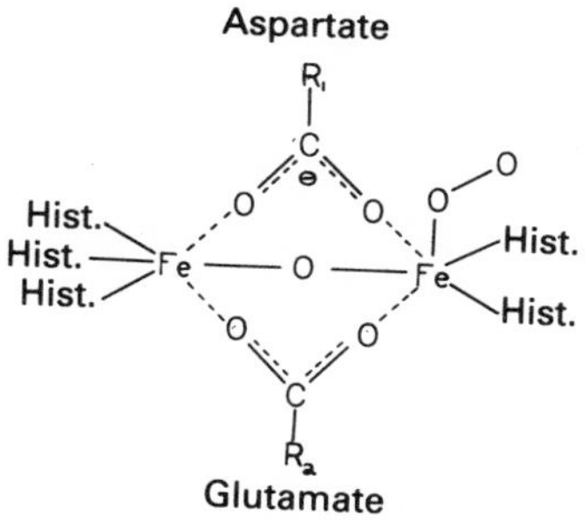

Figure 9-8. Structure of oxyhemerythrin.

There is enough donation of electrons from iron to dioxygen to give an O–O stretch frequency about equivalent to that of peroxide. The frequency is 844 cm^{-1} (Dunn *et al.*, 1973). Many of the properties of oxyhemerythrin are similar to those of methemerythrin (Loehr and Loehr, 1979), so that the complex appears to be hydrogen peroxide dianion bound to two ferric ions.

A model has been made of hemerythrin in which a ferrous iron is chelated by a saturated cyclic ring containing four amino groups and one pyridyl group (Kimura *et al.*, 1982). Two of these chelates bind one dioxygen molecule. The color of the oxycomplex is violet-pink similar to that of oxyhemerythrin.

References

Alben, J. O., Fuchsman, W. H., Beaudreau, C. A., and Caughey, W. S., 1968. Substituted deuteroporphyrins. III. Iron(II) derivatives: Reactions with oxygen and preparations from chloro- and methoxyporphyrins, *Biochemistry* 7:624–635.

Almog, J., Baldwin, J. E., Dyer, R. L., and Peters, M., 1975a. Condensation of tetraaldehydes with pyrrole: Direct synthesis of "capped" porphyrins, *J. Am. Chem. Soc.* 97:226–227.

Almog, J., Baldwin, J. E., and Huff, J., 1975b. Reversible oxygenation and autooxidation of a "capped" porphyrin iron(II) complex, *J. Am. Chem. Soc.* 97:227–228.

Austin, R. H., Beeson, K. W., Eisenstern, L., Frauenfelder, H., and Gunsalus, I. C., 1975. Dynamics of ligand binding to myoglobin, *Biochemistry* 14:5355–5373.

Baldwin, J. E., and Huff, J., 1973. Binding of dioxygen to iron(II): Reversible behavior in solution, *J. Am. Chem. Soc.* 95:5757–5759.

Barlow, C. H., Maxwell, J. C., Wallace, W. J., and Caughey, W. S., 1973. Elucidation of the mode of binding of oxygen to iron in oxyhemoglobin by infra-red spectroscopy, *Biochem. Biophys. Res. Commun.* 55:91–95.

Brown, C. B., 1970. Conformation of metalloporphyrins in solution, *J. Am. Chem. Soc.* 92:1423–1425.

Brunori, M., Saggese, U., Rotilio, G. C., Antonini, E., and Wyman, J., 1971. Redox equilibrium of sperm whale myoglobin, *Aplysia* myoglobin and *Chironomus thummi* hemoglobin, *Biochemistry* 10:1604–1609.

Brunori, M., Giardina, B., and Kuiper, H. A., 1982. Oxygen-transport proteins, *Inorg. Biochem.* 3:126–182.

Buchler, J. W., 1978. Hemoglobin—an inspiration for research in coordination chemistry, *Ang. Chem. Int. Ed. Engl.* 17:407–423.

Cerdonio, M., Congin-Castellano, A., Mogno, F., Pispisa, B., Romani, G. L., and Vitale, S., 1977. Magnetic properties of oxyhemoglobin, *Proc. Natl. Acad. Sci. U.S.A.* 74:398–400.

Cerdonio, M., Congin-Castellano, A., Calabresse, L., Morante, S., Pispisa, B., and Vitale, S., 1978. Room-temperature magnetic properties of oxy- and carbonmonoxyhemoglobin, *Proc. Natl. Acad. Sci. U.S.A.* 75:4916–4919.

Chang, C. K., and Traylor, T. G., 1973a. Synthesis of the myoglobin active site, *Proc. Natl. Acad. Sci. U.S.A.* 70:2647–2650.

Chang, C. K., and Traylor, T. G., 1973b. Proximal base influence on the binding of oxygen and carbon monoxide to heme, *J. Am. Chem. Soc.* 95:8477–8479.

Cohen, I. A., 1969. The dimeric nature of hemin hydroxides. *J. Am. Chem. Soc.* 91:1980–1983.

Collins, D. M., Countryman, R., and Hoard, L. L., 1972. Stereochemistry of low-spin iron porphyrins. I. Bis(imidazole)-$\alpha,\beta,\gamma,\delta$-tetra-phenylporphinatoiron(III) chloride, *J. Am. Chem. Soc.* 94:2066–2072.

Collman, J. P., Gagne, R. R., Reed, C. A., Robinson, W. T., and Rodley, G. A., 1974. Structure of an iron(II) dioxygen complex: A model for oxygen carrying heme proteins, *Proc. Natl. Acad. Sci. U.S.A.* 71:1326–1329.

Collman, J. P., Gagne, R. R., Reed, C. A., Talburt, T. R., Lang, C., and Robinson, W. T., 1975a. Picket fence porphyrins: Synthetic models of oxygen binding heme proteins, *J. Am. Chem. Soc.* 97:1427–1439.

Collman, J. P., Brauman, J. I., and Suslick, K. S., 1975b. Oxygen binding to iron porphyrins, *J. Am. Chem. Soc.* 97:7185–7186.

Collman, J. P., Brauman, J. I., Halbert, T. R., and Suslick, K. S., 1976. Nature of O_2 and CO binding to metalloporphyrins and heme proteins. *Proc. Natl. Acad. Sci. U.S.A.* 73:3333–3337.

Collman, J. P., Brauman, J. I., and Doxsee, K. M., 1979. Carbon monoxide binding to iron porphyrins, *Proc. Natl. Acad. Sci. U.S.A.* 76:6035–6039.

Coryell, C. D., Pauling, L., and Dodson, R. W., 1939. The magnetic properties of intermediates in the reactions of hemoglobin, *J. Phys. Chem.* 43:825–839.

Countryman, R., Collins, D. M., and Hoard, J. L., 1969. Stereochemistry of the low-spin porphyrin: Bis(imidazole)-$\alpha,\beta,\gamma,\delta$-tetraphenylporphinatoiron(III) chloride, *J. Am. Chem. Soc.* 91:5166–5167.

Dawson, J. W., Gray, H. B., Hoenig, H. E., Rossman, G. R., Schredder, J. M., and Wang, R.-H., 1972. A magnetic susceptibility study of hemerythrin using an ultrasensitive magnetometer, *Biochemistry* 11:461–465.

Drago, R. S., and Corden, B. B., 1980. Spin-pairing model of dioxygen binding and its application to various transition-metal systems as well as hemoglobin cooperativity, *Acc. Chem. Res.* 13:353–359.

Dunn, J. B. R., Shriver, D. F., and Klotz, I. M., 1973. Resonance Raman studies of the electronic state of oxygen in hemerythrin, *Proc. Natl. Acad. Sci. U.S.A.* 70:2582–2584.

Elam, W. T., Stern, E. A., McCallum, J. D., and Sanders-Laeher, J., 1982. Structure of the binuclear iron center in hemerythrin by X-ray absorption spectroscopy, *J. Am. Chem. Soc.* 104:6369–6373.

Fleischer, E., and Srivastava, T. S., 1969. The structure and properties of μ-oxo-bis[tetraphenylporphineiron(III)] *J. Am. Chem. Soc.* 91:2403–2405.

Garbett, K., Darnall, D. W., Klotz, I. M., and Williams, R. J. P., 1969. Spectroscopy and structure of hemerythrin, *Arch. Biochem. Biophys.* 135:419–434.

Garbett, K., Johnson, C. E., Klotz, I. M., Okamura, M. Y., and Williams, R. J. P., 1971. Hemerythrin: Further studies of Mossbauer structure, *Arch. Biochem. Biophys.* 142:574–583.

Gay, R. R., and Solomon, E. I., 1978. Polarized single-crystal spectroscopic studies of oxyhemerythrin, *J. Am. Chem. Soc.* 100:1972–1973.

Gupta, R. K., Mildvan, A. S., Yonetani, T., and Srivastava, T. S., 1975. EPR study of ^{17}O nuclear hyperfine interaction in cobalt–oxyhemoglobin: Conformation of bound oxygen, *Biochem. Biophys. Res. Commun.* 67:1005–1012.

Heidner, E. J., Ladner, R. C., and Perutz, M. F., 1976. Structure of horse carbonmonoxyhemoglobin, *J. Mol. Biol.* 104:707–722.

Hendrickson, W. A., Klippenstein, G. L., and Ward, K. B., 1975. Tertiary structure of myohemerythrin at low resolution, *Proc. Natl. Acad. Sci. U.S.A.* 72:2160–2164.

Hendrickson, W. A., Co, M. S., Smith, J. L., Hodgson, K. O., and Klippenstein, G. L., 1982. X-ray absorption spectroscopy of the dimeric iron site in azidomethemerythrin from *Phascolopsis gouldin, Proc. Natl. Acad. Sci. U.S.A.* 79:6255–6259.

Herman, Z. S., and Loew, G. H. 1980. A theoretical investigation of the magnetic ground state properties of model oxyhemoglobin complexes, *J. Am. Chem. Soc.* 102:1815–1821.

Hoard, J. L., Hamor, M. J., Hamor, T. A., and Caughey, W. S., 1965. The crystal structure and molecular stereochemistry of methoxyiron(III) mesoporphyrin-IX dimethyl ester, *J. Am. Chem. Soc.* 87:2312–2319.

Huber, R., Epp, O., and Formanek, H., 1970. Structures of deoxy- and carbonmonoxyerythrocruorin, *J. Mol. Biol.* 52:349–354.

Jameson, G. B., Rodley, G. A., Robinson, W. T., Gagne, R. R., Reed, C. A., and Collman, J. P., 1978. Structure of a dioxygen adduct of (1-methylimidazole)-meso-tetrakis ($\alpha,\alpha,\alpha,\alpha,O$-pival-amidophenyl) porphinatoiron-(II): An iron dioxygen model for the heme component of oxymyoglobin, *Inorg. Chem.* 17:850–857.

Jameson, G. B., Molinaro, F. S., Ibers, J. A., Collman, J. P., Braumann, J. L., Rose, E., and Suslick, K. S., 1980. Models for the active site of oxygen-binding hemoproteins: Dioxygen binding properties and the structures of (2-methyl-imidazole)-meso-tetra ($\alpha,\alpha,\alpha,\alpha,O$-pival-amidophenyl)-porphyrinatoiron-(II)ethanol and its dioxygen adduct, *J. Am. Chem. Soc.* 102:3224–3237.

Jones, R. D., Summerville, D. A., and Basolo, F., 1979. Synthetic oxygen carriers related to biological systems, *Chem. Rev.* 79:139–179.

Kimura, E., Kodama, M., Machida, R., and Ishizu, K., 1982. A new pyridyl-containing pentaaza macrocyclic ligand: Stabilization in aqueous solutions of the iron(II) complex and its dioxygen complex, *Inorg. Chem.* 21:595–602.

Klippenstein, G. L., Cote, J. L., and Ludlan, S. E., 1976. The primary structure of myohemerythrin, *Biochem.* 15:1128–1136.

Klotz, I. M., and Kurtz, Jr., D. M., 1984. Binuclear oxygen carriers: Hemerythrin, *Acc. Chem. Res.* 17:16–22.

Klotz, I. M., Klippenstein, G. L., and Hendrickson, W. A., 1976. Hemerythrin: Alternative oxygen carrier, *Science* 192:335–344.

Kurtz, D. M., Jr., Shriver, D. F., and Klotz, I. M., 1976. Resonance Raman spectroscopy with unsymmetrically isotopic ligands: Differentiation of possible structures of hemerythrin complexes, *J. Am. Chem. Soc.* 98:5033–5035.

Kurtz, D. M., Jr., Shriver, D. F., and Klotz, I. M., 1977. Structural chemistry of hemerythrin, *Coord. Chem. Rev.* 24:145–178.

Linard, J. E., Ellis, P. E., Jr., Budge, J. R., Jones, R. D., and Basolo, F., 1980. Oxygenation of iron(II) and cobalt(II) capped porphyrins, *J. Am. Chem. Soc.* 102:1896–1904.

Loehr, J. S., and Loehr, T. M., 1979. Hemerythrin: A review of structural and spectroscopic properties, in *Advances in Organic Biochemistry I*, G. L. Eichorn and L. G. Marzilla (eds.), Elsevier, New York, pp. 235–252.

Maxwell, J. C., Volpe, J. A., Barlow, C. H., and Caughey, W. S., 1974. Infrared evidence for the mode of binding of oxygen to iron of myoglobin from heart muscle, *Biochem. Biophys. Res. Commun.* 58:166–171.

Norvell, J. C., Nunes, A. D., and Schoenborn, B. P., 1975. Neutron diffraction analysis of myoglobin: Structure of the carbon monoxide derivative, *Science* 190:568–570.

Pauling, L., and Coryell, C. D., 1936. The magnetic properties and structure of hemoglobin: Oxyhemoglobin and carbonmonoxyhemoglobin, *Proc. Natl. Acad. Sci. U.S.A.* 22:210–216.

Peng, S.-M., and Ibers, J. A., 1976. Stereochemistry of carbonyl metalloporphyrins: The structure of (pyridine) (carbonyl) (5,10,15,20-tetraphenyl-porphinato) iron(III), *J. Am. Chem. Soc.* 98:8032–9036.

Perutz, M. F., 1970. Stereochemistry of cooperative effects in hemoglobin, *Nature (London)* 228:726–734.

Perutz, M. F., 1976. Structure and mechanism of hemoglobin, *Br. Med. Bull.* 32:195–208.

Perutz, M. F., 1979. Regulations of oxygen affinity of hemoglobin, *Annu. Rev. Biochem.* 48:327–386.

Perutz, M. F., Kilmartin, J. V., Nagai, K., Szabo, A., and Simon, S. R., 1976. Influence of globin structures on the state of the heme ferrous low spin derivatives, *Biochemistry* 15:378–387.

Phillips, S. E. V., 1978. Structure of oxymyoglobin, *Nature (London)* 273:247–248.

Sadasivan, N., Eberspaecher, H. I., Fuchsman, W. H., and Caughey, W. S., 1969, Substituted deuterporphryins. VI. Ligand-exchange and dimerization reactions of deuterohemins, *Biochemistry* 8:534–541.

Shaanan, B., 1982. The iron–oxygen bond in human oxyhemoglobin, *Nature (London)* 296:683–684.

Stenkamp, R. E., and Jensen, L. H., 1979. Hemerythrin and myohemerythrin: A review of models based on X-ray crystallographic data, in *Advances in Inorganic Biochemistry I*, G. L. Eichron and L. G. Marzilli (eds.), Elsevier, New York, pp. 219–234.

Stenkamp, R. E., Sieker, L. C., and Jensen, L. H., 1977. Structure of the iron complex in methemerythrin, *Proc. Natl. Acad. Sci. U.S.A.* 73:349–351.

Stenkamp, R. E., Sieker, L. C., Jensen, L. H., and Sanders-Loehr, J., 1981. Structure of the binuclear complex in metaazidohemerythrin from *Themiste dyscritum* at 2.2 Å resolution, *Nature (London)* 291:263–264.

Stenkamp, R. E., Sieker, L. C., and Jensen, L. H., 1982. Restrained least squares refinement of *Themiste dyscritum* methydroxahemerythrin at 2.0 Å resolution, *Acta Chrystallogr.* B38:784–792.

Taylor, D. S., and Coryell, C. D., 1938. The magnetic susceptibility of iron ferrohemoglobin, *J. Am. Chem. Soc.* 60:1177–1181.

Ten Eyck, L. F., 1979. Hemoglobin and myoglobin, in *The Porphyrins*, Vol. VII, D. Dolphin (ed.), Academic Press, New York, pp. 445–472.

Traylor, T. G., 1981. Synthetic model compounds for hemoproteins, *Acc. Chem. Res.* 14:102–109.

Wade, R. S., and Castro, D. E., 1973a. Oxidation of heme proteins by alkyl halides, *J. Am. Chem. Soc.* 95:231.

Wade, R. S., and Castro, C. E., 1973b. Oxidation of iron(III) porphyrins by alkyl halides, *J. Am. Chem. Soc.* 95:226–230.

Weber, E., Steigemann, W., Jones, T. A., and Huber, R., 1978. The structure of oxy-erythrocruorin at 1.4 Å resolution, *J. Mol. Biol.* 120:327–336.

York, J. S., and Bearden, A. J., 1970. Active site hemerythrin: Iron electronic states and the binding of oxygen, *Biochemistry* 9:4549–4554.

Hemocyanin: A Biological Copper Dioxygen Carrier

10.1 General Properties

Hemocyanin is a dioxygen-carrying copper protein found in crabs, snails, spiders, octopi, and certain other related animals (for a review, see Brunori *et al.*, 1982). There are two copper atoms at the oxidation level of cuprous ion at each reaction center. The structure of hemocyanin has been derived from X-ray studies.

The copper atoms are each bound by two histidines and are 5.6 Å apart (Co and Hodgson, 1981; Torensma and Phillips, 1983). It is surprising that there are only two ligands, but two ligands on a cuprous copper are not unknown. For example, $Cu(imidazole)_2ClO_4$ is a known compound. With the copper atoms at 5.6 Å, hemocyanin binds dioxygen with a low affinity. Alternatively, the copper atoms may move together to a distance of 3.55 Å and become bridged by a low-atomic-number atom in addition to a dioxygen bridge, as shown in Figure 10-1 (Co *et al.*, 1981; Torensma and Phillips, 1983). This latter form of hemocyanin has a high affinity for dioxygen.

The protein is normally colorless but becomes blue when it binds dioxygen. This band is relatively intense, having an extinction coefficient of 500 at about 600 nm. This is in the wavelength region of the cupric d–d transitions, but these normally have extinction coefficients of only 100 because the transition is forbidden by symmetry. The high extinction coefficient in oxyhemocyanin is the result of lack of symmetry around the copper plus some charge-transfer character, so that the band is not the result of pure d–d transition (Van Holde, 1967). The O–O frequencies of 744 and 749 cm^{-1} (Freedman *et al.*, 1976) found in oxyhemocyanin are much lower than those of the dioxygen-carrying iron proteins. These values are even below the 802 cm^{-1} frequency of peroxides.

Both oxygenated and deoxygenated hemocyanin are diamagnetic (Moss *et al.*, 1973). The diamagnetism of the oxyhemocyanin means that the two copper

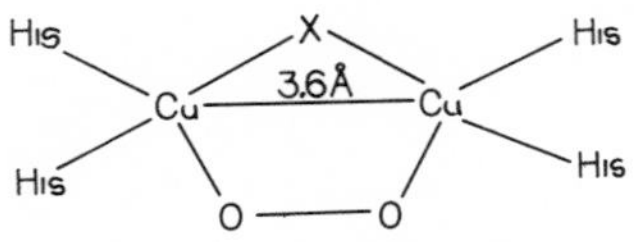

Figure 10-1. A structure proposed for oxyhemocyanin.

ions must be strongly coupled through the dioxygen bridge (Moss *et al.*, 1973). A dinuclear cupric complex has been prepared in which the copper atoms are bridged by both an alkoxide bridge and an azido group. Coupling between the two copper atoms causes the compound to be diamagnetic as in azidomethe-mocyanin (McKee *et al.*, 1981).

The resonance Raman spectrum of oxyhemocyanin in the frequency range corresponding to an oxygen–oxygen stretch has only one band for O^{16}–O^{18}, which indicates that the two oxygens are equivalent. The most likely structure is that of a peroxide bound by two cupric ions (Thamann *et al.*, 1977). This structure is analogous to the iron–dioxygen–iron intermediate formed in the aerobic oxidation of ferrous ion to ferric ion. It is not clear why this structure is not also on the pathway to the oxidation of cuprous ion to cupric ion. Why doesn't this complex dissociate to methemocyanin and to hydrogen peroxide? Possibly the unavailability of protons or the low dielectric constant in the protein is a contributing factor for stability.

10.2 Hemocyanin Models

Several models have been made to mimic hemocyanin. Bulkowski *et al.* (1977) prepared a model that has two copper atoms with a space between to accommodate a dioxygen molecule (Figure 10-2). The compound has been called "earmuff." Adding Cu(MeCN)$_4$ to the chelating agent in methyl ethyl ketone probably forms the chelate with the copper atoms positioned as shown. The

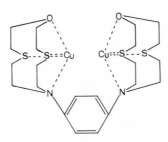

Figure 10-2. A model for hemocyanin. This copper chelate will reversibly bind dioxygen.

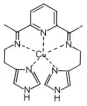

Figure 10-3. A copper chelate that will reversibly bind dioxygen in DMSO solution.

chelate precipitates out as a white powder that will combine with either carbon monoxide or dioxygen. The reaction with dioxygen is reversible. The dioxygen adduct is pale green.

Simmons and co-workers (Simmons and Wilson, 1978; Simmons *et al.*, 1980) have prepared a hemocyanin model that chelates only one copper atom. The chelate shown in Figure 10-3 will reversibly combine with dioxygen in dimethylsulfoxide (DMSO) solution. The chelate is red in DMSO, but turns green when dioxygen is added. Two chelates are required to bind one dioxygen so that the structure is probably Cu–O–O–Cu.

McKee *et al.* (1981) have prepared a copper chelate (Figure 10-4) with only 3.6 Å between the two copper atoms, which is much closer to the actual distances in hemocyanin (3.55 Å in oxoform) than the earlier binuclear "earmuff" model. In addition, there is a built-in oxo bridge, as is found in hemocyanin. The azido complex of the copper II chelate, in which the azide ion is represented by X, was found to have spectral and magnetic properties very similar to those of azidomethemocyanin. There is no report of a dioxygen complex.

The complex formed by two molecules of α,α,α′-bis(3,5-dimethypyrazol)-*m*-xylene and two atoms of Cu(I) forms a binuclear chelate (Sorrell and Jameson, 1982) that will bind one and only one molecule of CO per two copper atoms. This is the same stoichiometry between CO and copper that is found in hemocyanin.

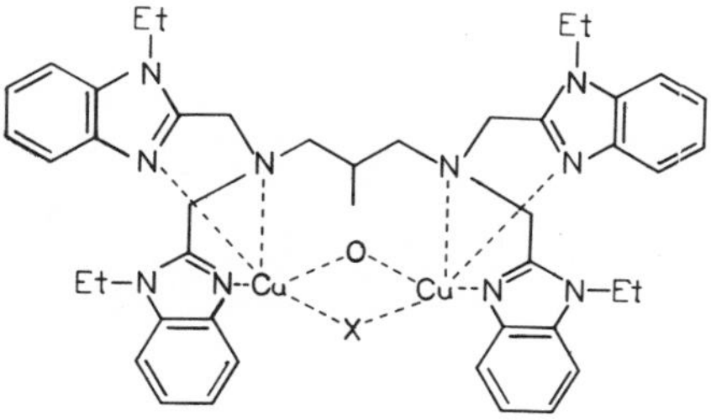

Figure 10-4. A hemocyanin model with an oxo bridge and a 3.6-Å Cu–Cu spacing.

References

Brunori, M., Giardina, B., and Kuiper, H. A., 1982. Oxygen-transport proteins, *Inorg. Biochem.* 3:126–182.

Bulkowski, J. E., Burk, P. L., Ludmann, M.-F., and Osborn, J. A., 1977. Two metal substrate interactions: The reversible reaction of a Cu^I–Cu^I complex with CO and O_2, *J. Chem. Soc. Chem. Commun.* 1977:498–499.

Co, M. S., and Hodgson, K. O., 1981. Copper site of deoxyhemocyanin: Structural evidence from X-ray absorption spectroscopy, *J. Am. Chem. Soc.* 103:3200–3201.

Co, M. S., Hodgson, X. O., Eccles, T. K., and Lontie, R., 1981. Copper site of molluscan oxyhemocyanins: Structural evidence from X-ray absorption spectroscopy, *J. Am. Chem. Soc.* 103:984–986.

Freedman, T. B., Loehr, J. S., and Loehr, T. M., 1976. A resonance Raman study of the copper protein hemocyanin: New evidence for the structure of the oxygen-binding site, *J. Am. Chem. Soc.* 98:2809–2815.

McKee, V., Dagdigian, J. V., Bau, R., and Reed, C. A., 1981. Copper(II) hemocyanin models, *J. Am. Chem. Soc.* 103:7000–7001.

Moss, T. H., Gould, D. C., Ehrenborg, A., Loehy, J. S., and Mason, H. S., 1973. Magnetic properties of *Cancer magister* hemocyanin, *J. Biol. Chem.* 12:2444–2449.

Simmons, M. G., and Wilson, L. J., 1978. A haemocyanin model: A synthetic copper(I) complex having imidazole ligands and reversible dioxygen activity, *J. Chem. Soc. Chem. Commun.* 1978:634.

Simmons, M. G., Merrill, C. L., Wilson, L. J., Bottomly, L. A., and Kadish, K. M., 1980. The {bis-2,6-[1-(2-imidazol-4-ylethylimino)ethyl]pyridine} copper(I) cation: A synthetic Cu oxygen carrier in solution as a potential model for oxy-hemocyanin, *J. Chem. Soc. Dalton Trans.* 1980:1827–1837.

Sorrell, T. N., and Jameson, D. L., 1982. An explanation for the observed stoichiometry of carbon monoxide binding to hemocyanin, *J. Am. Chem. Soc.* 104:2053–2054.

Thamann, T. J., Loehr, J. S., and Loehr, T. M., 1977. Resonance Raman study of oxyhemoglobin with unsymmetrically labeled oxygen, *J. Am. Chem. Soc.* 99:4187–4189.

Torensma, R., and Phillips, R. C., 1983. Oxygen binding by *Helix pomatra* α-haemocyanin studied by X-ray absorption spectroscopy, *Biochem. J.* 209:373–377.

Van Holde, K. E., 1967. Physical studies of hemocyanins. III. Circular dichroism and absorption spectra, *Biochemistry* 6:93–99.

Oxidases

11.1 Introduction

The oxidases are a group of enzymes that catalyze the acceptance of electrons by dioxygen. Labeled dioxygen is not incorporated into the product of the oxidation. These enzymes contain either copper or flavins. The discussion that follows considers first the copper oxidases and then the flavin oxidases.

The coppers in the copper oxidases are commonly classified into three different types according to their spectroscopic behavior (Powers *et al.*, 1979). Type I copper is called "blue copper" because of its absorption at 600 nm. It also has a band at 450 nm and shows an electron paramagnetic resonance (EPR) signal. Type II copper is called "nonblue copper" because of the very weak absorption at 600–700 nm. There is an EPR signal. Type III copper is called "silent copper" because it has no EPR signal. These are Cu^{2+} atoms that are close enough to be spin-paired.

11.2 Laccase

Laccase is a copper enzyme found in certain fungi and in the lacquer tree, *Rhus vernicifera,* which catalyzes the aerobic oxidation of aromatic 1,4-diamines and 1,4-diphenols. The enzyme contains four coppers per molecule. The amino acid content is only about 55%, which means the enzyme has only 10.4% nitrogen. The rest of the molecule is probably carbohydrate (Reinhammer, 1970). Only half the copper atoms give electron spin resonance (ESR) signals. A study of the electronic spectra has shown three types of copper atoms: one blue ESR-active copper called type I, one nonblue ESR-active copper called type II (Malmström *et al.*, 1968), and two ESR-inactive coppers called type III (Malmström *et. al.*, 1970). The type I copper has an intense band at 614 nm with an extinction coefficient of $\varepsilon = 5700$. The type III copper site has an absorption with $\varepsilon = 4500$ at 330 nm (Malkin *et al.*, 1969). This site appears to be a spin-paired couple (Malmström *et al.*, 1970) that can be reduced to a binuclear cuprous site (Lu Bien *et al.*, 1981). This binuclear site probably binds dioxygen. The type I and II coppers appear to be related mechanistically in that they are reduced at similar rates by hydroquinone (Holwerda and Gray, 1974). The redox level of type III copper affects the structure of copper site I (Lu Bien *et al.*, 1981). The type II coppers are also reduced during the catalytic cycle, but at a different rate (Branden and Reinhammer, 1975).

Exchange of a Co(II) for the type I Cu(II) gives a product containing cobalt with an intense electronic absorption band characteristic of a cobalt–sulfur charge transfer. The electronic spectrum is characteristic of a tetrahedral field (Larrabee and Spiro, 1979) with a tetragonal distortion (Dooley *et al.*, 1979; Siiman *et al.*, 1976). Copper types II and III are not in a tetrahedral field, but could be in a tetragonal or square planar field. From resonance Raman studies, there appear to be one cysteine and three nitrogens around the copper I (Siiman *et al.*, 1976). One of the nitrogens is apparently an amide nitrogen. Dioxygen combines with all reduction products of laccase to form various types of oxygenated enzyme species (Farver *et al.*, 1980).

11.3   Ceruloplasmin

Ceruloplasmin, a copper-containing protein in human blood plasma, has several diverse activities (for a review, see Frieden, 1980). It acts as an oxidase toward easily oxidizable substrates like hydroquinone and ascorbic acid (Dawson *et al.*, 1975) and may be the genetic precursor to ascorbic acid oxidase and laccase because of amino acid sequence similarities to these enzymes (Dawson *et al.*, 1975).

The oxidase activities of ceruloplasmin have been divided into three types by Frieden (1980): (1) ferrous iron, (2) many aromatic amines and phenols, and (3) substrates in which the oxidation is mediated by a type 1 or type 2 substrate. Ceruloplasmin may also function as a copper transport or in other cell functions (Frieden, 1980). Ceruloplasmin accounts for 90–95% of the serum copper. Ceruloplasmin contains five copper ions in three spectroscopically distinguishable states and one copper more, as required by elemental analysis, for a total of six copper ions (Frieden, 1980). The type I (blue copper) has a spectrum characteristic of cupric ion in a slightly flattened tetrahedral field (Dawson *et al.*, 1979; Siiman *et al.*, 1976), chelated by one sulfur and three nitrogens (Siiman *et al.*, 1976). On binding of azide or thiocyanate ion, there occur structural changes that make the environment less tetrahedral (Dawson *et al.*, 1979). The type II copper is believed to be in a tetragonal field.

11.4 Ascorbic Acid Oxidase

Ascorbic acid oxidase (Dawson *et al.*, 1975) is another blue-copper oxidase. It is oxidized by dioxygen and reduced by ascorbic acid. The enzyme is isolated from squash or cucumbers. The enzyme has a molecular weight of 140,000 composed of four subunits of two 38,000 A chains and two 28,000 B chains. The A and B chains appear to be bound by disulfide bonds and the AB pairs to be bound together by noncovalent forces (Strothkamp and Dawson, 1974). Separation of the AB pairs by sodium dodecyl sulfate causes loss of the 8–10 coppers per mole of enzyme. The spectral data are in agreement with a low-symmetry field around the copper (Lee and Dawson, 1973), probably again in a structure distorted between tetrahedral and planar, chelated by three nitrogens and one sulfur (Siiman *et al.*, 1976).

11.5 Amine Oxidases

Amine oxidases catalyze the oxidative deamination of amines by dioxygen (McEwen *et al.*, 1966). There are two types of monoamine oxidases: those that contain copper and those that contain flavins. In addition, there are diamine oxidases.

11.5.1 Monoamine Oxidase—Copper Enzyme

In this section, we will discuss the copper-containing monoamine oxidases (Yamada *et al.*, 1963, 1965; Malmström *et al.*, 1975). These enzymes are found in beef plasma (Yasunobu and Smith, 1971) and in *Aspergillus* (Yamada and Adachi, 1971).

Monoamine oxidase oxidizes primary amines attached to a methylene group to the corresponding aldehyde ($H_2O + RCH_2NH_2 + O_2 \rightarrow RCHO + NH_3 + H_2O_2$). The best substrates are the phenylethylamines. The copper is of type II, i.e., nonblue with an ESR signal (Suzuki *et al.*, 1980). The copper has at least two imidazole ligands (Suzuki *et al.*, 1982).

The enzymes are pink to pinkish yellow in the oxidized form when highly purified. The monamine oxidase from plasma has an absorption peak at 480 nm (Yamada and Yasunobu, 1962). The copper monoamine oxidase has been reported to contain pyridoxal (Yamada and Yasunobu, 1963; McEwen *et al.*, 1966). It is interesting to note that pyridoxal, manganous ion, and dioxygen will oxidize alamine to pyruvic acid (Hamilton and Revesz, 1966). Two mechanisms have been proposed for the reaction. In the first (Taylor *et al.*, 1972; Yadav and Knowles, 1981), pyridoxal forms a Schiff base with the amine, this step being followed by proton transfer and hydrolysis to form the aldehyde and pyridoxamines. The pyridoxamine is subsequently oxidized to pyridoxal and ammonium ion. The second mechanism (Suva and Abeles, 1978; Berg and Abeles, 1980) does not involve pyridoxal. The amine is directly oxidized to an imine that hydrolyzes to aldehyde and ammonium ion. The observation (Berg and Abeles, 1980) that aldehyde and ammonium ion appear as first products is consistent with this mechanism, but inconsistent with the pyridoxal mechanism.

11.5.2   Monoamine Oxidase—Flavin Enzyme

The enzyme that oxidizes monoamines to aldehydes found in animal mitochondria (Yasunobu and Gomes, 1971) contains both copper and flavin (Nara *et al.*, 1966; Erwin and Hellerman, 1967). The flavin is in the form of FAD and is covalently attached to the enzyme (Igave *et al.*, 1967; Gomes *et al.*, 1969) through the 8-α-methyl group of the flavin (Kearney *et al.*, 1971). The substrate N-oxide is not an intermediate in this reaction. The intermediate appears, instead, to be an imino derivative (Smith *et al.*, 1962). Thus, the enzyme is an oxidase rather than an oxygenase.

11.5.3   Diamine Oxidase

The diamine oxidases oxidize primary amines attached to a methylene carbon in diamines. They have slightly different properties as well as differences in specificity. One of the amino groups can be substituted by a group that has, or can develop, a positive charge. The remaining amino group is oxidized to an aldehyde. Examples are a dimethyl sulfonium group, a quaternary ammonium group, an imidazolium group, an isothiouronium group, and a pyridinium (Bardsley *et al.*, 1970). The enzyme must have an oxidizing center separated from an anionic binding center by about 6–9 Å of hydrophobic area (Bardsley and Hill, 1970). Positively charged organic ions as described above, without an amino group, act as inhibitors for the enzyme (Crabbe and Bardsley, 1973). These enzymes are obtained from *Aspergillus niger*, pea seedlings, bovine and pig plasma, and pig kidney.

Resting diamine oxidase is pink, with absorption peaks at 480 nm (Mondavi *et al.*, 1967a) for the pig kidney enzyme and at about 500 nm for the pea seedling enzyme (Hill and Mann, 1964). These absorption peaks disappear and the enzyme becomes yellow on the anaerobic addition of substrate.

The enzymes contain between one and three atoms of copper per enzyme molecule, and at least part, possibly all, of the copper can be detected by EPR (Van Heuvelen, 1965). The EPR signal is typical of a tetragonal Cu(II) (Mondavi et al., 1967a). Not surprisingly, the copper does not appear to be reduced during the catalytic cycle (Kluetz et al., 1980). The enzyme appears to contain no flavin (Mondavi et al., 1967a), but does contain pyridoxal phosphate (Mondavi et al., 1967b).

11.6 Galactose Oxidase

Galactose oxidase is a copper enzyme that catalyzes the dioxygen oxidation of the 6-CH_2OH group of galactose to an aldehyde. The other product of the reaction is hydrogen peroxide (Avigad et al., 1962). There is one copper atom per molecule of enzyme, which cannot be dialyzed out even in the presence of KCN. The copper of the resting enzyme has an EPR signal typical of the pseudo-square planar Cu(II) complex. The x and y directions are anisotropic (Giordano and Bereman, 1974). The copper ligands include at least two imidazoles and one exchangeable water (Bereman and Kosmin, 1977).

The enzyme has an essential disulfide bond. Reduction by β-mercaptoethanol completely eliminates all activity, but reoxidation restores activity. There is at least one essential tryptophan that interacts with the copper(II) (Weiner et al., 1977). A proposal has been made that the enzyme has copper in the 3+ state (Hamilton et al., 1973) (Figure 11-1) on the basis that the optical spectra and ESR spectra change on the addition of ferricyanide ion. However, the N-bromosuccinimide-treated enzyme shows no decrease in ESR signal under conditions that the optical spectra is modified with ferricyanide (Winkler and Bere-

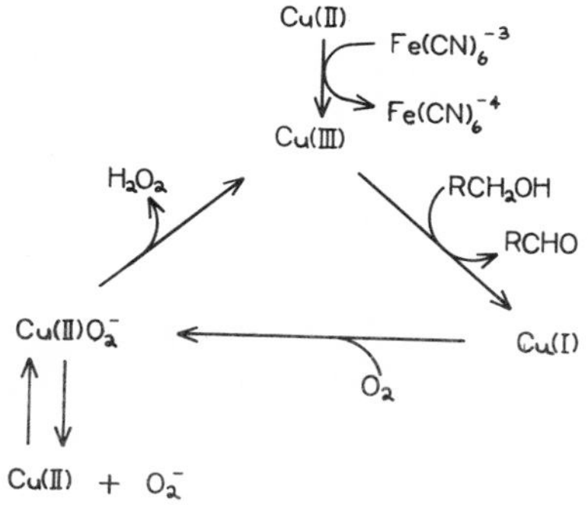

Figure 11-1. A mechanism of reaction proposed for galactose oxidase.

man, 1980). Thus, the change in the optical spectrum cannot be the result of the presence of Cu(III) in the enzyme.

11.7 Cytochrome Oxidase

Cytochrome oxidase, the terminal oxidase of the electron-transport chain, is an extremely important enzyme (for reviews, see Malmström, 1974; Caughey et al., 1976; Chance et al., 1982). Estimates are that about 90% of all biological oxygen is consumed by this enzyme. The reaction oxidizes four molecules of cytochrome c with one molecule of dioxygen. Cytochrome oxidase binds to both substrates, dioxygen and cytochrome c (Volpe and Caughey, 1974). It binds to the oxidized form of cytochrome c more strongly than to the reduced form. The dioxygen binding is irreversible (Greenwood et al., 1974) and has several intermediates of varying oxidation state and protein structure (Orii and King, 1972; Bruding et al., 1981; Brunori and Wilson, 1982).

Cytochrome oxidase is a complex protein consisting of seven subunits and varying amounts of phospholipid, depending on the source (Azzi and Casey, 1979; Azzi, 1980). Bovine cytochrome c has about 20% phospholipid. There are two heme groups and two copper atoms (Griffiths and Wharton, 1961) per molecule of enzyme. The fully oxidized form undergoes a four-electron reduction, so the irons must be in the ferric state and the coppers in the cupric state when fully oxidized.

The two heme groups are called cytochrome a and a_3. Cytochrome a is in a low-spin ferric state and cytochrome a_3 in a high-spin ferric state. When cytochrome a is partially reduced, the EPR signal decreases and a high-spin EPR signal appears. Both signals disappear on complete reduction (Erecinska and Wilson, 1978). The potentials of the two hemes differ considerably. The low-potential heme has a potential of $+210$ mV and the high-potential heme is at $+350$–375 mV. Cytochrome a_3 is oxidized more rapidly with dioxygen than with cytochrome a. Cytochrome a is in rapid equilibrium with cytochromes c and c' (Erecinska and Chance, 1972). Both hemes combine with carbon monoxide in an endwise fashion. There appears to be no CO bridging between the irons or between copper and iron (Volpe et al., 1975).

Cytochrome a has been isolated and its structure has been determined (Caughey et al., 1975). Two side chains are of interest. One is a formyl group that withdraws electrons from the heme, making it more difficult to oxidize the iron. The other is a farnesyl group that may be used to bind the heme to a hydrophobic area. Caughey et al. (1975) have suggested that the farnesyl group may orient in a position to transport electrons via the double bonds to the heme iron. Such electron transport requires a very special group and environment (Waleh and Ingraham, 1976).

Each of the cytochrome irons is associated with one of the coppers. Cytochrome a is 3.75 ± 0.5 Å (Powers $et\ al.$, 1981) from a type I blue copper. The iron and copper are bridged by a sulfur atom (Scott $et\ al.$, 1981). The two metal ions are spin-coupled (Powers $et\ al.$, 1979), so the complex is EPR-silent. The a_3 iron is liganded by four heme nitrogens, a proximal nitrogen, and the bridging sulfur (Powers $et\ al.$, 1981, 1982). The a_3 copper is liganded by two nitrogens and the bridging sulfur. Cytochrome a iron is surrounded by six equidistant nitrogens. The cytochrome a copper is highly covalently liganded by one or two nitrogens and three or four sulfurs (Powers $et\ al.$, 1981). This copper has an EPR signal at $g = 2.0$.

The addition of dioxygen to cytochrome a_3 probably forms a dioxygen bridge between the copper a_3 and iron a_3 (Karlsson $et\ al.$, 1979; Petty $et\ al.$, 1980). This is similar to the transient Fe–O–O–Fe intermediates detected in the monoxidation of ferrous compounds of dioxygen. Cytochrome oxidase also forms a complex with carbon monoxide that has a C≡O stretch frequency of 1963.5 cm^{-1} (Caughey $et\ al.$, 1976). This is very similar to "end-on" iron–carbon monoxide complexes in myoglobin and hemoglobin. One might expect by analogy that the iron would also combine with dioxygen. A Soret band at 426–428 nm has been assigned to the dioxygen–iron complex. There are several dioxygen complexes that may be observed by flash photolysis of the CO adduct in the presence of dioxygen (Chance $et\ al.$, 1975). The most reasonable explanation for these complexes (Clore $et\ al.$, 1980) is that the cytochrome a_3–copper–dioxygen complex accepts electrons one at a time from the cytochrome a–copper complex as the dioxygen is reduced to the level of water.

11.8 Xanthine Oxidase

Certain oxidases are quite complex, containing FAD, iron, sulfur, and molybdenum (for reviews, see Massey, 1973; Bray, 1975). The best known of these is xanthine oxidase, which hydroxylates xanthine to produce uric acid and hydroxylates many other substrates. It is most commonly isolated from milk. Xanthine oxidase is found in the highest concentration in the intestinal mucosa and in the liver, although it is present in detectable amounts in nearly all tissues. A very similar enzyme is aldehyde oxidase, which oxidizes many of the same substrates but also oxidizes aldehydes to carboxylic acids (Krenitsky $et\ al.$, 1972).

Xanthine oxidase (the oxygen-reducing form of the enzyme, or type "O") is not the native form of the enzyme. It is synthesized as xanthine dehydrogenase (an NAD$^+$-reducing enzyme, or type "D"). For every molecule of hypoxanthine or xanthine that is catabolically hydroxylated, NADH is produced. The dehydrogenase is converted to the oxidase via limited proteolysis or by sulfhydryl oxidation. These conversions usually occur during the isolation of the enzyme, but may occur *in vivo* under pathological conditions such as ischemia, giving rise to a pathological source of superoxide radicals (Battelli *et al.*, 1972; Roy and McCord, 1983).

The oxygen of the hydroxyl group is derived from water, so the enzyme is an oxidase rather than a monooxygenase. The enzyme appears to be oxidized by dioxygen in two phases that differ in rate by a factor of about 10. Only the first, faster phase is catalytically functional (Olson *et al.*, 1974a). The first phase requires five electrons and the second, slower phase only one electron. The products of dioxygen reduction are both hydrogen peroxide and superoxide ion (Fridovich, 1970). The six-electron oxidation of the enzyme by dioxygen occurs in four steps. The first two steps are two-electron oxidations producing hydrogen peroxide, the last two steps one-electron steps producing superoxide (Hille and Massey, 1981).

The superoxide ion in turn inactivates the enzyme. It has been suggested that the superoxide ion reacts with the hydrogen peroxide by the Haber–Weiss reaction to form singlet dioxygen (Lynch and Fridovich, 1979). However, note in the discussion in Chapter 3 that this reaction is slow compared with other reactions of superoxide ion. Luminescence has been observed in the xanthine oxidase reaction, but this has been ascribed to dimers of the carbon dioxide radical anion, since the luminescence is proportional to the square of the concentration of added carbon dioxide (Hodgson and Fridovich, 1976). The xanthine oxidase is protected by superoxide dismutase. Similarly, *p*-benzoquinone will oxidize xanthine oxidase by both one-electron and two-electron processes (Nakamura and Yamazaki, 1969).

The enzyme appears to be a dimer with a molecular weight of about 283,000. There are two FADs per mole. There is a ratio of one molybdenum and four iron per FAD. The enzyme has an essential sulfur that is believed to be part of a persulfide. The sulfur is lost very easily to form an inactive desulfo enzyme. Cyanide ion will convert the active enzyme to the desulfo enzyme and thiocyanate (Edmondson *et al.*, 1971). Cyanide ion reacts with persulfides. Conversion to the desulfo enzyme is inhibited by EDTA and by salicylate.

Partially reduced enzyme exhibits an EPR signal characteristic of a flavin semiquinone. The resting enzyme probably contains Mo(VI) because there is no EPR signal. Other forms of the enzyme show an EPR signal attributed to Mo(V).

A mechanism for the hydroxylation reaction (Figure 11-2) has been proposed (Olson *et al.*, 1974b; Edmondson *et al.*, 1971) in which the persulfide attacks the 2-position of the imidazole portion of the xanthine. Two electrons are withdrawn from the ring by the Mo(VI), which then becomes Mo(IV) as the proton at the 2-position is lost. This is followed by a displacement of the persulfide by water. In aldehyde oxidase, the persulfide would add to the carboxyl group of the aldehyde followed by oxidation.

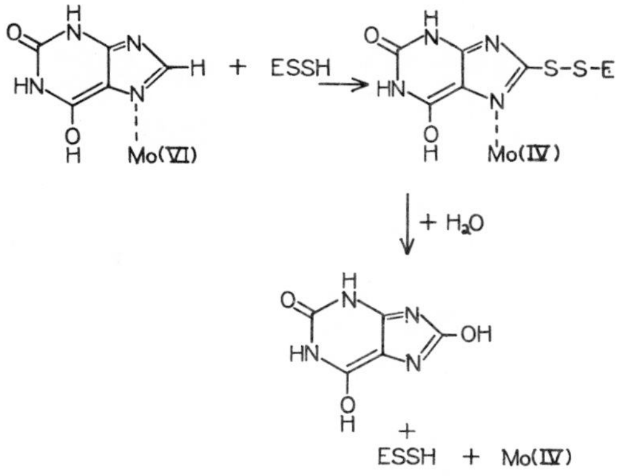

Figure 11-2. A mechanism proposed for the hydroxylation reaction of xanthine oxidase.

This proposed mechanism fits with kinetic data, but two questions arise: Why is a persulfide ion needed for the initial attack? Would not a sulfide be adequate? Also, one wonders about the displacement of a vinyl sulfur. Possibly the water could add to the carbon–nitrogen double bond catalyzed by the metal bound to the nitrogen. This would be followed by the elimination of persulfide ion. Aldehyde oxidase will hydroxylate N-methyl nicotinamide to the 2- and 4-pyridones. The mechanism has been proposed to be pseudo-base formation by the addition of hydroxide ion at the 2- or 4-position followed by oxidation (Felsted et al., 1973).

11.9   D-Amino Acid Oxidase

D-Amino acid oxidase oxidatively deaminates amino acids to form keto acids. The enzyme contains flavin as a prosthetic group. The first step in the reaction appears to be formation of an α-carbanion. There is an isotopic rate effect for the α-hydrogen, and false substrates give products derived from a carbanion (Walsh et al., 1971).

Porter et al. (1972) found that the D-amino acid oxidase will oxidize ethyl nitrate to acetaldehyde and nitrite ion. The reaction is inhibited by cyanide ion (Porter et al., 1973). Porter and Bright (1977) have proposed a mechanism (Figure 11-3) in which an α-carbanion condenses with the N^5 position of flavin. The UV spectrum of the cyanide-inhibited product corresponds to that expected for a 5-cyanomethyl flavin. The normal reaction (Figure 11-4) is not affected by cyanide ion, so the reactive Schiff base must not be formed.

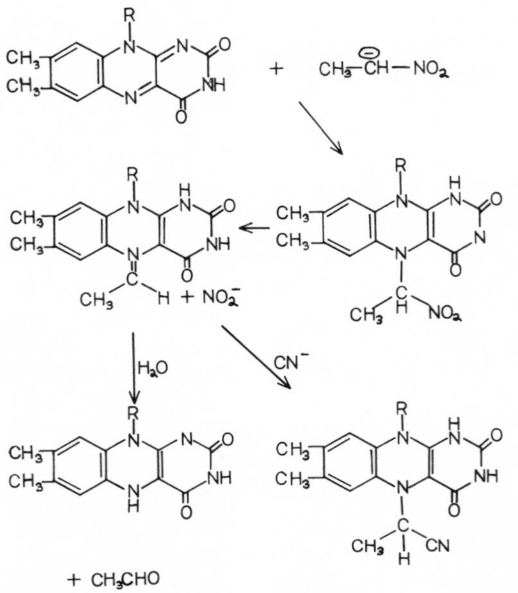

Figure 11-3. A mechanism proposed for the oxidation of nitroethane catalyzed by D-amino acid oxidase.

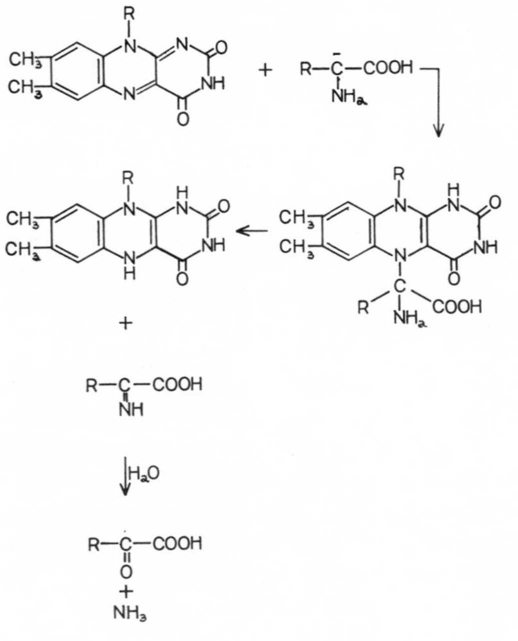

Figure 11-4. A mechanism proposed for the oxidation of an amino acid catalyzed by D-amino acid oxidase.

11.10   Glucose Oxidase

Glucose oxidase is a flavin enzyme that oxidizes glucose. Kinetic studies show that the mechanism is a Ping-Pong one in which the enzyme is first reduced by glucose to form reduced enzyme and D-gluconolactone. The reduced enzyme is then oxidized by dioxygen, forming hydrogen peroxide (Gibson *et al.*, 1964; Bright and Appleby, 1969).

Glucose oxidase also catalyzes the oxidation of nitroethane to acetaldehyde and nitrite (Porter and Bright, 1977), similar to the oxidation catalyzed by D-amino acid oxidase. The reaction with nitroethane is not clear. There is a small proportion of products that may be derived from radicals i.e., nitrate and dinitroethane. There has been considerable controversy over whether the reaction is a two-electron or a one-electron oxidation of glucose. Glucose oxidase will oxidize some α-hydroxy ketones by a one-electron process and others by a two-electron process (Chan and Bruice, 1977).

11.11   Lactate Oxidase

Lactate oxidase is a similar enzyme that oxidizes lactic acid to pyruvic acid. Williams and Bruice (1976) have considered all the redox potentials involved and conclude that this reaction could be an oxidation of the α-carbanion of lactic acid to the α-radical followed by loss of H to the flavin. Bruice believes that the preceding two enzymes could also proceed by this mechanism (Bruice, 1980).

However, recent studies of the oxidation of glycollate ion to glyoxylate ion catalyzed by lactate oxidase have shown that two stable adducts are formed between the glycollate ion and the flavin at the N^5 position of the flavin (Figure 11-5). These adducts, corresponding to the two prochiral positions of glycollate ion, can be isolated (Ghisla and Massey, 1980; Massey *et al.*, 1980). Unfortunately, this adduct does not proceed to reduced flavin and glyoxylate ion, but slowly decays to glycollate ion and oxidized flavin. This same adduct can be formed by the action of light on the lactate oxidase–tartronic acid complex (Ghisla

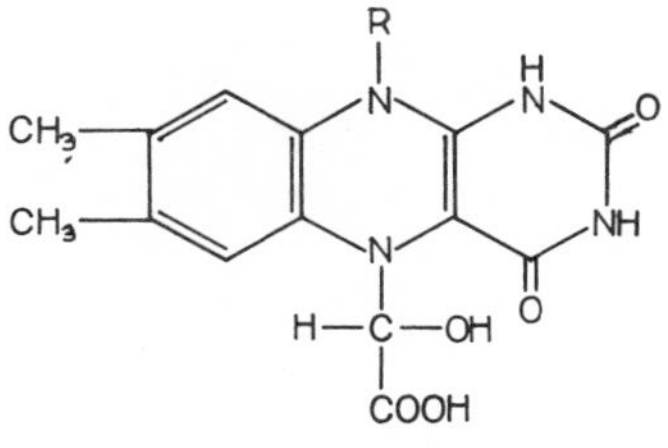

Figure 11-5. An intermediate found in the lactate-oxidase-catalyzed oxidation of glycollate ion to glyoxylate ion.

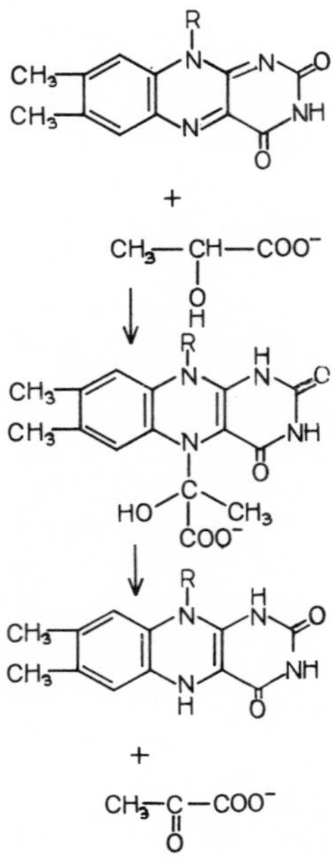

Figure 11-6. A reaction mechanism proposed for the lactate-oxidase-catalyzed oxidation of lactate ion to pyruvate ion.

et al., 1979). The other product of the reaction is carbon dioxide. This result allows a mechanism (Figure 11-6) similar to that proposed by Porter and Bright for D-amino acid oxidase.

References

Avigad, G., Amaral, D., Asension, C., and Horecker, B. L., 1962. The galactose oxidase of *Polyporus circinatus, J. Biol. Chem.* 237:2736–2743.

Azzi, A., 1980. Cytochrome C-oxidase: Towards a clarification of its structure, interactions and mechanism, *Biochim. Biophys. Acta* 594:231–252.

Azzi, A., and Casey, R. P., 1979. Molecular aspects of cytochrome *c* oxidase: Structure and dynamics, *Mol. Cell. Biochem.* 28:169–184.

Bardsley, W. G., and Hill, G. M., 1970. A study of the substrate binding site in hog kidney diamine oxidase, *Biochem. Biophys. Res. Commun.* 41:1068–1071.

Bardsley, W. G., Hill, G. M., and Lobley, R. W., 1970. A reinvestigation of the substrate specificity of pig kidney diamine oxidase, *Biochem. J.* 117:169–176.

Battelli, M. G., Corte, E. D., and Stirpe, F., 1972. Xanthine oxidase type D (dehydrogenase) in the intestine and other organs of the rat, *Biochem. J.* 126:747–749.

Bereman, R. D., and Kosman, D. J., 1977. Stereoelectronic properties of metalloenzymes. 5. Identification and assignment of ligand hyperfine splittings in the electron spin resonance spectrum of galactose oxidase, *J. Am. Chem. Soc.* 99:7322–7325.

Berg, K. A., and Abeles, R. H., 1980. Mechanism of action of plasma amine oxidase products released under anaerobic conditions, *Biochemistry* 19:3186–3189.

Branden, R., and Reinhammer, B., 1975. EPR studies of the aerobic reduction of fungal laccase: Evidence for participation of type 2 copper in the reduction mechanism, *Biochim. Biophys. Acta* 405:236–242.

Bray, R. C., 1975. Molybdenum iron–sulfur flavin hydroxylases and related enzymes, in *The Enzymes*, Vol. 12, P. D. Boyer (ed.), Academic Press, New York, pp. 229–419.

Bright, H. H., and Appleby, M., 1969. The pH dependence of the individual steps in the glucose oxidase reaction, *J. Biol. Chem.* 244:3625–3634.

Bruding, G. W., Stevens, T. H., Morse, R. H., and Chan, S. I., 1981. Conformations of oxidized cytochrome *c* oxidase, *Biochemistry* 20:3912–3921.

Bruice, T. C., 1980. Mechanisms of flavin catalysis, *Acc. Chem. Res.* 13:256–262.

Brunori, M., and Wilson, M. T., 1982. Cytochrome oxidase, *Trends Biochem. Sci.* 7:295–299.

Caughey, W. S., Smythe, G. A., O'Keefe, D. H., Maskasky, J. E., and Smith, M. L., 1975. Heme a of cytochrome *c* oxidase—Structure and properties: Comparison with hemes b, c, and s and derivatives, *J. Biol. Chem.* 250:7602–7622.

Caughey, W. A., Wallace, W. J., Volpe, J. A., and Yoshikawa, S., 1976. Cytochrome *c* oxidase, in *The Enzymes*, Vol. XIII, Part C, P. D. Boyer (ed.), Academic Press, New York, pp. 299–344.

Chan, T. W., and Bruice, T. C., 1977. One and two electron transfer reactions of glucose oxidase, *J. Am. Chem. Soc.* 99:2387–2389.

Chance, B., Saronio, C., and Leigh, J. S., Jr., 1975. Functional intermediates in reaction of cytochrome oxidase with oxygen, *Proc. Natl. Acad. Sci. U.S.A.* 72:1635–1640.

Chance, B., Powers, L., and Ching, X., 1982. Structure and function of the redox site of cytochrome oxidase, *Adv. Exp. Med. Biol.* 148:95–109.

Clore, M., Andreasson, L.-E., Karlsson, B., Aasa, R., and Malmström, B. G., 1980. Characterization of the low temperature intermediates of the reaction of fully reduced soluble cytochrome oxidase with oxygen by electron paramagnetic resonance and optical spectroscopy, *Biochem. J.* 185:139–154.

Crabbe, M. J. C., and Bardsley, W. G., 1973. The inhibition of human placental diamine oxidase by substrate analogues, *Biochem. J.* 139:183–189.

Dawson, C. R., Strothkamp, D. G., and Krul, K. G., 1975. Ascorbic oxidase and related copper proteins, *Ann. N. Y. Acad. Sci.* 258:209–220.

Dawson, J. H., Dooley, D. M., Clark, R., Stephens, P. F., and Gray, H. B., 1979. Spectroscopic studies of ceruloplasmin: Electronic structures of the copper sites, *J. Am. Chem. Soc.* 101:5046–5053.

Dooley, D. M., Rawlings, J., Dawson, J. H., Stephens, P. J., Andreasson, L.-E., Malmström, B. G., and Gray, H. B., 1979. Spectroscopic studies of *Rhus vernicifera* and *Polyporus versicolor* laccase: Electronic structure of the copper sites, *J. Am. Chem. Soc.* 101:5038–5046.

Edmondson, D., Massey, V., Palmer, G., Beachan, L. M., II, and Ellon, G. B., 1971. The resolution of active and inactive xanthine oxidase by affinity chromatography, *J. Biol. Chem.* 247:1597–1604.

Erecinska, M., and Chance, B., 1972. Studies of the electron transport chain at subzero temperatures: Electron transport at site III, *Arch. Biochem. Biophys.* 151:304–315.

Erecinska, M., and Wilson, D. F., 1978. Cytochrome oxidase, *Arch. Biochem. Biophys.* 188:1–14.

Erwin, V. G., and Hellerman, L., 1967. Mitochrondrial monoamine oxidase. I. Purification and characterization of the bovine kidney enzyme, *J. Biol. Chem.* 242:4230–4238.

Farver, O., Goldberg, M., and Pecht, I., 1980. A circular dichroism study of the reactions of *Rhus* laccase with dioxygen, *Eur. J. Biochem.* 104:71–77.

Felsted, R. L., Chu, A. E.-Y., and Chaykin, S., 1973. Purification and properties of the aldehyde oxidases from hog and rabbit livers, *J. Biol. Chem.* 248:2580–2687.

Fridovich, I., 1970. Quantitative aspects of the production of superoxide anion radical by milk xanthine oxidase, *J. Biol. Chem.* 245:4053–4057.

Frieden, E., 1980. Caeruloplasmin: A multi-functional metalloprotein of vertebrate plasma—biological roles of copper, *Ciba Found. Symp.* 79:93–124.

Ghisla, S., and Massey, V., 1980. Studies on the catalytic mechanism of lactate oxidase, *J. Biol. Chem.* 255:5688–5696.

Ghisla, S., Massey, V., and Choong, Y. S., 1979. Covalent adducts of lactate oxidase, *J. Biol. Chem.* 254:10662–10669.

Gibson, O., Bennett, E. P., Swoboda, E. P., and Massey, V., 1964. Kinetics and mechanism of action of glucose oxidase, *J. Biol. Chem.* 239:3927–3934.

Giordano, R. S., and Bereman, R. D., 1974. Stereoelectronic properties of metallo-enzymes. I. A comparison of the coordination of copper(II) in galactose oxidase and a model system, *N,N'*-ethylenebis(trifluoroacetylacetonimato) copper(II), *J. Am. Chem. Soc.* 96:1019–1023.

Gomes, B., Igave, I., Kloepfer, H. G., and Yasunobu, K. T., 1969. Amine oxidase. XIV. Isolation and characterization of the multiple beef liver amino oxidase components, *Arch. Biochem. Biophys.* 132:16–27.

Greenwood, C., Wilson, M. T., and Brunori, M., 1974. Studies on partially reduced mammalian cytochrome oxidase: Reactions with carbon monoxide and oxygen, *Biochem. J.* 137:205–215.

Griffiths, D. E., and Wharton, D. C., 1961. Studies of the electron transport system. XXXVI. Properties of copper in cytochrome oxidase, *J. Biol. Chem.* 236:1857–1862.

Hamilton, G. A., and Revesz, A., 1966. Oxidation by molecular oxygen IV: A possible model reaction for some amine oxidases, *J. Am. Chem. Soc.* 88:2069–2070.

Hamilton, G. A., Libby, R. D., and Hartzell, G. R., 1973. The valence of copper and the role of superoxide in the D-galactose oxidase catalyzed reaction, *Biochem. Biophys. Res. Commun.* 55:333–340.

Hill, J. M., and Mann, P. J. G., 1964. Further properties of the diamine oxidase of pea seedlings, *Biochem. J.* 91:171–182.

Hille, R., and Massey, V., 1981. Studies on the oxidative half-reactions of xanthine oxidase, *J. Biol. Chem.* 256:9090–9095.

Hodgson, E. K., and Fridovich, I., 1976. The mechanism of the activity-dependent luminescence of xanthine oxidase, *Arch. Biochem. Biophys.* 172:202–205.

Holwerda, R. A., and Gray, G. B., 1974. Mechanistic studies of the reduction of *Rhus vernicifera* laccase by hydroquinone, *J. Am. Chem. Soc.* 96:6008–6022.

Igave, I., Gomes, B., and Yasunobu, K. T., 1967. Beef mitochondrial monoamine oxidase, a flavin dinucleotide enzyme, *Biochem. Biophys. Res. Commun.* 29:562–570.

Karlsson, B., Andreasson, L.-E., Aasa, R., Malmström, B. G., and Clore, G. M., 1979. Studies of the reaction of cytochrome *c* oxidase with oxygen at low temperature, *Acta Chem. Scand. Ser. B* 33:615–618.

Kearney, E. B., Salach, J. I., Walker, W. H., Seng, R., and Singer, T. P., 1971. Structure of the covalently bound flavin of monoamine oxidase, *Biochem. Biophys. Res. Commun.* 42:490–496.

Kluetz, M. D., Adamsons, K., and Flynn, J. E., Jr., 1980. Cryoenzymology and spectrophotometry of pea seedling diamine oxidase, *Biochemistry* 19:1617–1621.

Krenitsky, T. A., Neil, A. M., Elion, G. B., and Hitchings, G. H., 1972. A comparison of the specificities of xanthine oxidase and aldehyde oxidase, *Arch. Biochem. Biophys.* 150:585–599.

Larrabee, J. A., and Spiro, T. G., 1979. Cobalt II substitution in the Type I site of the multi-copper oxidase *Rhus* laccase, *Biochem. Biophys. Res. Commun.* 88:753–760.

Lee, M. H., and Dawson, C. R., 1973. Ascorbate oxidase: Spectral characteristics of the enzyme, *J. Biol. Chem.* 248:6603–6609.

Lu Bien, C. D., Winkler, M. E., Thamann, T. J., Scott, R. A., Co, M. S., Hodgson, K. O., and Solomon, E. I., 1981. Chemical and spectroscopic properties of the binuclear copper active site in laccase: Direct confirmation of a reduced binuclear type III copper site in type II depleted laccase and intramolecular coupling of the type III to the type I and type II copper sites, *J. Am. Chem. Soc.* 103:7014–7016.

Lynch, R. E., and Fridovich, I., 1979. Autoinactivation of xanthine oxidase: The role of superoxide radical and hydrogen peroxide, *Biochim. Biophys. Acta* 571:195–200.

Malkin, R., Malmström, B. G., and Vanngard, T., 1969. Spectroscopic differentiation of the electron-accepting sites in fungal laccase: Association of a near ultraviolet band with a two electron accepting unit, *Eur. J. Biochem.* 10:324–329.

Malmström, B. G., 1974. Cytochrome *c* oxidase: Some current biochemical and biophysical problems, *Q. Rev. Biophys.* 6:389–431.

Malmström, B. G., Reinhammer, B., and Vanngard, T., 1968. Two forms of copper(II) in fungal laccase, *Biochim. Biophys. Acta* 156:67–76.

Malmström, B. G., Reinhammer, B., and Vanngard, T., 1970. The state of copper in stellacyanin and laccase from the lacquer tree *Rhus vernicifera*, *Biochim. Biophys. Acta* 205:48–57.

Malmström, B. G. Andreasson, L.-E., and Reinhammer, B., 1975. Copper-containing oxidases and superoxide dismutase, in *The Enzymes*, Vol. XII, Part B., P. D. Boyer (ed.), Academic Press, New York, pp. 507–579.

Massey, V., 1973. Iron–sulfur flavoprotein hydroxylases, in *Iron–Sulfur Proteins*, Vol. I, W. Lovenberg (ed.), Academic Press, New York, pp. 225–300.

Massey, V., Ghisla, S., and Kieschke, K., 1980. Studies on the reaction mechanism of lactate oxidase: Formation of two covalent flavin–substrate adducts on reaction with glycollate, *J. Biol. Chem.* 255:2796–2806.

McEwen, C. M., Jr., Cullen, K. T., and Sober, A. J., 1966. Rabbit serum monoamine oxidase: Purification and characterization, *J. Biol. Chem.* 241:4544–4556.

Mondavi, B., Rotilio, G., Costa, M. T., Finazzi-Agro, A., Chancone, E., and Hansen, R. D., 1967a. Diamine oxidase from pig kidney: Improved purification and properties, *J. Biol. Chem.* 242:1160–1167.

Mondavi, B., Costa, M. T., Finazzi-Argo, A., and Rotilio, G., 1967b. Pyridoxal phosphate as a prosthetic group of pig kidney diamine oxidase, *Arch. Biochem. Biophys.* 119:373–381.

Nakamura, S., and Yamazaki, I., 1969. One-electron transfers in biochemical systems. IV. A mixed mechanism in the reaction of milk xanthine oxidase with electron acceptors, *Biochim. Biophys. Acta* 189:29–37.

Nara, S., Igave, I., Gomes, B., and Yasunobu, K. T., 1966. The prosthetic groups of animal amine oxidases, *Biochem. Biophys. Res. Commun.* 23:324–328.

Olson, J. S., Ballou, D. P., Palmer, G., and Massey, V., 1974a. The reaction of xanthine oxidase with molecular oxygen, *J. Biol. Chem.* 249:4350–4362.

Olson, J. A., Ballou, D. P., Palmer, G., and Massey, V., 1974b. The mechanism of action of xanthine oxidase, *J. Biol. Chem.* 249:4363–4382.

Orii, Y., and King, T. E., 1972. New species of the oxygenated compound of cytochrome oxidase, *FEBS Lett.* 21:199–202.

Petty, R. H., Welch, B. R., Wilson, L. J., Bottomly, L. A., and Kadish, K. M., 1980. Cytochrome oxidase models: μ-Bipyramidyl mixed-metal complexes as synthetic models for the Fe/Cu binuclear active site of cytochrome oxidase, *J. Am. Chem. Soc.* 102:611–620.

Porter, D. J. T., and Bright, H. J., 1977. Mechanism of oxidation of nitroethane by glucose oxidase, *J. Biol. Chem.* 252:4361–4370.

Porter, D. J. T., Voegt, J. G., and Bright, H. J., 1972. Nitromethane: A novel substrate of D-amino acid oxidase, *J. Biol. Chem.* 241:1951–1953.

Porter, D. J. T., Voegt, J. G., and Bright, H. J., 1973. Direct evidence for carbanions and covalent N^5-flavin–carbanion adducts as catalytic intermediates in the oxidation of nitroethane by D-amino acid oxidase, *J. Biol. Chem.* 248:4400–4416.

Powers, L., Blumberg, W.-E., Chance, B., Barlow, C. H., Leigh, I. S., Smith, J., Yonetani, T., Vik, S., and Peisach, J., 1979. The nature of the copper atoms of cytochrome *c* oxidase as studied by optical and X-ray absorption edge spectroscopy, *Biochim. Biophys. Acta* 546:520–538.

Powers, L., Chance, B., Ching, Y., and Angiolillio, P., 1981. Structural features and the reaction mechanism of cytochrome oxidase, *Biophys. J.* 34:465–498.

Powers, L., Chance, B., Ching, Y., Muhoberac, B., Weintraub, S. T., and Wharton, D. C., 1982. Structural features of the copper-depleted cytochrome oxidase from beef heart: Iron EXAFS, *FEBS Lett.* 138:245–248.

Reinhammer, B., 1970. Purification and properties of laccase and stellacyanin from *Rhus vernicifera*, *Biochim. Biophys. Acta* 205:35–47.

Roy, R. S., and McCord, J. M., 1983. Superoxide and ischemia: Conversion of xanthine dehydrogenase to xanthine oxidase, in *Oxy Radicals and Their Scavenger Systems*, Vol. II, *Cellular and Molecular Aspects*, R. A. Greenwald and G. Cohen (eds.), Elsevier, New York, pp. 145–153.

Scott, R. A., Cramer, S. P., Shaw, R. W., Beinert, H., and Gray, H. B., 1981. Extended X-ray absorption fine structure of copper in cytochrome *c* oxidase: Direct evidence for copper–sulfur ligation, *Proc. Natl. Acad. Sci. U.S.A.* 78:664–667.

Siiman, O., Young, N. M., and Carey, P. R., 1976. Resonance Raman spectra of "blue" copper proteins and the nature of their copper sites, *J. Am. Chem. Soc.* 98:744–748.

Smith, T. E., Weissbach, H., and Udenfriend, S., 1962. Studies on the mechanism of action of monoamine oxidase: Metabolism of *N,N*-dimethyl-tryptamine and *N,N*-dimethyl-tryptamine-*N*-oxide, *Biochemistry* 1:137–143.

Strothkamp, K. G., and Dawson, C. R., 1974. Concerning the quarternary structure of ascorbate oxidase, *Biochemistry* 13:434–440.

Suva, R. A., and Abeles, R. H., 1978. Studies on the mechanism of action of plasma amine oxidase, *Biochemistry* 17:3538–3545.

Suzuki, S., Sakurai, T., Nakahara, A., Oda, O., Manabe, T., and Okuyama, T., 1980. Spectroscopic aspects of copper binding site in bovine serum amine oxidase, *FEBS Lett.* 116:17–20.

Suzuki, S., Sakurai, T., Nakahara, A., Oda, O., Manabe, T., and Okuyama, T., 1982. Copper binding site in serum amine oxidase treated with sodium diethylthiocarbamate, *Chem. Lett.* 487–490.

Taylor, C. E., Taylor, R. S., Rasmussen, C., and Knowles, P. F., 1972. A catalytic mechanism for the enzyme benzylamine oxidase from fig plasma, *Biochem. J.* 130:713–728.

Van Heuvelen, A., 1965. Electron spin resonance spectrum of Cu^{++} in diamine oxidase, *Nature (London)* 208:888–889.

Volpe, J. A., and Caughey, W. S., 1974. Preparation and properties of bovine heart cytochrome *c* oxidase with high solubility in detergent free media, *Biochem. Biophys. Res. Commun.* 61:502–509.

Volpe, J. A., O'Toole, M. C., and Caughey, W. S., 1975. Quantitative infra-red spectroscopy of CO complexes of cytochrome *c* oxidase, hemoglobin and myoglobin: Evidence for one CO per heme, *Biochem. Biophys. Res. Commun.* 62:48–53.

Waleh, A., and Ingraham, L. L., 1976. Electron transport in cytochrome *c*, in *Environmental Effects on Molecular Structure and Properties*, B. Pullman (ed.), Reidel, Dordrecht Holland, pp. 505–515.

Walsh, C. T., Schonbrunn, A., and Abeles, R. H., 1971. Studies on the mechanism of action of D-amino acid oxidase: Evidence for removal of substrate α-hydrogen as a proton, *J. Biol. Chem.* 246:6855–6866.

Weiner, R. E., Ettinger, M. E., and Kosman, D. J., 1977. Fluorescence properties of copper enzyme galactose oxidase and its tryptophan-modified derivatives, *Biochemistry* 16:1602–1606.

Williams, R. F., and Bruice, T. C., 1976. The kinetics and mechanisms of 1,5-D-hydroflavin reduction of carbonyl compounds and flavin oxidation of alcohols. 2. Ethyl pyruvate, pyruvamide and pyruvic acid, *J. Am. Chem. Soc.* 98:7752–7768.

Winkler, M. E., and Bereman, R. D., 1980. Stereoelectronic properties of metalloenzymes. 6. Effects of anions and ferricyanide on the copper(II) site of the histidine and the tryptophan modified forms of galactose oxidase, *J. Am. Chem. Soc.* 102:6244–6247.

Yadav, K. D. S., and Knowles, P. F., 1981. A catalytic mechanism for benzylamine oxidase from pig plasma, *Eur. J. Biochem.* 114:139–144.

Yamada, H., and Adachi, I., 1971. Amine oxidase (*Aspergillus niger*), *Methods Enzymol.* 17B:705–709.

Yamada, H., and Yasunobu, K. T., 1962. Monamine oxidase, I. Purification, crystallization and properties of plasma monoamine oxidase, *J. Biol. Chem.* 237:1511–1516.

Yamada, H., and Yasunobu, K., 1963. Monoamine oxidase. IV. Nature of the second prosthetic group of plasma monoamine oxidase, *J. Biol. Chem.* 238:2669–2675.

Yamada, H., Yasunobu, K., Yamano, T., and Mason, H. S., 1963. Copper in plasma amine oxidase, *Nature (London)* 198:1092–1093.

Yamada, H., Adachi, O., and Ogata, K., 1965. Amine oxidase of microorganisms. IV. Further properties of amine oxidase of *Aspergillus niger*, *Agric. Biol. Chem.* 29:912–917.

Yasunobu, K. T., and Gomes, B., 1971. Mitochondrial amine oxidase (monoamine oxidase) (beef liver), *Methods Enzymol.* 17B:708–717.

Yasunobu, K. T., and Smith, R. A., 1971. Amine oxidase (beef plasma), *Methods Enzymol.* 17B:698–704.

Flavin Monooxygenases

12

12.1 Introduction

Flavins and pteridines act as catalysts in many hydroxylation reactions, particularly those in bacterial systems. One atom of dioxygen is incorporated in the hydroxyl group and the other is reduced to water, so these enzymes are monooxygenases. Flavin monooxygenases (for a review, see Ballou, 1982) perform aromatic hydroxylations, oxidative decarboxylations, and oxidation of amines. Flavin monooxygenases are not able to hydroxylate aliphatic hydrocarbons, and most, if not all, aromatic substrates are fairly well activated by hydroxy or amino groups.

The flavin monooxygenases may be divided into those that require an external two-electron reductant, called *external enzymes,* and those that utilize the substrate itself for the source of two electrons, called *internal enzymes.*

The external enzymes act on aromatic substrates and require either NADH or NADPH as the external reductant. The substrates act both as effectors and as substrates. The rate of reduction of the flavin prosthetic group by NADH or NADPH may be increased by 10,000 times when the substrate is in high enough concentration to act as an effector. The first reaction of the reductant with dioxygen produces hydrogen peroxide. When substrate is present, the total reaction to give hydroxylated product produces water. During the hydroxylation, an intermediate called I is formed. This intermediate is believed to be the 4a-hydroperoxyflavin. This point is discussed further in Section 12.13.

12.2 Salicylate Hydroxylase

Salicylate hydroxylase catalyzes the oxidative decarboxylation of salicylic acid (Massey and Hemmerich, 1975; Kamin *et al.,* 1978). The hydroxylation occurs on the same carbon that was bonded to the carboxyl group (Hamzah and Tu, 1981). The product of the reaction is CO_2, not H_2CO_3, so the carboxyl group must be decarboxylated in the anionic form (Suzuki and Katagiri, 1981). The enzyme utilizes the NADH for the external reductant. The reducing step is at least partially rate-determining as found from isotopic rate studies (Wang et al., 1982).

Many derivatives of salicylic acid also act as substrate, including 2,3-dihydroxybenzoic acid, 2,4-dihydroxybenzoic acid, 2,5-dihydroxybenzoic acid, 2,6-dihydroxybenzoic acid, p-aminosalicylic acid, 1-hydroxy-2-naphthoic acid, and 3-methylsalicylic acid. The usual source of the enzyme is Pseudomonas putida. This enzyme has 1 mole of FAD per mole of enzyme, but enzymes from other sources have 2 moles of FAD per mole of enzyme. There appears to be no metal present. The oxidation of NADH by dioxygen will take place in the absence of substrate to be hydroxylated, but the reaction rate is increased by the presence of the hydroxylation substrate (Takemori et al., 1972) or by certain pseudo-substrates such as benzoic acid (White-Stevens and Kamin, 1972). Both substrates and pseudo-susbtrates perturb the spectrum of the enzyme, making a convenient method to measure enzyme binding (White-Stevens and Kamin, 1972).

12.3   m-Hydroxybenzoate Hydroxylase

m-Hydroxybenzoate hydroxylase forms gentisic acid from m-hydroxybenzoic acid (Figure 12-1) (Massey and Hemmerich, 1975). This is a rather unusual reaction, since hydroxylations are usually ortho to the existing hydroxyl group. The enzyme is obtained from Pseudomonas aeruginosa and requires either NADH or NADPH as an external reductant.

12.4   m-Hydoxybenzoate 4-Hydroxylase

m-Hydroxybenzoate 4-hydroxylase (Premkumar et al., 1969) from Aspergillus niger hydroxylates m-benzoic acid ortho to the existing hydroxy group (Figure 12-2). Surprisingly, this enzyme is inhibited by superoxide dismutase

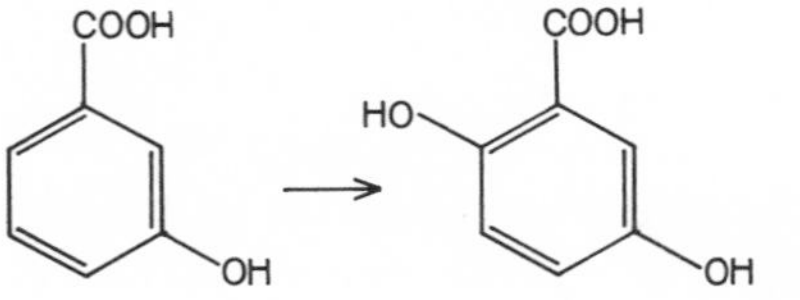

Figure 12-1. Reaction catalyzed by m-hydroxybenzoate hydroxylase.

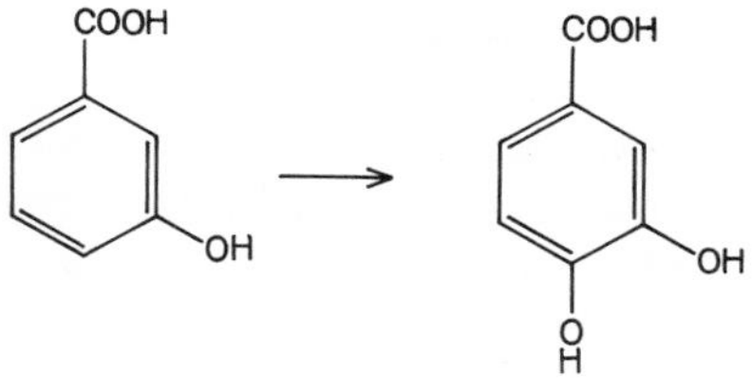

Figure 12-2. Reaction catalyzed by *m*-hydroxyben-
zoate 4-hydroxylase.

(Kumar *et al.*, 1972). The authors have taken this for evidence that superoxide anion is the hydroxylating agent.

12.5 *p*-Hydroxybenzoate Hydroxylase

p-Hydroxybenzoate hydroxylase hydroxylates *p*-hydroxybenzoic acid and derivatives to 3,4-dihydroxybenzoic acid (Spector and Massey, 1972c; Entsche *et al.*, 1976). The enzyme is isolated from *Pseudomonas putida, Ps. desmolytica,* and *Ps. fluorescens* (Husain *et al.*, 1978). The enzyme contains 1 mole of FAD per mole of enzyme (Howell *et al.*, 1972) and requires NADPH as an external reductant. There is an essential arginine at the substrate-binding site (Shoun *et al.*, 1980) and an essential histidine at the nicotinamide-binding site (Shoun and Beppu, 1982).

The substrates serve as effectors, but not all effectors are substrates. Both 3,4-dihydroxybenzoate and 2,4-dihydroxybenzoate are effectors for this enzyme. The 2,4-dihydroxybenzoate is hydroxylated in the 3-position, but surprisingly, 3,4-dihydroxybenzoate is not hydroxylated in the 5-position (Spector and Massey, 1972b). 6-Hydroxynicotinate is an effector for the enzyme, but is not hydroxylated (Howell and Massey, 1970). Both chloride ion and iodide ion are inhibitors of the enzyme (Steenis *et al.*, 1973). Wierenga *et al.* (1979) have done an X-ray crystallographic study of this enzyme. The crystals were formed in the presence of substrate, so that the orientation of substrate to isoalloxazine in the active site could be determined. Both substrate and flavin are seen to be buried deep in the enzyme with little room for movement, and the substrate is clearly adjacent to the C^{4a}–N^5 edge of flavin, strongly supporting the theory that the reactive oxygenated flavin has an electrophilic oxygen moiety in this region. NADH and dioxygen apparently bind in the same site (in agreement with kinetic data), which is on the opposite side of the flavin from the substrate-binding site.

That the reaction occurs without major perturbation of the enzyme structure (i.e., as would be required for substrate to align with some other portion of the flavin) is supported by the demonstration of catalytic activity in the crystallized enzyme.

12.6   Melilotic Hydroxylase

Melilotic monooxygenase hydroxylates *o*-hydroxyphenyl pyruvic acid (melilotic acid) to 2,3-dihydroxyphenyl pyruvic acid (Strickland and Massey, 1973a, b). This enzyme contains 1 mole of FAD per mole and requires NADH as an external reductant.

12.7   Phenol Hydroxylase

Phenol hydroxylase (Massey and Hemmerich, 1975) occurs in a yeast. It catalyzes the hydroxylation of phenol and many derivatives of phenol to catechols and on to pyrogallols. There is 1 FAD per mole of enzyme, and the external reductant is NADPH (Neujahr and Gaal, 1973). The enzyme appears to have an essential thiol group. Intermediates in the reaction have spectra corresponding to 4a-hydroperoxyflavin and 4a-hydroxyflavin (Delmar *et al.*, 1982). The significance of these intermediates is discussed in Section 12.13.

12.8   Orcinol Hydroxylase

Orcinol hydroxylase (Massey and Hemmerich, 1975) is found in *Ps. putida*. The external reducing agent is NADH, but NADPH also shows some activity. There is 1 FAD per mole of enyzme. Orcinol is hydroxylated to produce 2,3,5-trihydroxytoluene (Ortha and Ribbons, 1970).

12.9   Kynurenine 3-Hydroxylase

Kynurenine 3-hydroxylase (Massey and Hemmerich, 1975) hydroxylates kynurenine at the 3-position (Figure 12-3). The enzyme occurs in rat liver mi-

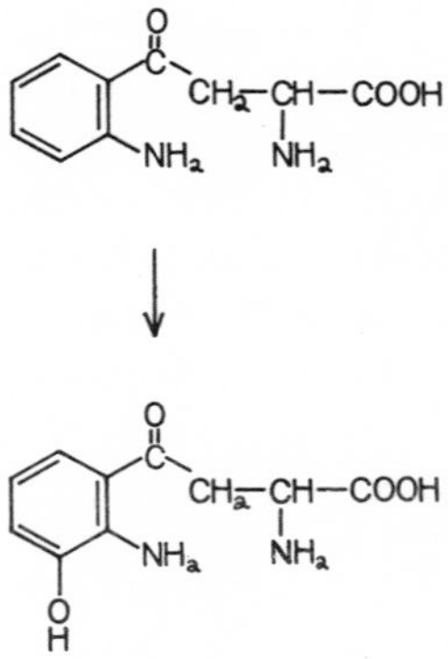

Figure 12-3. Reaction catalyzed by kynurenine 3-hydroxylase.

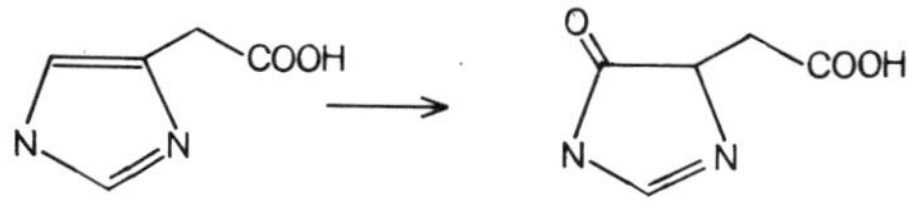

Figure 12-4. Reaction catalyzed by imidazolylacetate monooxygenase.

tochondria and is believed to have an FAD prosthetic group. The enzyme requires
either NADPH or NADH as an external reductant, although NADPH is preferred.

12.10   Imidazolylacetate Monooxygenase

Imidazolylacetate monooxygenase (Massey and Hemmerich, 1975) is a
flavoprotein (Maki *et al.*, 1966) that catalyzes the oxidation of imidazolylacetate
to imidazolonylacetate (Figure 12-4). One atom of dioxygen is transferred to the
product (Rothberg and Hayaishi, 1957). This reaction is the first step in the
pathway between imidazolylacetate and formyl aspartic acid. The imidazolonylacetate is converted to *N*-formimino-L-aspartic acid and then to formyl aspartic
acid (Hayaishi *et al.*, 1957). The enzyme contains 1 mole of FAD (Maki *et al.*,
1969). Either NADH or NADPH can be used as an external reductant, but
NADPH functions the better. The enzyme contains two sulfhydryl groups, of
which one is essential for catalytic activity of the enzyme. The sulfhydryl group
is essential for substrate binding, but does not appear to have any function in
the hydroxylation reaction (Okamoto *et al.*, 1968).

12.11   Amine Oxidase

Amine oxidase (Massey and Hemmerich, 1975) from liver microsomes and
from hog kidney will oxidize secondary amines to the hydroxylamines and tertiary
amines to the *N*-oxides. With a few exceptions, there is no action on primary
amines, although the presence of primary amines activates the enzyme toward
secondary and tertiary amines (Ziegler and Mitchell, 1972). These enzymes will
also oxidize thio ethers to the sulfoxide (Paulsen and Ziegler, 1979; Hajjar and
Hodgson, 1980). They are probably responsible for the degradation in mammals
of many of the sulfur-containing pesticides (Hajjar and Hodgson, 1980). The
external reductant is NADPH (Ziegler and Mitchell, 1972). Amine oxidases are
inhibited by certain amino acetylenes (Figure 12-5). The inhibitor adds to the
N^5 position of the flavin as shown in Figure 12-6. Both pargyline [*N*-methyl-

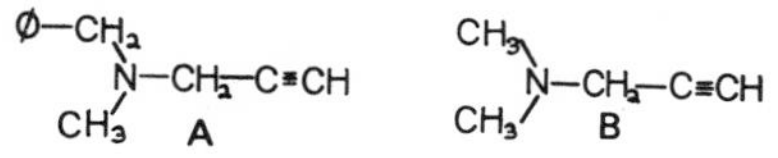

Figure 12-5. Inhibitors for amine oxidase. (A) Pargyline [*N*-methyl-*N*-(2-propynyl)-benzylamine]; (B)
3-dimethyl-amino-1-propyne.

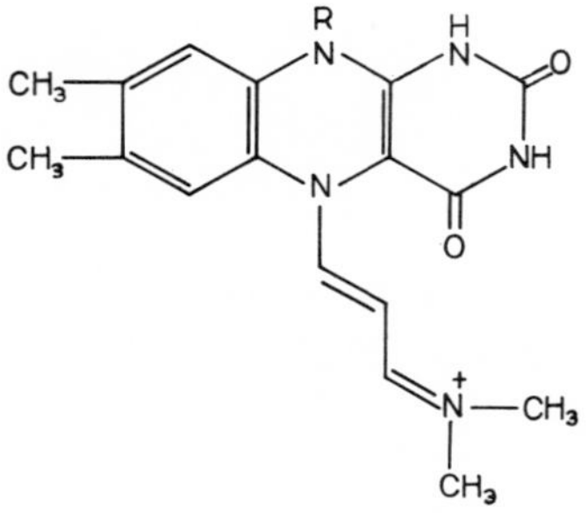

Figure 12-6. Adduct formed when dimethylpropynyl amine is added to amine oxidase.

N(2-propynyl)-benzylamine] (A in Figure 12-5) (Hellerman and Erwin, 1968; Chuang *et al.*, 1974) and 3-dimethylamino-1-propyne (B) (Maycock *et al.*, 1976) act in this manner.

These enzymes proceed through a relatively stable intermediate I that absorbs at 375 nm corresponding to a 4a-hydroperoxy flavin (Paulsen and Ziegler, 1979). No intermediate II has been detected. Intermediate I reacts with substrates to produce a new intermediate that has a spectrum corresponding to a 4a-hydroxy flavin (Beaty and Ballou, 1980, 1981).

12.12   *p*-Cresol Methyl Hydroxylase

This is an aliphatic hydroxylating enzyme that produces *p*-hydroxybenzyl-alcohol from *p*-cresol. This enzyme utilizes a flavin derivative for a coenzyme, 8-α-O-tyrosyl-FAD (McIntire *et al.*, 1980, 1981).

12.13   Mechanisms of Hydroxylation

There appear to be two different kinds of hydroxylation mechanisms: One type, typified by the amine oxidase, oxidizes easily oxidizable nucleophiles; the other type hydroxylates the aromatic rings. We will first discuss the aromatic hydroxylations. Most studies of hydroxylation have been on the enzyme *p*-hydroxybenzoate hydroxylase.

The reaction sequence for the *p*-hydroxybenzoate hydroxylase (Entsche *et al.* (1976), and probably for the other aromatic flavin monooxygenases as well, starts with the reduced flavin on the enzyme reacting with dioxygen in a second-order reaction to give an intermediate absorbing at 380–390 nm that is designated I (Spector and Massey, 1972a; Entsche *et al.*, 1976). Intermediate I can decompose with poor substrates to give the oxidized flavin and hydrogen peroxide or, with good substrates, to give the new intermediate II absorbing at 390–420 nm.

The rate of product formation has been correlated with the rate of conversion of intermediate I to II (Entsche *et al.*, 1974), so that this conversion is the rate-determining step. Intermediate II then forms an intermediate very similar to I. This intermediate, designated III, absorbs at 380–385 nm. Intermediates I and III have also been detected in the hydroxylation of 2-hydroxycinnamate ion by melilotate hydroxylase (Schopfer and Massey, 1980). Intermediate I is believed to be a 4a-hydroperoxy adduct of flavin, and either this or a kinetically related species is believed to be the hydroxylating species (Entsche *et al.*, 1976).

Although there is a deuterium isotope effect of about 8% in the hydroxylation reaction of a deuterated substrate (melilotate), deuterium on the aromatic ring does not affect the rate of formation of I or the reduction of flavin (Strickland *et al.*, 1975). No deuterium migration, characteristic of an NIH shift [see Chapter 13 (Section 13.2)], has been detected during the hydroxylation. The overall deuterium rate effect, utilizing deuterated nicotinamides, is 1.8–3.5 for these enzymes even though the reduction step has an isotope rate effect of 10 ± 2. The individual steps in the reaction must have quite similar energy barriers so that no step is definitely rate-determining (Ryerson *et al.*, 1982).

1-Deazaflavin can be added to the apoenzyme to produce an enzyme similar to a normal flavin enzyme. The reconstituted enzyme can oxidize NADPH with dioxygen to form $NADP^+$ and hydrogen peroxide. When substrate is added as an effector, the 4a-hydroperoxide is formed, but the reaction does not proceed further (Entsche *et al.*, 1980). This result indicates that the nitrogen at the 1-position is necessary for the hydroxylation reaction (or at least for the decay of intermediate I).

A mechanism has been proposed (Entsche *et al.*, 1976) in which the center ring opens during the hydroxylation (Figure 12-7). Intermediate II would then be a flavin derivative (C) in which cleavage has occurred between the 4a- and 5-positions. However, Wessiak and Bruice (1981) have synthesized (C) and found that its spectral maximum is at 342 nm instead of 390–420 nm. Unless the protein causes a large spectral shift in this compound, this makes the open form of flavin an unlikely candidate for intermediate II.

Intermediate II then proceeds to form 4a-hydroxyflavin (D), which has been assigned as the structure of intermediate III. The absorption spectra of intermediates I and III on the enzyme are very similar to that of 5-ethyl-4a-hydroxy flavin when absorbed on the enzyme (Entsche *et al.*, 1976). The 4a-hydroxy flavin then eliminates water to produce the oxidized flavin, which is reduced by a pyridine nucleotide. The hydroxylated substrate is not released until the oxidized flavin is formed.

Hamilton (1971) has proposed a very similar mechanism differing only in the detail that the center ring of the 4a-hydroperoxide opens before the hydroxylation (Figure 12-8). The open intermediate (B) is a carbonyl oxide that has a low electron density on the terminal oxygen. However, as discussed above, the present data for intermediate II do not correspond to the open flavin (C).

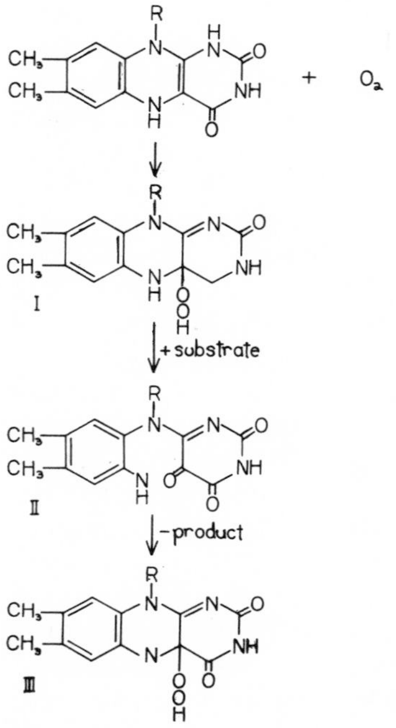

Figure 12-7. A reaction mechanism proposed for aromatic hydroxylation by flavin monooxygenases.

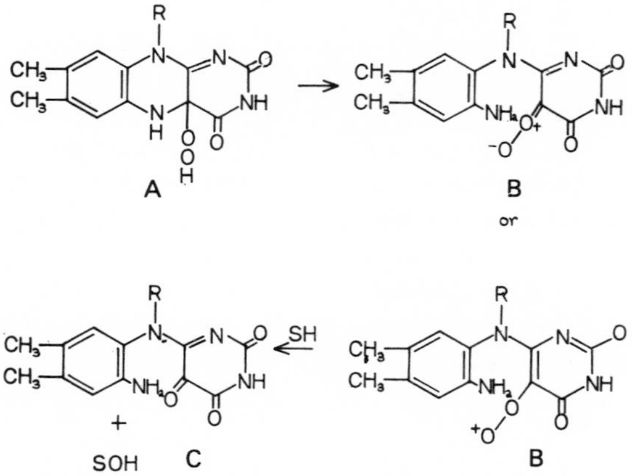

Figure 12-8. A proposed mechanism for hydroxylation by flavin monooxygenase that proceeds through a carbonyl oxide.

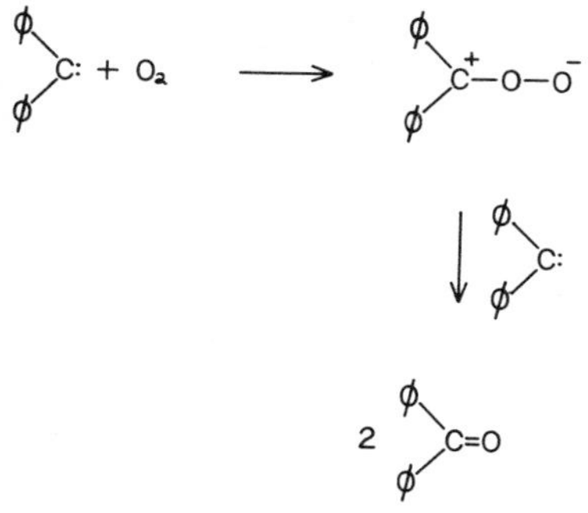

$$R_2CN_2 \xrightarrow{h\nu} R_2C: \; + \; N_2$$

$$R_2C: \; + \; O_2 \longrightarrow R_2CO_2$$

Figure 12-9. Carbonyl oxide intermediate in the carbene–dioxygen reaction.

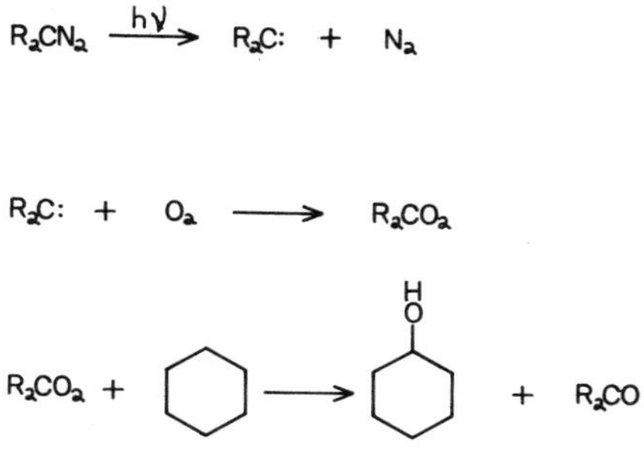

Figure 12-10. Hydroxylation of cyclohexane saturated with dioxygen by the photolysis of a diazo compound.

Carbonyl oxides occur as intermediates in the reaction of carbenes with dioxygen (Figure 12-9). The product of the reaction is usually the ketone formed by the reaction of the carbonyl oxide with another carbene (Kirmse *et al.*, 1958). Unlike flavin monooxygenases, which require electron-rich substrates, carbonyl oxides will hydroxylate hydrocarbons. The photolysis of a diazo compound to form a carbene in a solution of oxygenated cyclohexane will form cyclohexanol (Figure 12-10) (Hamilton and Giacin, 1966).

The hydroxylation of 2-methyl-butane by a carbonyl oxide produces a ratio of primary to secondary to tertiary alcohols of 1 : 15 : 140. This result would be expected for a highly selective free radical. Carbonyl oxides can be drawn as free radicals as well as ionic species (Figure 12-11). Carbonyl oxides will dimerize to form cyclic diperoxides (Figure 12-12). This reaction of carbonyl oxides can be rationalized in terms of either the diradical or the ionic resonance forms (Bartlett and Traylor, 1962).

α-Carbonyl groups make carbonyl oxides better hydroxylating agents. Benzoylphenyldiazomethane gives a carbonyl oxide that will produce about 500 times as much dimethylsulfoxide from dimethylsulfide as does diphenyl diazomethane because the terminal oxygen is more electrophilic or less nucleophilic (Figure 12-13) (Ando *et al.*, 1979).

Hamilton (1971) has suggested that the salicylate hydroxylase reaction may

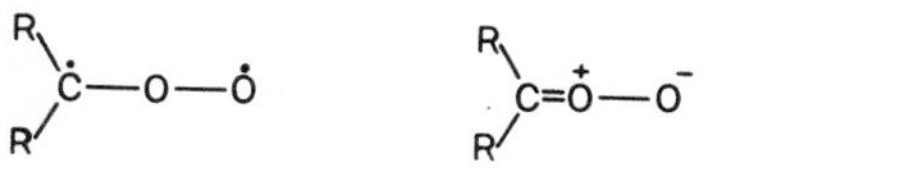

Figure 12-11. Resonance forms of a carbonyl oxide.

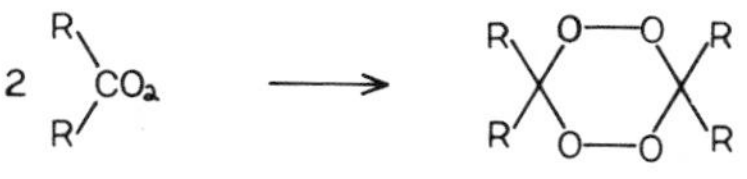

Figure 12-12. Dimerization of carbonyl oxide.

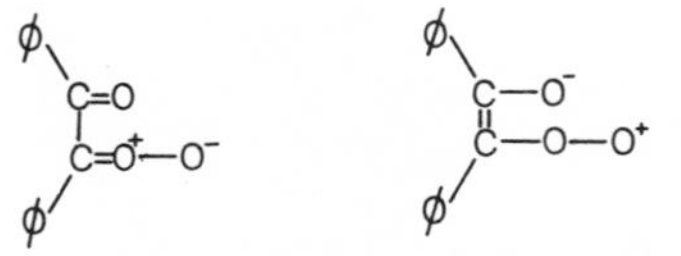

Figure 12-13. Resonance forms showing why α-carbonyl oxides are more electrophilic than unsubstituted carbonyl oxides.

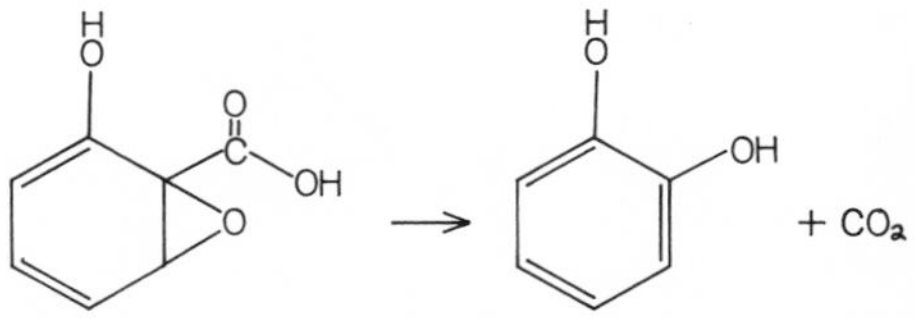

Figure 12-14. A proposed intermediate epoxide in the enzymatic hydroxylation of salicylic acid.

proceed through an epoxide that subsequently rearranges and cleaves to give catechol and carbon dioxide (Figure 12-14). 1,2-Epoxides have been found in hydroxylations of aromatic compounds by cytochrome P-450 (cf. Chapter 15). Orf and Dolphin (1974) have proposed (Figure 12-15) that the actual hydroxylating agent in the flavin monooxygenases is an oxaziridine formed from the 4a-hydroperoxide and that the oxaziridine opens to an N^5-oxide with a very low electron density on the oxygen. Flavin-N^5-oxides have been synthesized and found to be effective hydroxylating agents in the excited state (Rastetter *et al.*, 1979). The hydroxylating agent has been shown to be a nitroxyl radical. The

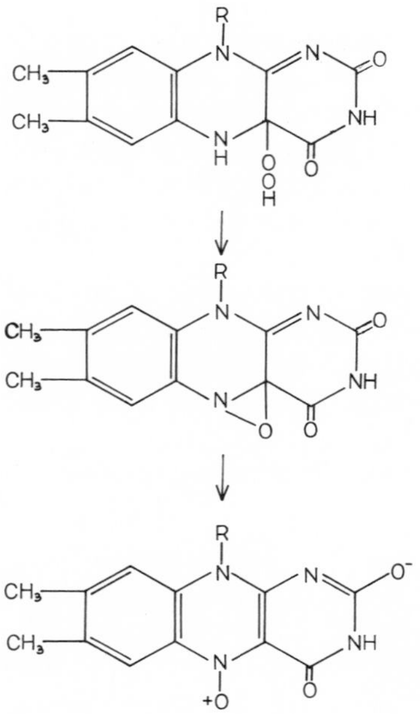

Figure 12-15. A proposed mechanism for hydroxylation by flavin monooxygenases that proceeds through an oxaziridine.

nitroxyl radical could be formed from the oxaziridine or alternatively from a hydroxyl amine intermediate (Frost and Rastetter, 1981). Tokumura *et al.* (1980) report loss of oxygen from an aromatic oxaziridine on warming from 77°K. Presumably solvent is hydroxylated in the process.

The formation of the 4a-hydroperoxyflavin from dioxygen and reduced flavin is an interesting reaction because the overall reaction requires a change in spin. Kemal *et al.* (1977) have proposed that a triplet flavin–dioxygen complex is formed that decays to a singlet complex:

$$FH^- + O_2 \rightarrow {}^3[FH \cdot O_2]^-$$

$${}^3[FH \cdot O_2]^- \rightarrow {}^1[FH \cdot O_2]^-$$

$$H^+ + {}^1[FH \cdot O_2]^- \rightarrow FHOOH$$

The other type of flavin hydroxylation reaction is that with an easily oxidizable nucleophile. Amine oxidase oxidizes amines to N-oxide and thiol ethers to the sulfoxides. In these reactions, there is no indication of an intermediate II. The hydroxylation is performed by intermediate I, the 4a-hydroperoxide of flavin, which directly forms intermediate III.

A chemical model has been found for this reaction. The 4a-hydroperoxide of 5-ethylflavin has been prepared (Kemal and Bruice, 1976) and found to hydroxylate certain easily hydroxylated compounds, such as amines, without an enzyme. 4a-Hydroperoxy-5-ethylflavin will oxidize N,N-dimethylaniline and many other amines to the N-oxides (Ball and Bruice, 1979, 1980). It will also oxidize thioxane to the sulfoxide. The 4a-hydroperoxy-5-ethylflavin is reduced to a 4a-hydroxy-5-ethylflavin. In the enzymatic reaction where there is no alkyl group at the 5-position, the hydroxyflavin would eliminate water to re-form the oxidized flavin. The reaction does not proceed when the 4a-hydroperoxide-5-ethylflavin is in the anionic form, nor does the reaction occur with hydrogen peroxide. The mechanism is believed to be the attack of the amine or thio ether on the outer oxygen of the peroxide. The flavin hydroperoxide reacts very much faster than *t*-butyl hydroperoxide because of the electronegativity of the flavin. Substitution of a carbon for nitrogen-1 in the flavin reduces the oxidizing ability of the hydroperoxide (Ball and Bruice, 1981).

It is interesting to note that the 4a-hydroperoxide of N^5-ethyl-3-methyl lumiflavin can be prepared from the semiquinone radical and superoxide (Nanni *et al.*, 1981).

A mechanism has been proposed in which a flavin dioxygenase can also act as a flavinmonooxygenase. For a discussion of this mechanism, see Section 18.12 in Chapter 18.

12.14   Baeyer–Villiger Enzymes

A small number of flavin monooxygenase enzymes are known that catalyze Baeyer–Villiger reactions. This is a very well known organic reaction of ketones with peracids or peroxides to form esters. The reaction is characterized by migration of the more stable carbonium ion (of the two carbonyl substituents) and retention of configuration in the migrating group.

One example of this type of enzyme has been isolated from *Nocardia globerula* CL1 and *Acinetobacter* NCIB 9871 (Donoghue *et al.*, 1976) that allows these two organisms to survive with cyclohexanone as sole carbon source. The *N. globerula* enzyme has a molecular weight of 53,000 and that from *Acinetobacter* 59,000, and they show pH optima of 8.4 and 9.0, respectively. The reaction product is 1-oxa-2-oxocycloheptane (ε-caprolactone).

A study of the *Acinetobacter* enzyme was undertaken to establish a similarity between the biochemical and organic reaction paths (Schwab, 1981). Cyclohexanone stereospecifically labeled with deuterium at the α-carbons was synthesized enzymatically. Labeled substrate was subjected to the enzymatic oxidation, then derivatized, eventually yielding a single camphanate ester isomer. Deuterium nuclear magnetic resonance showed that the reaction had proceeded with retention of configuration of the migrating group. Hence, the enzymatic reaction appears to follow the uncatalyzed reaction stereochemically, even in approximately symmetrical substrates.

12.15   Lactic Monooxygenase

Lactic monooxygenase catalyzes the oxidative decarboxylation of lactic acid to produce carbon dioxide and acetic acid (Massey and Hemmerich, 1975). This enzyme is called an internal monooxygenase because no extraneous two-electron reductant is utilized. The substrate itself furnishes the two electrons for reducing the flavin. The enzyme is rather general for several α-hydroxy acids. The acetic acid produced in the oxidative decarboxylation of lactic acid is labeled when labeled dioxygen is used in the reaction mixture, but not when oxygen-labeled water is used (Hayaishi and Sutton, 1957).

The enzyme is inhibited by several anions including phosphate ion (Lockridge *et al.*, 1972). The phosphate ion appears to be competitive for the lactate ion. A kinetic study of the enzyme has shown that there are two steps in the oxidation (Lockridge *et al.*, 1972). In the first reaction, oxidized flavin oxidizes the α-hydroxy acid to an α-keto acid. This reaction will occur anaerobically if oxidized FMN is used as a substrate (Katagiri and Takenori, 1971). The second step requires dioxygen (Yamauchi *et al.*, 1973). In this reaction, reduced $FMNH_2$,

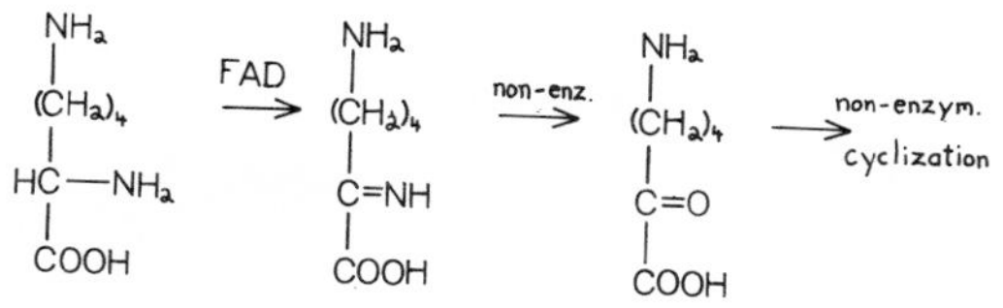

Figure 12-16. Anaerobic action of lysine monooxygenases on lysine.

dioxygen, and the α-keto acid produce acetic acid, carbon dioxide, and oxidized FMN.

12.16 Lysine Monooxygenase

Another flavin monooxygenase, lysine monooxygenase (Massey and Hemmerich, 1975), catalyzes the aerobic oxidative decarboxylation of lysine to produce δ-amino valeramide (Takeda et al., 1969).

This enzyme will also act on L-arginine, although at a much slower rate. This is another internal oxygenase in which the substrate furnishes the two electrons for reduction of the flavin. The flavin is in the form of FAD. The enzyme has four subunits, and there is one FAD molecule per subunit (Flashner and Massey, 1974). It is an oxygenase because labeled dioxygen produces labeled amide (Itada et al., 1961). There appear to be no metals present. The enzyme acts on L-ornithine (Nakazawa et al., 1972) or anaerobically on L-lysine (Yamamoto et al., 1972) to form the corresponding α-keto acid. The enzyme acts as a dehydrogenase to form the α-imino acid, which hydrolyzes nonenzymatically to the α-keto acid. The α-keto acid subsequently cyclizes to α-piperidine-2-carboxylic acid (Figure 12-16).

12.17 Arginine Monooxygenase

Arginine monooxygenase is similar to lysine monooxygenase (Massey and Hemmerich, 1975). The enzyme catalyzes the oxidation of arginine to the aldehyde (Figure 12-17). This again is an internal flavin monooxygenase, and the

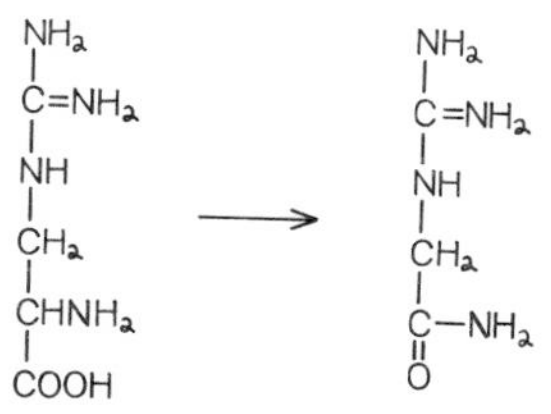

Figure 12-17. Reaction catalyzed by arginine monooxygenase.

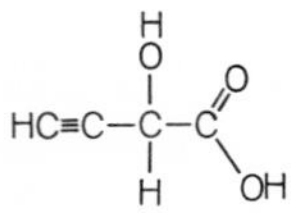

HC≡C−C−C
       | ‖O
       H   OH   Figure 12-18. An inhibitor for lactate oxidase.

(with H, O, O, OH substituents on the carbon chain)

prosthetic group is FAD. There is an essential lysine at the active center that is believed to abstract the α-proton of the arginine. The enzyme is a monooxygenase because one atom of dioxygen is formed in the product (Pho *et al.*, 1966).

12.18   Oxalic Acid Oxidase

Oxalic acid oxidase, found in barley seedlings (Chiriboga, 1966), catalyzes the oxidation of oxalic acid to carbon dioxide. The dioxygen is reduced to hydrogen peroxide. Because the enzyme is activated by flavins, we mention it here with the flavin monooxygenases.

12.19   Mechanism of Oxidative Decarboxylation

Although the detailed mechanism of these reactions is not known, it is clear that the α-amino acids proceed through α-imino acids and the α-hydroxy acids proceed through α-keto acids. The formation of the α-keto acid from the α-hydroxy acid appears to proceed through an α-carbanion.

Studies with inhibitors and unnatural substrates both indicate that the first step is the formation of an α-carbanion. β-Chlorolactic acid is an unnatural substrate for lactic monooxygenase. The product of the reaction, pyruvic acid, can be rationalized by the initial formation of α-carbanion, followed by elimination of chloride ion to produce the enol of pyruvic acid (Walsh *et al.*, 1973).

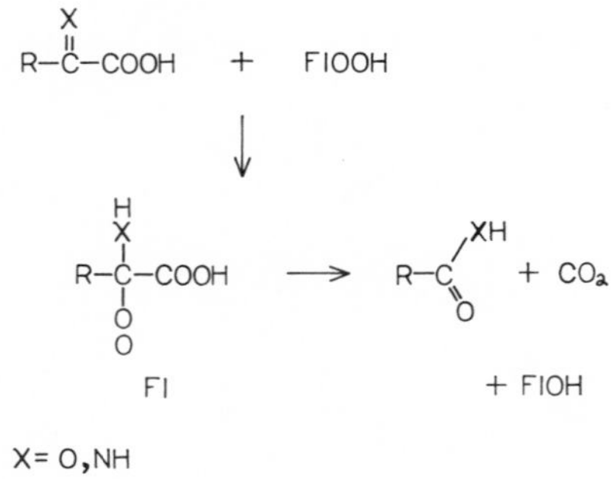

$$
\begin{array}{c}
\overset{X}{\underset{\|}{R-C}}-COOH \quad + \quad FlOOH \\[2em]
\downarrow \\[2em]
\underset{\underset{\underset{Fl}{O}}{O}}{\overset{\overset{\overset{H}{X}}{|}}{R-C}}-COOH \quad \longrightarrow \quad R-C\overset{XH}{\underset{O}{}} \quad + \quad CO_2 \\[1em]
\hspace{6em} + \; FlOH
\end{array}
$$

X= O,NH

Figure 12-19. A proposed mechanism for oxidative decarboxylation of flavin enzymes.

The acetylenic hydroxy acid 1-hydroxy-2-butynoic acid (Figure 12-18) is an inhibitor for lactate oxidase (Walsh *et al.*, 1972). The α-hydrogen is lost, but the γ-hydrogen is retained, in the inhibition reaction. A flavin adduct of unknown structure is formed with the inhibitor (Walsh *et al.*, 1972). Probably the α-carbanion of the inhibitor condenses with the flavin.

The most likely reaction following the formation of α-keto or α-imino acid is the addition of the 4a-hydroperoxy flavin to the imino or keto acids (Hamilton, 1971) followed by ionic cleavage of the hydroperoxide (Figure 12-19).

References

Ando, W., Miyazaki, H., and Kohmoto, S., 1979. Oxygen atom transfer by an intermediate in the photosensitized oxygenation of diazo compounds, *Tetrahedron Lett.* 1317–1320.

Ball, S., and Bruice, T. C., 1979. 4a-Hydroperoxyflavin *N*-oxidation of teriary amines, *J. Am. Chem. Soc.* 101:4017–4019.

Ball, S., and Bruice, T. C., 1980. Oxidation of amines by 4a-hydroperoxyflavin, *J. Am. Chem. Soc.* 102:6498–6503.

Ball, S., and Bruice, T. C., 1981. The chemistry of 1-carba-1-deaza-N^5-methyl lumiflavins: Influence of the N^1 upon the reactivity of flavin 4a-hydroperoxides, *J. Am. Chem. Soc.* 103:5494–5503.

Ballou, D. P., 1982. Flavin monooxygenases, *Dev. Biochem.* 21:301–310.

Bartlett, P. D., and Traylor, T. G., 1962. Reaction of diphenyldiazomethane with oxygen: The Criegee carbonyl oxide, *J. Am. Chem. Soc.* 84:3488–3409.

Beaty, N. B., and Ballou, D. P., 1980. Transient kinetic study of liver microsomal FAD-containing monooxygenase, *J. Biol. Chem.*, 255:3817–3819.

Beaty, N. B., and Ballou, D. P., 1981. The oxidative half-reaction of liver microsomal FAD-containing monooxygenase, *J. Biol. Chem.* 256:4619–4625.

Chiriboga, J., 1966. Purification and properties of oxalic acid oxidase, *Arch. Biochem. Biophys.* 116:516–523.

Chuang, H. Y. K., Patek, D. R., and Hellerman, L., 1974. Mitochondrial monoamine oxidase inactivation by pargyline: Adduct formation, *J. Biol. Chem.* 249:2381.

Detmar, K., Massey, V., Ballou, D. P., and Neujahr, H. Y., 1982. Steady state and rapid reaction studies on phenol hydroxylases, *Dev. Biochem.* 21:334–338.

Donoghue, N. A., Norris, D. B., and Trudgill, P. W., 1976. The purification and properties of cyclohexanone oxygenase from *Nocardia globerula* CL1 and *Acinetobacter* NCIB 9871, *Eur. J. Biochem.* 63:175–192.

Entsche, B., Massey, V., and Ballou, D. P., 1974. Intermediates in flavoprotein catalyzed hydroxylation, *Biochem. Biophys. Res. Commun.* 57:1018–1026.

Entsche, B., Ballou, D. P., and Massey, V., 1976. Flavin–oxygen derivatives involved in hydroxylation by *p*-hydroxybenzoate hydroxylase, *J. Biol. Chem.* 251:2550–2563.

Entsche, B., Hussain, M., Ballou, D. P., Massey, V., and Walsh, C., 1980. Oxygen reactivity of *p*-hydroxybenzoate hydroxylase containing 1-deazaflavin, *J. Biol. Chem.* 255:1420–1429.

Flashner, M. I. S., and Massey, V., 1974. Purification and properties of L-lysine monooxygenase from *Pseudomonas fluorescens*, *J. Biol.*, *Chem.* 249:2579–2586.

Frost, J. W., and Rastetter, W. H., 1981. Flavoprotein monooxygenase: A chemical model, *J. Am. Chem. Soc.* 103:5242–5245.

Hajjar, N. P., and Hodgson, E., 1980. Flavin adenine dinucleotide-dependent monooxygenase: Its role in the sulfoxidation of pesticides in mammals, *Science* 209:1134–1136.

Hamilton, G. A., 1971. The proton in biological redox reactions, *Prog. Bioorg. Chem.* 1:83–157.

Hamilton, G. A., and Giacin, J. R., 1966. Oxidations by molecular oxygen. III. Oxidation of saturated hydrocarbons by an intermediate in the reaction of some carbenes with oxygen, *J. Am. Chem. Soc.* 88:1584–1585.

Hamzah, R. Y., and Tu, S.-C., 1981. Determination of the position of monooxygenation in the formation of catechol catalyzed by salicylate hydroxylase, *J. Biol. Chem.* 256:6392–6394.

Hayaishi, O., and Sutton, W. B., 1957. Enzymatic oxygen fixation into acetate concomitant with the enzymatic decarboxylation of L-lactate, *J. Am. Chem. Soc.* 79:4809–4810.

Hayaishi, O., Tabor, H., and Hayaishi, T., 1957. N-Formimino-L-aspartic acid as an intermediate in the enzymatic conversion of imidazole-acetic acid to formylaspartic acid, *J. Biol. Chem.* 227:161–180.

Hellerman, L., and Erwin, V. G., 1968. Mitochondrial amine oxidase. II. Action of various inhibitors for the bovine kidney enzyme catalytic mechanism, *J. Biol. Chem.* 243:5234–5243.

Howell, L. G., and Massey, V., 1970. A non-substrate effector of *p*-hydroxybenzoate hydroxylase, *Biochem. Biophys. Res. Commun.* 40:887–893.

Howell, L. G., Spector, T., and Massey, V., 1972. Purification and properties of *p*-hydroxybenzoate hydroxylase from *Pseudomonas fluorescens*, *J. Biol. Chem.* 247:4340–4350.

Husain, M., Schapfer, L. M., and Massey, V., 1978. *p*-Hydroxybenzoate hydroxylase and melilotate hydroxylase, *Methods Enzymol.* 53:543–518.

Itada, N., Ichihara, A., Makita, T., Hayaishi, O., Suda, M., and Sasaki, N., 1961. L-Lysine oxidase, a new oxygenase, *J. Biochem.* 50:118–121.

Kamin, H., White-Stevens, R. H., and Presswood, R. P., 1978. Salicylate hydroxylase, *Methods Enzymol.* 53:527–543.

Katagiri, M., and Takenori, S., 1971. The Reaction mechanism of flavin-containing enzymes, in *Flavins and Flavoproteins*, H. Kamin (ed.), University Park Press, Baltimore, Maryland, pp. 447–462.

Kemal, C., and Bruice, T. C., 1976. Simple synthesis of a 4a-hydroperoxy adduct of a 1,5-dihydroflavin: Preliminary studies of a model for bacterial luciferase, *Proc. Natl. Acad. Sci. U.S.A.* 73:995–999.

Kemal, C., Chan, T. W., and Bruice, T. C., 1977. Reaction of $^3O_2$ with dihydroflavins. I. $N^{3,5}$-Dimethyl-1,5-dihydrolumiflavin and 1,5-dihydroisoalloxazines, *J. Am. Chem. Soc.* 99:7272–7286.

Kirmse, W., Norner, L., and Hoffman, H., 1958. Umsetzungen photochemisch erzeugter Carbene, *Annalen* 614:19–30.

Kumar, R. P., Ravindranath, S. D., Vaidyanathan, C. S., and Rao, N. A., 1972. Mechanism of hydroxylation of aromatic compound. II. Evidence for the involvement of superoxide anions in enzymatic hydroxylations, *Biochem. Biophys. Res. Commun.* 49:1422–1426.

Lockridge, O., Massey, V., and Sullivan, P. A., 1972. Mechanism of action of the flavoenzyme lactate oxidase, *J. Biol. Chem.* 247:8098–8106.

Maki, Y., Yamamoto, S., Nozaki, M., and Hayaishi, O., 1966. Crystallization of imidazoleacetate monooxygenase and its characterization as a flavoprotein, *Biochem. Biophys. Res. Commun.* 25:609–613.

Maki, T., Yamamoto, S., Nozaki, M., and Hayaishi, O., 1969. Studies of monooxygenase. II. Crystallization and some properties of imidazole and acetate monooxygenase, *J. Biol. Chem.* 244:2942–2950.

Massey, V., and Hemmerich, P., 1975. Flavin and pteridine monooxygenases, in *The Enzymes*, Vol. 12, P. D. Boyer (ed.), Academic Press, New York, pp. 191–252.

Maycock, A. L., Abeles, R. H., Salach, J. I., and Singer, R. P., 1976. The structure of the covalent adduct formed by the interaction of 3-dimethylamino-1-propyne and the flavine of mitochondrial amine oxidase, *Biochemistry* 15:114–125.

McIntire, W., Edmondson, D. F., and Singer, T. P., 1980. 8α-O-Tryosyl-FAD: A new form of covalently bound flavin from *p*-cresol methyl hydroxylase, *J. Biol. Chem.* 255:6553–6555.

McIntire, W., Edmondson, D. E., Hopper, D. J., and Singer, T. P., 1981. 8α-(O-Tyrosyl) flavin adenine dinucleotide, the prosthetic group of bacterial p-cresol methylhydroxylase, *Biochemistry* 20:3068–3075.

Nakazawa, T., Hori, K., and Hayaishi, O., 1972. Studies on monooxygenases. V. Manifestation of amino acid oxidase activity by L-lysine monooxygenase, *J. Biol. Chem.* 247:3439–3444.

Nanni, E. J., Jr., Sawyer, D. T., Ball, S. S., and Bruice, T. C., 1981. Redox chemistry of N^5-ethyl-3-methyl-lumiflavinium cation and N^5-ethyl-4a-hydroperoxy-3-methyllumiflavin in dimethylformamide: Evidence for the formation of the N^5-ethyl-4a-hydroperoxy-3-methyllumiflavin anion via radical–radical coupling with superoxide ion, *J. Am. Chem. Soc.* 103:2797–2802.

Neujahr, H. Y., and Gaal, A., 1973. Phenol hydroxylase from yeast: Purification and properties of the enzyme from *Trichosporon cutaneum*, *Ev. J. Biochem.* 35:386–400.

Okamoto, H., Nozaki, M., and Hayaishi, O., 1968. A role of sulfhydryl groups in imidazoleacetate monooxygenase, *Biochem. Biophys. Res. Commun.* 32:30–36.

Orf, H. W., and Dolphin, D., 1974. Oxaziridines as possible intermediates in flavin monooxygenases, *Proc. Natl. Acad. Sci. U.S.A.* 71:2646–2650.

Ortha, Y., and Ribbons, D. W., 1970. Crystallization of orcinol hydroxylase from *Pseudomonas putida*, *FEBS Lett.* 2:189–192.

Paulsen, L. L., and Ziegler, D. M., 1979. The liver microsomal FAD-containing monooxygenase, *J. Biol. Chem.* 254:6449–6455.

Pho, D. B., Olomucki, A., and Thoai, N. V., 1966. L-Arginine oxygenase decarboxylante. IV. Incorporation de ^{18}O dans la γ-guanidino-butryamide, *Biochim. Biophys. Acta* 118:311–315.

Premkumar, R., Roa, R. V. S., Sreeleela, N. S., and Vaidyanathan, C. S., 1969. m-Hydroxybenzoic acid 4-hydroxylase from *Asperigillus niger*, *Can. J. Biochem.* 47:825–827.

Rastetter, W. H., Gadek, T. R., Tane, J. P., and Frost, J. W., 1979. Oxidations and oxygen transfers effected by a flavin N(5)-oxide: A mode for flavin-dependent monooxygenases, *J. Am. Chem. Soc.* 101:2228–2231.

Rothberg, S., and Hayaishi, O., 1957. Studies on oxygenases: Enzymatic oxidation of imidazoleacetic acid, *J. Biol. Chem.* 229:897–903.

Ryerson, R. R., Ballou, D. P., and Walsh, C. 1982. Kinetic isotope effects in the oxidation of isotopically labeled NAD(P)H by bacterial flavoprotein monooxygenases, *Biochemistry* 21:1144–1151.

Schopfer, L. M., and Massey, V., 1980. Kinetic and mechanistic studies on the oxidation of the melilotate hydroxylase 2-OH-cinnamate complex by molecular oxygen, *J. Biol. Chem.* 255:5355–5363.

Schwab, J. M., 1981. Stereochemistry of an enzymatic Baeyer–Villiger reaction: Application of deuterium NMR, *J. Am. Chem. Soc.* 103:1876–1879.

Shoun, H., and Beppu, T., 1982. A histidine residue in p-hydroxybenzoate hydroxylase essential for binding of reduced nicotinamide adenine dinucleotide phosphate, *J. Biol. Chem.* 257:3422–3428.

Shoun, H., Beppu, T., and Arima, K., 1980. An essential arginine residue at the substrate-binding site of p-hydroxybenzoate hydroxylase, *J. Biol. Chem.* 255:9319–9324.

Spector, T., and Massey, V., 1972a. p-Hydroxybenzoate hydroxylase from *Pseudomonas fluorescens*: Evidence for an oxygenated flavin intermediate, *J. Biol. Chem.* 247:5632–5636.

Spector, T., and Massey, V., 1972b. Studies on the effector specificity of p-hydroxybenzoate hydroxylase from *Pseudomonas fluorescens*, *J. Biol. Chem.* 247:6479–6487.

Spector, T., and Massey, V., 1972c. p-Hydroxybenzoate hydroxylase from *Pseudomonas fluorescens*: Reactivity with oxygen, *J. Biol. Chem.* 247:7123–7127.

Steenis, P. G., Cordes, M. M., Hilkens, G.-H., and Muller, F., 1973. On the interaction of para-hydroxybenzoate hydroxylase from *Pseudomonas fluorescens* with halogen ions, *FEBS Lett.* 36:177–180.

Strickland, S., and Massey, V., 1973. The purification and properties of the flavoprotein melilotate hydroxylase, *J. Biol. Chem.* 248:2944–2952.

Strickland, S., and Massey, V., 1973b. The mechanism of action of the flavoprotein melilotate hydroxylase, *J. Biol. Chem.* 248:2953–2962.

Strickland, S., Schopfer, L. M., and Massey, M., 1975. Kinetics and mechanistic studies on the reaction of melilotate hydroxylase with deuterated melilotate, *Biochemistry* 14:2230–2235.

Suzuki, K., and Katagiri, M., 1981. Mechanism of salicylate hydroxylase-catalyzed decarboxylation, *Biochim. Biophys. Acta* 657:530–534.

Takeda, H., Yamamoto, S., Kojima, V., and Hayaishi, O., 1969. Studies on monooxygenases. I. General properties of crystalline L-lysine monooxygenase, *J. Biol. Chem.* 244:2935–2941.

Takemori, S., Nakamura, M., Suzuki, K., Katagiri, M., and Nakamura, T., 1972. Mechanism of the salicylate hydroxylase reaction. V. Kinetic analysis, *Biochim. Biophys. Acta* 284:382–393.

Tokumura, K., Goto, H., Kashiwabara, H., Kaneko, C., and Itoh, H., 1980. Formation and reaction of oxaziridine intermediate in the photochemical reaction of 6-cyano phenanthridine 5-oxide at low temperature, *J. Am. Chem. Soc.* 102:5643–5647.

Walsh, C. T., Schonbrun, A., Lockridge, O., Massey, V., and Abeles, R. H., 1972. Inactivation of a flavoprotein lactate oxidase by an acetylenic substrate, *J. Biol. Chem.* 247:6004–6006.

Walsh, C. T., Lockridge, O., Massey, V., and Abeles, R. H., 1973. Studies on the mechanism of action of the flavoenzyme lactic oxidase: Oxidation and elimination with β-chloroacetate, *J. Biol. Chem.* 248:7049–7054.

Wang, L.-H., Hamzah, R. H., and Tu, S. C., 1982. On the mechanism of salicylate hydroxylase: Studies using deuterated substrates, *Dev. Biochem.* 21:346–349.

Wessiak, A., and Bruice, T. C., 1981. On the nature of the intermediate between 4a-hydroperoxy-flavin and 4a-hydroxyflavin in the hydroxylation reaction of *p*-hydroxybenzoate hydroxylase: Synthesis of 6-aminopyrimidine-2,4,5(34)-triones and the mechanism of aromatic hydroxylation by flavin monooxygenases, *J. Am. Chem. Soc.* 103:6996–6998.

White-Stevens, R. H., and Kamin, H., 1972. Studies of a flavoprotein: Salicylate hydroxylase. II. Enzyme mechanism, *J. Biol. Chem.* 247:2371–2381.

Wierenga, R. K., de Jong, R. J., Kalk, K. H., Hol, W. G. J., and Drenth, J., 1979. Crystal structure of *p*-hydroxybenzoate hydroxylase, *J. Mol. Biol.* 131:55–73.

Yamauchi, T., Yamamoto, S., and Hayaishi, O., 1973. Reversible conversion of lysine monoox-ygenase to an oxidase by modification of sulfhydryl groups, *J. Biol. Chem.* 248:3750–3752.

Yamamoto, S., Nakazawa, T., and Hayaishi, O., 1972. Studies on monooxygenases. IV. Anaerobic formation of an α-keto acid by L-lysine monooxygenase, *J. Biol. Chem.* 247:3434–3438.

Ziegler, D. M., and Mitchell, C. H., 1972. Microsomal oxidase. IV. Properties of a mixed-function amine oxidase isolated from pig liver microsomes, *Arch. Biochem. Biophys.* 150:116–125.

Pterin Monooxygenases

13

13.1 Introduction

Pterins are very similar to flavins, and the reactions catalyzed by pterin monooxygenases (for a review, see Massey and Hemmerich, 1975) are very similar to those catalyzed by the flavin monooxygenases.

The pterin monooxygenases differ from the flavin monooxygenases in that they contain iron and require two enzymes for the total reaction. One enzyme catalyzes the reaction between tetrahydropterin, oxygen, and substrate to produce hydroxylated substrate, dihydropterin, and water, and another enzyme is needed to reduce the dihydropterin back to tetrahydropterin with NADH or NADPH. The pterin is bound much less tightly the flavin cofactors, so the kinetics show an extra substrate rather than a prosthetic group on the enzyme.

13.2 Phenylalanine Hydroxylase

Phenylalanine hydroxylase is a monooxygenase. One atom of the dioxygen molecule produces water, the other is found in the hydroxyl group of tryosine (Kaufman *et al.*, 1962).

The full reaction is tetrahydrobiopterin (A in Figure 13-1), phenylalanine, and dioxygen to produce tyrosine, 6,7-dihydrobiopterin, and water. Like all pterin monooxygenases, another enzyme is required to reduce the 6,7-dihydro-biopterin to tetrahydrobiopterin with NADH. A third enzyme will reduce the commonly isolated 7,8-dihydrobiopterin (C) to tetrahydrobiopterin. This latter reaction requires NADPH as a cofactor.

Oxidized phenylalanine hydroxylase contains high-spin ferric ion that becomes electron-paramagnetic-resonance-inactive on the addition of phenylalanine

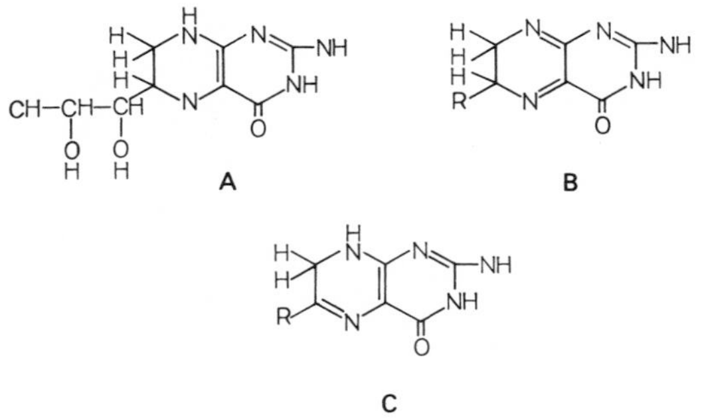

Figure 13-1. Some oxidation states of biopterin. (A) Tetrahydrobiopterin; (B) 6,7-dihydrobiopterin; (C) 7,8-dihydrobiopterin.

and dimethyltetrahydropterin (Fisher *et al.*, 1972). The enzyme exists as two isozymes of almost equal molecular weight, around 55,000. They combine to form dimers of about 110,000 and tetramers of about 210,000 (Kaufman and Fisher, 1970).

The tetrahydropterin can be oxidized by dioxygen if the enzyme is altered by adding lysolecithin or α-chymotrypsin in the presence of tyrosine. The dioxygen is reduced to hydrogen peroxide under these conditions (Fisher and Kaufman, 1973).

Hydroxylation of *p*-deuterophenylalanine results in tyrosine with little loss of deuterium (Guroff *et al.*, 1966a). The reason for this result is that phenylalanine hydroxylase causes a simultaneous migration of substituents from the *para* position to the *meta* position as the *para* position is hydroxylated. This migration is called an NIH shift because the research on the migration was performed at the National Institutes of Health. 4-Chlorophenylalanine is hydroxylated to form 3-chloro-4-hydroxyphenylalanine (chlorotyrosine) (Guroff *et al.*, 1966b), and 4-tritiophenylalanine is hydroxylated to form 3-tritio-4-hydroxyphenylalanine (tritiotyrosine) (Guroff *et al.*, 1966c). Also, 4-methylphenylalanine is converted to 3-methyltyrosine by this enzyme (Daly and Guroff, 1968). The simplest mechanism consistent with these results involves a tritium migration with a pair of electrons (Figure 13-2) (Udenfriend *et al.*, 1967).

The driving force for the rearrangement is that cation B in Figure 13-2 is a much more stable cation than A. The product contains about 85% tritium. The high percentage of tritium product is the result of a tritium isotope rate effect for the loss of tritium.

This type of rearrangement occurs to a small extent in certain nonenzymatic hydroxylations. Some tritium migration occurs in the trifluoroperactic acid hydroxylation of *p*-tritioacetanilide (Jerina *et al.*, 1967).

Phenylalanine hydroxylase will cause the hydroxyl substitution of 4-fluo-

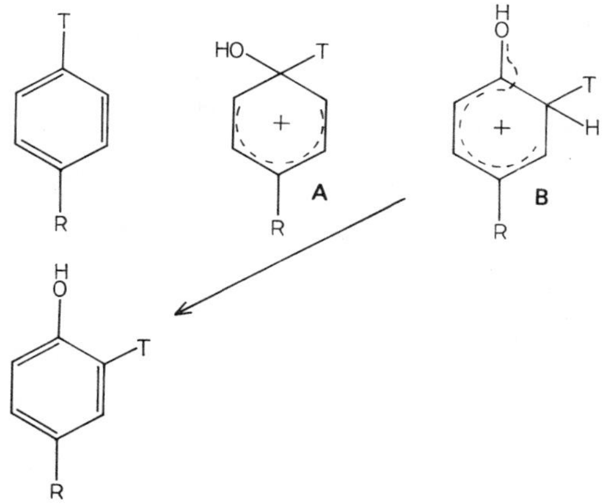

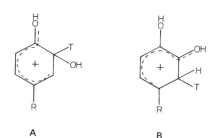

Figure 13-2. A proposed mechanism for the tritium shift in the enzymatic hydroxylation of 4-tritiophenylalanine.

rophenylalanine (Kaufman, 1961). The product is tyrosine. This reaction is not a simple hydrolysis because it requires dioxygen, NADPH, and tetrahydropterin.

13.3 Tyrosine Hydroxylase

Tyrosine hydroxylase, which produces dihydroxyphenylalanine, utilizes tetrahydropteridine (Brenneman and Kaufman, 1964; Nagatsu *et al.*, 1964; Ellenbogen *et al.*, 1965), as does also tryptophan hydroxylase. This latter enzyme forms 5-hydroxytryptophan.

Tyrosine hydroxylase and tryptophan hydroxylase both cause the NIH shift when hydroxylating phenylalanine and tryptophan (Guroff *et al.*, 1967; Renson *et al.*, 1966). However, tyrosine hydroxylase does not cause an NIH shift when acting on tyrosine. The hydroxylation of tyrosine has no driving force for rearrangement because cations A and B (Figure 13-3) have very similar stabilities.

Figure 13-3. Cations with very similar stabilities required for migration in tyrosine hydroxylation.

A

B

References

Brenneman, A. R., and Kaufman, S., 1964. The role of tetrahydropterins in the enzymatic conversion of tyrosine to 3,4-dihydroxyphenylalanine, *Biochem. Biophys. Res. Commun.* 17:177–183.

Daly, J., and Guroff, G., 1968. Production of *m*-methyltyrosine and *p*-hydroxyphenylalanine from *o*-methylphenylalanine by phenylalanine hydroxylase, *Arch. Biochem. Biophys.* 125:136–141.

Ellenbogen, L., Taylor, R. J., Jr., and Brundage, G. B., 1965. On the role of pteridines as cofactors for tyrosine hydroxylase, *Biochem. Biophys. Res. Commun.* 19:708–715.

Fisher, D. B., and Kaufman, S., 1973. Tetrahydropterin oxidation without hydroxylation catalyzed by rat liver phenylalanine hydroxylase, *J. Biol. Chem.* 248:4300–4304.

Fisher, D. B., Kirkwood, R., and Kaufman, S., 1972. Rat liver phenylalanine hydroxylase, an iron enzyme, *J. Biol. Chem.* 247:5161–5167.

Guroff, G., Reifsnyder, C. A., and Daly, J., 1966a. Retention of deuterium in *p*-tryosine formed enzymatically from *p*-deuterophenylalanine, *Biochem. Biophys, Res. Commun.* 24:720–724.

Guroff, G., Kondo, K., and Daly, J., 1966b. The production of *meta*-chlorotyrosine from *para*-chlorotyrosine by phenylalanine hydroxylase, *Biochem. Biophys. Res. Commun.* 25:623–628.

Guroff, G., Levitt, M., Daly, J., and Udenfriend, S., 1966c. The production of *meta*-tritiotyrosine from *p*-tritiophenylalanine by phenylalanine hydroxylase, *Biochem. Biophys. Res. Commun.* 25:253–259.

Guroff, G., Daly, J. W., Jerina, D. M., Renson, J., Witkop, B., and Udenfriend, S., 1967. Hydroxylation induced migration: The NIH shift, *Science* 157:1524–1530.

Jerina, D., Daly, J., Landis, W., Witkop, B., and Udenfriend, S., 1967. Intramolecular migration of deuterium during nonenzymic aromatic hydroxylation, *J. Am. Chem. Soc.* 89:3347–3349.

Kaufman, S., 1961. The enzymatic conversion of 4-fluorophenylalanine to tyrosine, *Biochim. Biophys. Acta* 51:619–621.

Kaufman, S., and Fisher, D. R., 1970. Purification and some physical properties of phenylalanine hydroxylase from rat liver, *J. Biol. Chem.* 245:4745–4750.

Kaufman, S., Bridgers, W. F., Eisenberg, F., and Friedman, S., 1962. The source of oxygen in the phenylalanine hydroxylase and the dopamine β-hydroxylase catalyzed reactions, *Biochem. Biophys. Res. Commun.* 9:497–502.

Massey, V., and Hemmerich, P., 1975. Flavin and pteridine monooxygenases, in *The Enzymes*, Vol. 12, P. D. Boyer (ed.), Academic Press, New York, pp. 191–252.

Nagatsu, T., Levitt, M., and Udenfriend, S., 1964. Tyrosine hydroxylase: The initial step in norepinephrine biosynthesis, *J. Biol. Chem.* 239:2910–2917.

Renson, J., Daly, J., Weisbach, H., Witkop, B., and Udenfriend, S., 1966. Enzymatic conversion of 5-nitro-tryptophan to 4-tritio-5-hydroxytryptophan, *Biochem. Biophys. Res. Commun.* 25:504–513.

Udenfriend, S., Zaltzman-Nireberg, P., Daly, J., Guroff, G., Chidsey, C., and Witkop, B., 1967. Intramolecular migration of deuterium and tritium during enzymatic hydroxylation of *p*-deuteroacetanilide and *p*-tritioacetanilide, *Arch. Biochem. Biophys.* 120:413–419.

Copper Hydroxylases

14

14.1 Tyrosinase

The copper-containing enzyme tyrosinase functions both as a hydroxylase for phenols and as an oxidase for catechols. Since cresol is usually used as a substrate for the hydroxylation reaction, this is called *cresolase activity*. The oxidase activity is called *catecholase activity*. These two activities appear to take place at different sites and have different properties. The catecholase exchanges copper with the solution during the catecholase reaction, but the cresolase activity does not (Doessler and Dawson, 1960). The hydroxylation reaction is always *ortho* to the existing hydroxyl group. The oxygen of the new hydroxyl group is derived from molecular oxygen (Mason *et al.*, 1955), so the enzyme is a mono-oxygenase.

The hydroxylation of tyrosine is the first step in the formation of melanin. Tyrosine is hydroxylated to dopa, which is oxidized to a dopaquinone and then cyclized through the amino group to leuko-dopaquinone. Oxidation to dopa-chrome, decarboxylation to 2,5-dihydroxyindole, and oxidation to the quinone of 2,5-dihydroxyindole are followed by polymerization to finally give melanin (Hearing *et al.*, 1980).

Tyrosinase in the cuprous state combines reversibly with dioxygen to form an oxygenated compound with a strong absorption of 345 nm and a weaker absorption at 600 nm, absorptions similar to those found in oxyhemocyanin. Tyrosinase also forms a carbon monoxide complex, as does hemocyanin. It is interesting to note that both these complexes are luminescent (Kuiper *et al.*, 1980) a property that indicates further similarities between tyrosinase and hemo-cyanin. Oxytyrosinase decomposes slowly to the cupric form and hydrogen peroxide in another reversible reaction (Jolley *et al.*, 1974).

There are two binding sites on the enzyme, an aromatic site competitively inhibited by benzoic acid and a dioxygen site competitively inhibited by cyanide.

The binding to these two sites appears to be sequential, the aromatic compound binding before dioxygen (Duckworth and Coleman, 1969). The rate of oxidation of catechols substituted with electron-withdrawing groups is slower than that of unsubstituted catechols.

Studies of visible and electron paramagnetic resonance spectra of tyrosinase and its complexes with inhibitors have led (Himmelwright et al., 1980; Winkler et al., 1981) to a structure of oxytyrosinase similar to that of oxyhemocyanin.

A proposed mechanism for hydroxylation by this type of an active center is shown in Figure 14-1. The oxygenated form of the enzyme (A) is composed of two cupric coppers bridged by a hydrogen peroxide. The phenol to be hydroxylated displaces a water (B), and a rearrangement of the copper structure occurs. The initial square planar arrangement (B) changes to a bipyramid (C). This latter structure of copper would bind ligands less tightly so that the peroxy would be labilized. This allows for ortho hydroxylation of the phenol (D), followed by release of the o-quinone. The resulting binuclear cuprous center (E) combines with dioxygen again to form oxytyrosinase (A). One would expect the hydroxylation to be an electrophilic displacement reaction, which is hard to rationalize with the attack of peroxy anion in this mechanism.

Electrophilic attacks on aromatic compounds involve first the addition of the electrophil followed by loss of a proton (Melander, 1949), as shown in Figure 14-2. In the nitration of toluene, the first step is rate-determining, whereas in the nitration of a phenol, the second step is rate-determining. Unexpectedly, the hydroxylation of 3,4-dimethlcresol by tyrosinase has a very low isotope rate

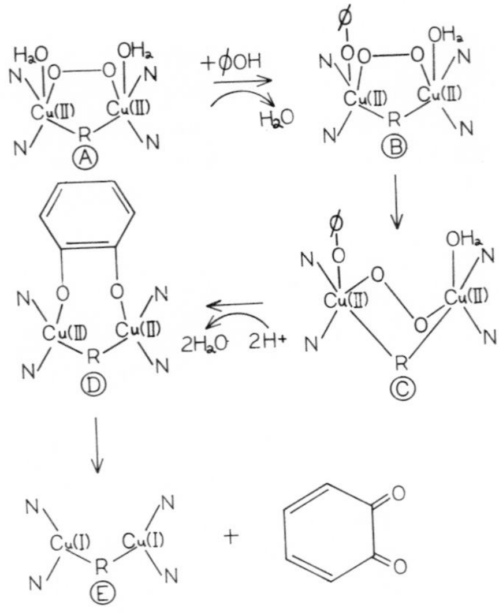

Figure 14-1. A proposed mechanism of action of tyrosinase.

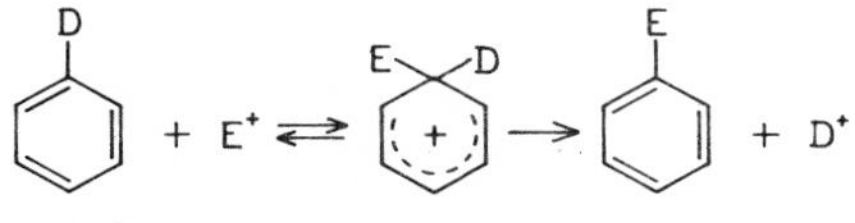

Figure 14-2. Mechanism of electrophilic aromatic substitution.

effect (k_H/k = 1.3) (Wood and Ingraham, 1962), indicating that the two steps are comparable in rate. This can be rationalized by a consideration of catalysis of the release of the proton by the enzyme. For example, the coupling of a diazonium compound to an α-naphthol has an isotope rate effect (Zollinger, 1955) of k_H/k_D = 6.45, which is reduced to 3.63 in the presence of pyridine. The lowering of the isotope rate effect shows that the second step is catalyzed by the pyridine pulling the deuteron off the intermediate cation. Similarly , the low isotope rate effect for tyrosinase must indicate that the enzyme aids in removal of the proton.

14.2   Dopamine-β-Hydroxylase

Dopamine-β-hydroxylase is of medical interest because it catalyzes a step on the important pathway between tyrosine and epinephrine (Figure 14-3) (Kaufman and Friedman, 1965). The reaction sequence is: tyrosine to dopa to dopamine to norepinephrine to epinephrine. Dopamine-β-hydroxylase produces norepinephrine from dopamine. The total reaction requires ascorbic acid, dopamine, and dioxygen to produce dehydroascorbic acid, water, and norepinephrine as

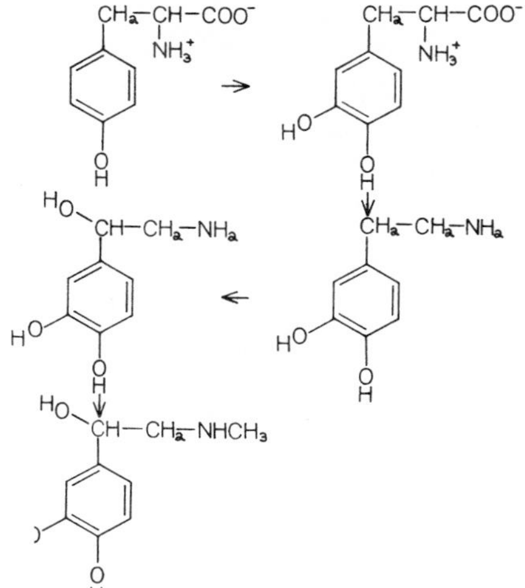

Figure 14-3. Metabolic sequence from tyrosine to epinephrine.

products. Fumarate ion is an activator for the enzyme. The enzyme is isolated from bovine adrenal medulla (Foldes *et al.*, 1972). The enzyme has a rather broad specificity. Benzyl amine is a good substrate.

One atom of dioxygen appears in the hydroxyl group of norepinephrine (Kaufman *et al.*, 1962), so the enzyme is a monooxygenase. The enzyme contains two equivalents of copper that have been shown to be involved in catalysis by electron spin resonance studies (Blumberg *et al.*, 1965). Both coppers are in the cupric state in the resting enzyme. Adding copper back to the apoenzyme, which has had the copper removed, gives activity proportional to the square of the copper added to the enzyme (Blackburn *et al.*, 1980), which indicates that two coppers are necessary for activity. However, it is not known whether the active site contains only one or two copper atoms (Villafranca, 1981). The copper is reduced to the cuprous state by the ascorbic acid by a Ping-Pong-type mechanism (Goldstein *et al.*, 1968) in which the dehydroascorbic acid leaves before the dioxygen binds.

The enzyme combines with dioxygen, and the resulting species hydroxylates dopamine, leaving the coppers in the cupric state (Friedman and Kaufman, 1965). In the resting state, the copper on the enzyme binds one water molecule per copper (Blumberg *et al.*, 1965). The copper does not change valence when dopamine is added to the enzyme, nor is there any indication that the dopamine is bound to the copper.

Villafranca (1981) has proposed a free-radical mechanism for the hydroxylation in which the ascorbic acid acts as a one-electron reductant and a benzyl radical is formed before the hydroxylation occurs.

There has been considerable interest in inhibitors for the enzyme because of their possible medical use. Copper chelates are inhibitors for the enzyme (Green, 1964). These include 8-hydroxyquinoline, 2,9-dimethyl-*o*-phenanthroline, 2,2-dipyridyl, and sodium dithiocarbamate. Other inhibitors include serotonin, epinephrine, and compounds similar to benzylamine such as benzyl hydrazine (Creveling *et al.*, 1962a) and benzyl hydroxylamine (Creveling *et al.*, 1962b).

REFERENCES

Blumberg, W. E., Goldstein, M., Lauber, E., and Peisach, J., 1965. Magnetic resonance studies on the mechanism of the enzymatic β-hydroxylation of 3,4-dihydroxyphenylalanine, *Biochim. Biophys. Acta* 99:187–190.

Blackburn, N. J., Mason, H. S., and Knowles, P. F., 1980. Dopamine hydroxylase: Evidence for bionuclear copper sites, *Biochem. Biophys. Res. Commun.* 95:1275–1281.

Creveling, C. F., Daly, J. W., Witkop, B., and Udenfriend, S., 1962a. Substrate and inhibitors of dopamine-β-oxidase, *Biochim. Biophys. Acta* 64:125–134.

Creveling, C. F., Van der Schoot, J. B., and Udenfriend, S., 1962b. Phenylethylamine isoesters as inhibitors of dopamine-β-oxidase, *Biochem. Biophys. Res. Commun.* 8:215–219.

Doessler, H., and Dawson, C. R., 1960. On the nature and mode of action of the copper protein, tyrosinase. II. Exchange experiments with radioactive copper and the functioning enzyme, *Biochim. Biophys. Acta* 45:515–524.

Duckworth, H. W., and Coleman, J. E., 1969. Physiochemical and kinetic properties of mushroom tyrosinase, *J. Biol. Chem.* 245:1613–1625.

Foldes, A., Jeffrey, P. L., Preston, B. N., and Austin. L., 1972. Dopa hydroxylase of bovine adrenal medullae, *Biochem. J.* 126:1209–1217.

Friedman, S., and Kaufman, S., 1965. 3,4-Dihydroxyphenolethylamine β-hydroxylase: Physical properties, copper content and role of copper in the catalytic activity, *J. Biol. Chem.* 241:2256–2259.

Goldstein, M., Joh, T. H., and Garvey, T. Q., III, 1968. Kinetic studies of the enzymatic dopamine β-hydroxylation reaction, *Biochemistry* 7:2724–2730.

Green, A. L., 1964. The inhibition of dopamine-β-oxidase by chelating agents, *Biochim. Biophys. Acta* 81:394–397.

Hearing, V. J., Jr., Ekel, T. M., Montague, P. A., and Nicholson, J. M., 1980. Mammalian tyrosinase: Stoichiometry and measurements of reaction products, *Biochim. Biophys. Acta* 611:251–268.

Himmelwright, R. S., Eickman, N. C., LuBien, C. D., Lerch, K., and Solomon, E. I., 1980. Chemical and spectrascopic studies of binuclear copper active site of *Neurospora* tyrosinase: Comparison to hemocyanins, *J. Am. Chem. Soc.* 102:7339–7344.

Jolley, R. L., Evans, L. H., Makino, H., and Mason, H. S., 1974. Oxytyrosinase, *J. Biol. Chem.* 249:335–345.

Kaufman, S., and Friedman, S., 1965. Dopamine-β-hydroxylase, *Pharm. Rev.* 17:71–100.

Kaufman, S., Bridgers, W. F., Eisenberg, F., and Friedman, S., 1962. The source of oxygen in the phenylalanine hydroxylase and the dopamine-β-hydroxylase catalyzed reactions, *Biochem. Biophys. Res. Commun.* 9:497–502.

Kuiper, H. A., Lerch, K., Brunori, M., and Agro, A. F., 1980. Luminescence of the copper–carbon monoxide complex of *Neurospora* tyrosinase, *FEBS Lett.* 111:232–234.

Mason, H. S., Fowlks, W. L., and Peterson, E., 1955. Oxygen transfer and electron transport by the phenolase complex, *J. Am. Chem. Soc.* 77:2914–2915.

Melander, L., 1949. Introduction and substituents in the aromatic nucleus: Exploration of a mechanism by means of isotopic hydrogen, *Acta Chem. Scand.* 3:95–96.

Villafranca, J. J., 1981. Dopamine-β-hydroxylase, in *Copper Proteins*, T. G. Spiro (ed.), John Wiley, New York, pp. 265–289.

Winkler, M. W., Lerch, K., and Solomon, E. I., 1981. Competitive inhibitor binding to binuclear copper active site in tyrosinase, *J. Am. Chem. Soc.* 103:7001–7003.

Wood, B. J. B., and Ingraham, L. L., 1962. The tritium rate effect in tyrosinase-catalyzed hydroxylation, *Arch. Biochem. Biophys.* 98:479–484.

Zollinger, H., 1955. Uber der Natur der Protonabspaltung bei Azokupplungen, *Helv. Chem. Acta* 38:1623–1631.

Cytochrome P-450

15

15.1 Properties

Cytochrome P-450 is a group of nonspecific hydroxylating systems that hydroxylate substrates of widely varying structure (Ullrich, 1979). Cytochrome P-450 is important in the degradative hydroxylation of drugs in the body (Gillette *et al.*, 1972). The enzyme can hydroxylate steroids and hydrocarbons. For example, *Corynebacterium* utilizes P-450 to oxidize *n*-octane to 1-octanol (Cardini and Justishuk, 1970), and fatty acids are hydroxylated at the ω and (ω-1) positions (Bjorkhem and Danielson, 1970; Ellin *et al.*, 1972; Ichihara *et al.*, 1979). Either cytochrome P-450 or similar iron-containing monooxygenases cleave inositol to D-glucuronic acid (Charalampous, 1959, 1960).

The reactions catalyzed by cytochrome P-450 have been reviewed and classified by Guengerich and McDonald (1984) into six different types: (1) hydroxylation of carbon compounds, (2) oxidative cleavage of a heteroatom, (3) heteroatom hydroxylation, (4) epoxidation of olefins or aromatic compounds, (5) oxidative rearrangements, and (6) destruction of the P-450 heme group by an intermediate or product of the reaction.

The hydroxylation reaction requires dioxygen and also two equivalents of reducing power (usually NADPH). The overall reaction is one new hydroxyl group per dioxygen consumed by the substrate, SH:

$$SH + O_2 + NADPH + H^+ \rightarrow SOH + H_2O + NADP^+$$

Thus, cytochrome P-450 is a monooxygenase because one oxygen of dioxygen appears in the product and the other appears in water.

The active center of cytochrome P-450 has a heme group. Hemes have a strong absorption at around 400 nm that moves to slightly longer wavelengths ($\sim$410 nm) when the ion of the heme binds CO. However, in cytochrome P-450, this band moves to 450 nm in the presence of carbon monoxide. This unique property of P-450 aided greatly in the isolation of the enzyme. The CO complex

185

is enzymatically inactive, but light at 450 nm will reverse the reaction and reactivate the enzyme (Cooper *et al.*, 1965; Wilson and Harding, 1970).

The resting enzyme contains predominantly low-spin ferric heme (Whysner *et al.*, 1970), although quantitatively this depends on the source and purity of the enzyme. In rat liver microsomes, the P-450 is about 48% high-spin, which decreases to 9.7% high-spin after purification (Cinti *et al.*, 1979). Similarly, the purified enzyme from the bacterium *Pseudomonas putida* is about 8% high-spin at room temperature (Sligar, 1976). In all cases, the low-spin form is more favored at lower temperatures. The low-spin form has a spin of $\frac{1}{2}$ and the high-spin form a spin of $\frac{5}{2}$, which means that the low-spin is a doublet whereas the high-spin is a sextet. The change in spin alone would increase the entropy by $R \ln \frac{6}{2} = -2$ e.u. from low-spin to high-spin. The observed decrease in entropy of -14 e.u. means that much more must be happening to the protein than merely a change in multiplicity of the iron (Sligar, 1976).

Considerable effort has been made to prepare various model hemes that will mimic some of the properties of P-450. Thiol adducts of heme produce both electron spin resonance (ESR) and UV spectra similar to those of cytochrome P-450 (Bayer *et al.*, 1969). The spectra of cytochrome P-450 can be interpreted in terms of a sulfur chelated to the iron. In the high-spin oxidized form of cytochrome P-450, there are bands at 323 and 567 nm that disappear on reduction. Molecular orbital [Iterative Extended Hückel (IEH)] calculations show that these bands are sulfur-to-iron charge-transfer transitions (Hanson *et al.*, 1977). The absorption at 450 nm of the CO adduct of the ferroenzyme also appears to be the result of a mercaptide bound as the fifth ligand to the heme iron. The CO complex has a band at 363 nm in addition to the band at 450 nm. The 363 nm band is primarily a charge-transfer transition from sulfur to the porphyrin ring. Calculations (IEH) and polarization studies have both shown that this band has the correct symmetry (Z polarization) to mix with the Soret band. The observed band at 450 nm is the result of mixing between the charge-transfer band and the Soret band (Hanson *et al.*, 1977).

Similar results are found by (INDO) calculations (Lowe and Rohmer, 1980); i.e., the Soret band is split by mixing with sulfur orbitals. It is interesting to note that the Soret band is also split in oxycytochrome P-450, but both bands are at shorter wavelengths. A thiol, CO, heme, and strong base in a dimethyl-sulfoxide–ethanol solution have absorption very close to that of cytochrome P-450. The absorption peaks are at 450 and 555 nm. In the absence of a thiol, the maxima are at 413, 535, and 566 nm, which is about normal for heme–CO adducts (Stern and Peisach, 1974).

The ESR spectrum of cytochrome P-450 has values of $g = 2.45, 2.26$, and 1.19. This is to be compared with the di-*N*-methylimidazole complex of heme, which has values of $g = 2.29$ and 1.57. However, when one of the ligands is bound to the iron by a sulfur as in the mixed *p*-nitro-thiophenol and *N*-methylimidizole complex, the values are $g = 2.42, 2.26$, and 1.91 (Koch *et al.*, 1975). These are very similar to those of the low-spin resting form of P-450. Similarly, the dithiophenol complex has values of $g = 2.45, 2.26$, and 1.91 (Collman *et al.*, 1975), which are identical to the values found in P-450. These results, combined with electron paramagnetic resonance studies (Peisach *et al.*, 1979), support an imidazole–heme–cysteine structure for P-450. The techniques of "extended X-ray absorption fine structure" have verified that the fifth and sixth ligands are a mercaptide and an imidazole (Cramer *et al.*, 1978). The imidazole ligand can be displaced from the ferri form of the enzyme by cyanide or guanidine or by amines. In the ferro form, the imidazole is displaced by carbon monoxide or nitrous oxide (Chevion *et al.*, 1977).

15.2   Reaction Mechanisms

On the addition of substrate (Figure 15-1), the spin mixture becomes predominantly high-spin (Tsai *et al.*, 1970). The addition of camphor will change the spin to about 60% high-spin and 40% low-spin ferric heme. Neither D- nor L-camphor will affect the last 40% of the low-spin ferric heme. During this change in spin, the *pK* of the enzyme changes from 6.5 in the high-spin form to 5.8 in the low-spin form (Sligar and Gunsalus, 1979). Again, the equilibrium depends on the purity and source of the enzyme. The bacterial enzyme (*Ps.*

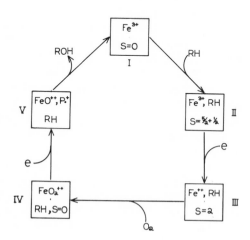

Figure 15-1. Mechanism of action of cytochrome P-450.

putida) becomes 94% high-spin in the presence of substrate (Sligar, 1976). As with the enzyme without substrate, the proportion of low-spin form increases as the temperature is lowered. The thiol group is still attached to the iron as a fifth ligand after the substrate is bound, as determined by magnetic circular dichroism (CD) studies (Dawson *et al.*, 1976).

The ferric heme (structure II) is reduced to the ferrous heme (structure III) by putidaredoxin, which occurs in *Ps. putida,* and is associated with the P-450 (Tyson *et al.*, 1972). The product of the reduction is a high-spin ferrous iron species (S = 2) (Sharrock *et al.*, 1976).

The putidaredoxin is in turn reduced by a flavin enzyme that is reduced by NADPH (Omura *et al.*, 1966). The COOH-terminus of the putidaredoxin appears to have a function in the cytochrome P-450 reduction, because if the first two C-terminal amino acids, tryptophan and glutamine, are hydrolyzed off by carboxypeptidase A, the acitivity of the putidaredoxin is reduced by a factor of about 50 (Sligar *et al.*, 1974a).

The high-spin ferrous heme in turn reacts with dioxygen to form an oxygenated species (IV) with a new electronic absorption band (Estabrook *et al.*, 1971; Ishimura *et al.*, 1971; Peterson *et al.*, 1972). Magnetic CD studies show that structure IV no longer has an axial histidine group (Dawson and Cramer, 1978). The dioxygen must have taken the position of the histidine. Structure IV has a Mossbauer spectrum (Sharrock *et al.*, 1973, 1976) similar to that of oxygenated hemoglobin. Structure IV has other similarities to oxygenated hemoglobin in that it is a low-spin complex that arises from the addition of dioxygen to a high-spin ferrous complex (structure III). Structure III will also add carbon monoxide to form an adduct similar to the carbon monoxide addition product of hemoglobin (Sharrock *et al.*, 1973). Structure IV is reduced by putidaredoxin (Tyson *et al.*, 1972) in the bacterial enzyme system and by cytochrome b_5 (Guengerich *et al.*, 1976) in the liver microsomal enzyme system. Both these reductants also act as effectors for the respective enzymes. In the absence of putidaredoxin, the iron of the bacterial enzyme is autoxidized (Lipscomb *et al.*, 1976). The product of the autoxidation is superoxide ion, which is claimed to disproportionate to hydrogen peroxide and singlet dioxygen. Chemiluminescence of the singlet dioxygen is reported (Sligar *et al.*, 1974b), this being inhibited by superoxide dismutase. However, this explanation becomes difficult to rationalize in light of results that show that the disproportionation of superoxide ion produces at most infinitesimal quantities of singlet dioxygen (Foote *et al.*, 1980).

Structure V is of considerable mechanistic interest because it is this intermediate that actually carries out the hydroxylation of the substrate. It is stoichiometrically equivalent to complex I of peroxidase and catalase. Thus, a likely structure would be FeO^{2+} plus a porphyrin cation radical.

The observations that P-450 mediates allylic rearrangements during hydroxylation combined with a large isotopic rate effect for the hydrogen displaced

(Groves and Subramanian, 1984) support a hydroxylation mechanism in which structure V withdraws a hydrogen atom from the substrate followed by hydroxylation of the resulting radical.

There is experimental evidence that argues against a dioxygen structure for intermediate V. Many oxidants—sodium periodate, sodium chlorite, hydrogen peroxide, alkyl hydroperoxides, acyl hydroperoxides, and iodosobenzenes (Kadlubar et al., 1973; Hrycay et al., 1975; Nordblom et al., 1976; Lichtenberger, 1976; Gustafsson et al., 1979; Blake and Coon, 1980) will oxidize P-450 to structure V and hydroxylate substrates. This reaction is not inhibited by carbon monoxide and does not require flavoproteins for reduction. The first two oxidants, sodium periodate and sodium chlorite, rule out any dioxygen structures. The peroxides lend support to the idea that structure V is comparable to compound I of catalase and peroxidase.

The reaction with iodosobenzene appears to be unique because the hydroxylated product contains labeled oxygen if the reaction is performed in labeled water. This does not occur with the other oxidants (Heimbrook and Sligar, 1981). Heimbrook and Sligar (1981) propose two mechanisms for P-450 hydroxylations, one in which the oxygen of the intermediate hydroxylating species does not exchange with the oxygen of water and another in the iodosobenzene reaction in which a different intermediate is formed in which there is facile oxygen exchange with solvent. The oxygen of iodosobenzene does not exchange with water.

Lipoic acid will react with structure IV to give structure V. When labeled dioxygen is used, one of the oxygens is found in the carboxyl group of the lipoic acid and the other in the hydroxylated product. A reasonable mechanism (Figure 15-2) for this observation would be the formation and decomposition of acylated heme-bound oxygen (Sligar et al., 1980).

Another possibility that appears unlikely but cannot be eliminated by these experiments is that the acylated dioxygen complex decomposes to a peracid that does the hydroxylation. However, peracids show a very low activity toward

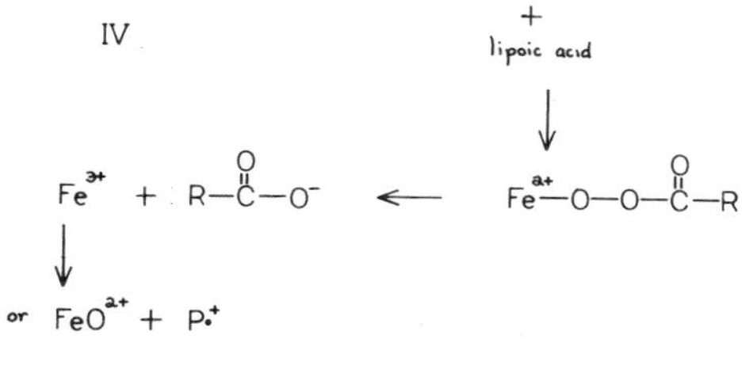

Figure 15-2. Formation of structure V by the reaction of lipoic acid with structure IV.

saturated hydrocarbons. There is probably a carboxyl group on the P-450 enzyme that functions like the lipoic acid.

15.3 Model Studies

Ferriporphyrins will catalyze the iodosylbenzene hydroxylation of both aromatic (Groves *et al.*, 1979) and aliphatic compounds (Groves and Nemo, 1983). A plausible intermediate in this reaction is FeO^{3+}, so the reaction could be analogous to P-450 hydroxylations.

Model porphyrin ferryl radical cations have been prepared (Groves *et al.*, 1981). The oxidation of tetramesityl ferriporphyrin with *m*-chloroperbenzoic acid produces a green Fe(IV) porphyrin radical cation of A_{2u} symmetry. In basic solution, the green species changes to a red species; in acidic solution, it reverts to the green species. The red species may be formed directly from the ferriporphyrin and iodosobenzene. Both species will oxidize norbornene to norbornene oxide. In the presence of oxygen-labeled water, the label appears in the norbornene oxide.

A compound at the oxidation level of structure V may be produced by the one-electron electrolytic reduction of ferrous octaethylporphyrin dioxygen complex (a model for structure IV) (Welborn *et al.*, 1981) or from Fe(I) octaethylporphyrin and dioxygen or from ferrous octaethylporphyrin and superoxide ion. However, this compound does not epoxidize styrene. Probably the oxygen–oxygen bond has not been broken.

Porphyrin ferryl complexes (FeO^{2+}) but without the π-cation radical may be formed by the addition of a base to the peroxo-bridged Fe(III) complexes described in Chapter 8 (Chin *et al.*, 1980a). These compounds have been shown to be Fe(IV) compounds by the Mossbauer spectra (Simmoneaux *et al.*, 1982). They will oxidize triphenyl phosphine to triphenyl phosphine oxide (Chin *et al.*, 1980b).

The structure proposed for V, FeO^{2+} plus a porphyrin radical, or FeO^{3+} would act as a strong electrophile or possibly as a radical. To test whether P-450 acts as an electrophile or a radical, Groves *et al.*, (1978) studies the P-450-catalyzed hydroxylation of norbornane. There is a rather large isotope rate effect ($k_H/k_D = 11.5$), which supports a radical mechanism, and also some epimerization between exo and endo, which would be expected for a radical hydroxylation. However, the percentage epimerization is relatively low, being only 14% for exo to endo and 18% for endo to exo. Similarly, the epoxidation of olefins by cytochrome P-450 will cause isomerizations to occur (May *et al.*, 1977). The epoxidation of *trans*-olefins produces 70% of *cis*-epoxide and 30% *trans*-epoxide.

Either of these structures for structure V could also be looked on as a source of atomic oxygen (Hamilton, 1964). Atomic oxygen can be considered an "oxene" because it is neutral but lacks two electrons for a closed shell, as does a carbene. The lower states of atomic oxygen are a singlet state (^{1}S), in which all electrons are paired, or a triplet state (^{3}P), in which two electrons are unpaired. Thus, an "oxene" or atomic oxygen could act as an electrophile (^{1}S) or as a radical (^{3}P). The ground state of atomic oxygen is the ^{3}P state.

Molecular orbital studies (Pudzianowski and Loew, 1980) by a modified intermediate neglect of differential overlap method predict that closed-shell ^{1}S oxygen atoms should hydroxylate with retention of configuration, whereas radical hydroxylations with ^{3}P oxygen atoms would cause loss of configuration. The ^{1}S closed-shell calculation gave a lower activation energy. Atomic oxygen in the ^{3}P state, prepared by the radiolysis of CO_2, has been found to hydroxylate anisole, toluene, chlorobenzene, and other aromatic compounds (Takamuku *et al.*, 1980). The NIH shift of deuterium [see Chapter 13 (Section 13.2)] is observed. Substituents that favor electrophilic substitution of the aromatic ring also favor the oxene hydroxylation. A dipolar intermediate (Figure 15-3) was proposed to explain these results.

15.4 Oxides and Amine Dealkylation

Microsomal hydroxylating systems, which are essentially crude P-450 preparations, as well as purified P-450 preparations, will produce arene oxides. Anthracene is first oxidized to anthracene-1,2-epoxide (Akhtar *et al.*, 1979), naphthalene to naphthalene-1,2-epoxide (Jerina *et al.*, 1968), and phenanthrene to the 9,10-epoxide (Rahimtula *et al.*, 1978). These arene oxides can be reduced to phenols, opened with glutathione, or hydrolyzed to a *trans*-glycol (Figure 15-4). The latter reaction is catalyzed by an enzyme called epoxide hydrolase (Pesch *et al.*, 1972). The 1,2-epoxide of naphthalene rearranges to about 95% α-naphthol and 5% β-naphthol. The proposed mechanism for this rearrangement proceeds through the keto form of the phenol (Figure 15-5) (Kasperek and Bruice, 1972).

If the hydroxylation position is substituted with tritium, the tritium will migrate (Udenfriend *et al.*, 1967). This results in an NIH shift. The retention of deuterium in the arrangement of [1 − ^{2}H]naphthalene-1,2-oxide varies from 60 to 85% depending on the pH of the solution (Boyd *et al.*, 1972). This range covers the results found in deuterium retention in enzymatic hydroxylations.

Figure 15-3. Proposed dipolar intermediate in the hydroxylation of aromatic compounds by ^{3}P dioxygen.

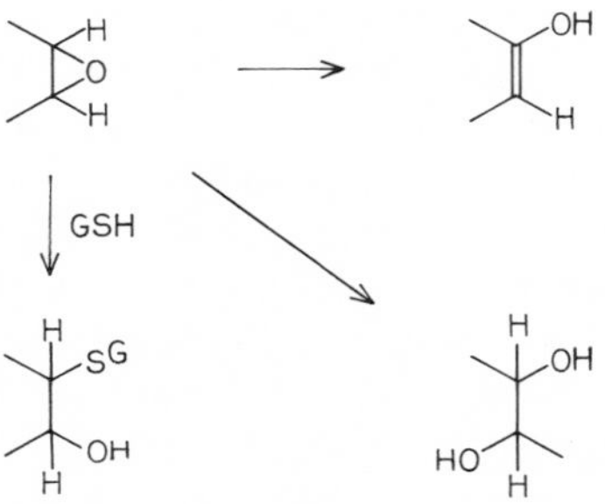

Figure 15-4. Reactions of an arene oxide.

In addition to the phenol products, *trans*-glycol and the glutathione adduct are found. Glutathione adds at the 1-position (Jerina *et al.*, 1970). Similarly, dibenz-(a,h)-anthracene, a potent carcinogen, is oxidized to an epoxide by microsomes (Selkirk *et al.*, 1971), and phenanthrene is oxidized to the 9,10-epoxide (Grover *et al.*, 1971). Certain olefins, *n*-1-octene, *n*-4-octene, and 3-ethyl-2-pentene, are oxidized directly to the corresponding glycols by microsomal enzymes that contain epoxide hydrolase (Mynert *et al.*, 1970).

The addition of an oxygen atom to a double bond to form an epoxide is analogous to a carbene reaction in which a carbon with six electrons (e.g., CH_2) adds to a double bond to form a cyclopropane. Accordingly, the oxygenating species in this reaction is believed to be an "oxene" (Jerina *et al.*, 1970).

Cytochrome P-450 enzymes also dealkylate amines (Gorrod, 1978). This same reaction has been accomplished by a P-450 model system (Shannon and Bruice, 1981). An oxidizing agent comparable to structure V of cytochrome P-450 is formed from dimethylaniline *N*-oxide and chlorotetraphenyl Fe(III) in anhydrous ethanol. This reagent will oxidize dimethylaniline to *N*-methyl aniline and formaldehyde. The mechanism proposed proceeds through a carbon radical (B in Figure 15-6), which can be further oxidized to the Schiff base (C). This

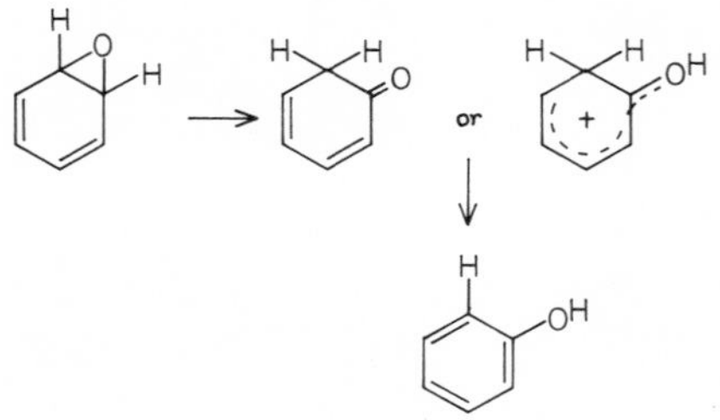

Figure 15-5. A proposed mechanism for the rearrangement of an arene oxide.

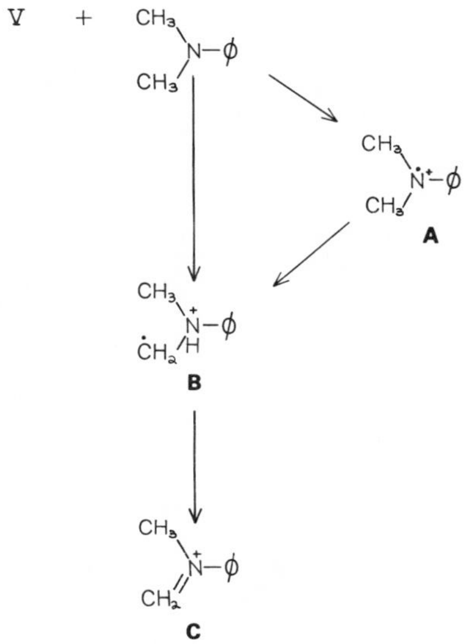

Figure 15-6. A mechanism for oxidative dealkylation of an amine.

would hydrolyze utilizing the hydroxyl attached to the iron. The carbon radical (B) can be formed directly or by an oxidation of the nitrogen followed by hydrogen-atom transfer.

References

Akhtar, N., Hamilton, J. G., Boyd, D. R., Brawnstein, A., Seilfried, H. E., and Jerina, D. M., 1979. Anthracene-1,2-oxide: Synthesis and role in the metabolism of anthracene by mammals, *J. Chem. Soc. Perkins Trans. 1* 1979:1442–1446.

Bayer, E., Hill, H. A. O., Roder, A., and Williams, R. J. P., 1969. The interaction between haem-iron and thiols, *Chem. Commun.* 1969:109.

Bjorkhem, I., and Danielsson, H., 1970. ω and (ω-1)-oxidation of fatty acids by rat liver microsome, *Eur. J. Biochem.* 17:450–459.

Blake, R. C., II, and Coon, M. J., 1980. On the mechanism of action of cytochrome P450, *J. Biol. Chem.* 255:4100–4111.

Boyd, D. R., Daly, J. W., and Jerina, D. M., 1972. Rearrangement of [1 − ^{2}H] and [2 − ^{2}H] naphthalene-1,2-oxides to 1-naphthol: Mechanisms of the NIH shift, *Biochemistry* 11:1961–1966.

Cardini, J., and Justishuk, P., 1970. The enzymatic hydroxylation of *n*-octane by *Corynebacterium* sp. strain 7EIC, *J. Biol. Chem.* 245:2789–2796.

Charalampous, F. C., 1959. Biochemical studies of inositol. V. Purification and properties of the enzyme that cleaves inositol to D-glucuronic acid, *J. Biol. Chem.* 234:220–227.

Charalampous, F. C., 1960. Biochemical studies on inositol. VI. Mechanism of cleavage of inositol to D-glucuronic acid, *J. Biol. Chem.* 235:1286–1291.

Chevion, M., Peisach, J., and Blumberg, W. E., 1977. Imidazole, the ligand *trans* to mercaptide in ferric cytochrome P450, *J. Biol. Chem.* 252:3637–3645.

Chin, D.-H., Balch, A. L., and LaMar, G. N., 1980a. Formation of porphyrin ferryl (FeO^{+2}) complexes through the addition of nitrogen bases to peroxo-bridged iron(III) porphyrins, *J. Am. Chem. Soc.* 102:1446–1448.

Chin, D.-H., LaMar, G. N., and Balch, A. L., 1980b. Role of ferryl (FeO^{++}) complexes in oxygen atom transfer reactions: Mechanism of iron(II) porphyrin catalyzed oxygenation of triphenyl-phosphine, *J. Am. Chem. Soc.* 102:5945–5947.

Cinti, D., Sligar, S. G.., Gibson, G. G., and Schenkman, J. B., 1979. Temperature-dependent spin equilibrium of microsomal and solubilized P450 from rat liver, *Biochemistry* 18:36–42.

Collman, J. P., Sorrell, T. N., and Hoffman, B. M., 1975. Models for cytochrome P450, *J. Am. Chem. Soc.* 97:913–914.

Copper, D. Y., Levin, S., Narasimhula, S., and Rosenthal, O., 1965. Photochemical action spectrum of the terminal oxidase of mixed function oxidase systems, *Science* 147:400–402.

Cramer, S. P., Dawson, J. H., Hodgson, K. O., and Hager, L. P., 1978. Studies on the ferric forms of cytochrome P450 and chloroperoxidase by extended X-ray absorption fine structure: Characterization of the Fe–N and Fe–S distances, *J. Am. Chem. Soc.* 100:7282–7290.

Dawson, J. H., and Cramer, S. P., 1978. Oxygenated cytochrome P450: Evidence against axial histidine ligation of iron, *FEBS Lett.* 88:127–130.

Dawson, J. H., Holm, R. H., Trudel, J. R., Barth, G., Linden, R. E., Bunnenberg, E., Djerassi, C., and Tang, S. C., 1976. Oxidized cytochrome P450: Magnetic circular dichroism evidence for thiolate ligation in the substrate-bound form—Implications for the catalytic mechanism, *J. Am. Chem. Soc.* 98:3707–3709.

Ellin, A., Jakobsson, S. V., Schenkman, J. B., and Orrenius, S., 1972. Cytochrome P450 of rat kidney cortex microsomes: Its involvement in fatty acid ω and (ω-1)-hydroxylation, *Arch. Biochem. Biophys.* 150:64–71.

Estabrook, R. W., Hildebrandt, A. G., Baron, J., Netter, N. J., and Leibaman, K., 1971. A new spectral intermediate associated with cytochrome P450 function in liver microsomes, *Biochem. Biophys. Res. Commun.* 42:132–139.

Foote, C. S., Shook, F. C., and Abakerli, R. A., 1980. Chemistry of superoxide ion. 4. Singlet oxygen is not a major product of dismutation, *J. Am. Chem. Soc.* 102:2503–2504.

Gillette, J. E., Davis, D. C., and Sasame, H. A., 1972. Cytochrome P450 and its role in drug metabolism, *Annu. Rev. Pharm.* 12:57–84.

Gorrod, J. W. (ed.), 1978. *Biological Oxidation of Nitrogen*, Elsevier North-Holland, New York.

Grover, A. L., Hewer, A., and Sims, P., 1971. Epoxides as microsomal metabolites of polycyclic hydrocarbons, *FEBS Lett.* 18:76–79.

Groves, J. T., and Nemo, T. E., 1983. Aliphatic hydroxylation catalyzed by iron porphyrin complex, *J. Am. Chem. Soc.* 105:6243–6248.

Groves, J. T., and Subramanian, D. V., 1984. Hydroxylation by cytochrome P450 and metallo-porphyrin models: Evidence for allylic rearrangements, *J. Am. Chem. Soc.* 106:2177–2181.

Groves, J. T., McClusky, G. A., White, R. E., and Coon, M. J., 1978. Aliphatic hydroxylation by highly purified liver microsomal cytochrome P450: Evidence for a carbon radical interme-diate, *Biochem. Biophys. Res. Commun.* 81:154–160.

Groves, J. T., Nemo, T. E., and Myers, R. S., 1979. Hydroxylation and epoxidation catalyzed by iron–porphine complexes: Oxygen transfer from iodosylbenzene, *J. Am. Chem. Soc.* 101:1032–1033.

Groves, J. T., Haushalter, R. C., Nakamura, M., Wemo, T. E., and Evans, B. J., 1981. High-valent iron porphyrin complexes related to peroxidase and cytochrome P450, *J. Am. Chem. Soc.* 103:2884–2886.

Guengerich, F. P., and McDonald, T. L., 1984. Chemical mechanism of catalysis by cytochromes P450: A unified view, *Acc. Chem. Res.* 17:9–16.

Guengerich, F. P., Ballou, D. P., and Coon, M. J., 1976. Spectral intermediates in the reaction of oxygen with purified liver microsomal cytochrome P450, *Biochem. Biophys. Res. Commun.* 70:951–956.

Gustafsson, J., Rondahl, L., and Bergmann, J., 1979. Iodosylbenzene derivatives as oxygen donors in cytochrome P450 catalyzed steroid hydroxylations, *Biochemistry* 18:865–870.

Hamilton, G. A., 1964. Oxidation by molecular oxygen. II. The oxygen atom transfer mechanism of mixed-function oxidases and the model for mixed function oxidases, *J. Am. Chem. Soc.* 86:3391–3392.

Hanson, L. K., Sligar, S. G., and Gunsalus, I. C., 1977. Electronic structure of cytochrome P450, *Croat. Chem. Acta* 49:237–250.

Heimbrook, E. C., and Sligar, S. G., 1981. Multiple mechanisms of cytochrome P450 catalyzed substrate hydroxylations, *Biochem. Biophys. Res. Commun.* 9:530–535.

Hrycay, E. G., Gustafsson, J. A., Ingleman-Sundberg, M., and Ernster, L., 1975. Sodium periodate, sodium chlorite, organic hydroperoxides and H_2O_2 as hydroxylating agents in steroid hydroxylation reactions catalyzed by partially purified cytochrome P-450, *Biochem. Biophys. Res. Commun.* 66:209–216.

Ichihara, K., Yamakawa, I., Kusinose, E., and Kusinose, M., 1979. Fatty acid ω and (ω-1)-hydroxylation in rabbit intestinal mucosa microsomes, *J. Biochem.* 86:139–146.

Ishimura, Y., Ullrich, V., and Peterson, J. A., 1971. Oxygenated cytochrome P450 and its possible role in enzymatic hydroxylation, *Biochem. Biophys. Res. Commun.* 42:140–146.

Jerina, D. M., Daly, J. W., Witkop, B., Zaltzman-Nirenberg, P., and Udenfriend, S., 1968. The role of arene oxide–oxepin systems in the metabolism of aromatic substrates. III. Formation of 1,2-naphthalene oxide from naphthalene by liver microsomes, *J. Am. Chem. Soc.* 90:6525–6527.

Jerina, D. M., Daly, J. W., Witkop, B., Zaltzman-Nirenberg, P., and Udenfriend, S., 1970. 1,2-Naphthalene oxide as an intermediate in the microsomal hydroxylation of naphthalene, *Biochemistry* 9:147–155.

Kadlubar, F. F., Morton, K. C., and Ziegler, D. M., 1973. Microsomal-catalyzed hydroperoxide-dependent C-oxidation of amines, *Biochem. Biophys. Res. Commun.* 54:1255–1261.

Kasperek, G. J., and Bruice, T. C., 1972. The mechanism of aromatization of arene oxides, *J. Am. Chem. Soc.* 94:198–202.

Koch, S., Tang, S. C., Holm, R. H., Frankel, R. B., and Ibers, J. A., 1975. Ferric porphyrin thiolates: Possible relationship to cytochrome P450 enzymes and the structure of (*p*-nitrobenzene-thiolate) iron(III) protoporphyrin IX dimethyl ester, *J. Am. Chem. Soc.* 97:916–918.

Lichtenberger, F., Nastainczyk, W., and Ullrich, V., 1976. Cytochrome P450 as an oxene transferase, *Biochem. Biophys. Res. Commun.* 70:939–946.

Lipscomb, J. D., Sligar, S. G., Namtvedt, M. J., and Gunsalus, I. C., 1976. Autoxidation and hydroxylation reactions of oxygenated P450, *J. Biol. Chem.* 251:1116–1124.

Loew, G. H., and Rohmer, M. M., 1980. Electronic spectra of model oxy, carboxy P450 and carboxy heme complexes, *J. Am. Chem. Soc.* 102:3655–3657.

May, S. W., Gordon, S. L., and Steltenkamp, M. S., 1977. Enzymatic epoxidation of *trans,trans* 1,8-dideutero-1,7-octadiene: Analysis using partially relaxed proton Fourier transform NMR, *J. Am. Chem. Soc.* 99:2017–2024.

Mynert, E. W., Foreman, R. L., and Watabe, T., 1970. Epoxides as obligatory intermediates in the metabolism of olefins to glycols, *J. Biol. Chem.* 245:5234–5238.

Nordblom, G. D., White, R. E., and Coon, M. J., 1976. Studies on the hydroperoxide-dependent substrate hydroxylation by purified rat liver microsomal cytochrome P450, *Arch. Biochem. Biophys.* 175:524–533.

Omura, T., Saunders, E., Estabrook, R. W., Cooper, D. V., and Rosenthal, O., 1966. Isolation from adrenal cortex of a nonheme iron protein and flavoprotein functional as a reduced triphosphopyridine nucleotide cytochrome P450 reductase, *Arch. Biochem. Biophys.* 117:660–673.

Peisach, J., Mims, W. B., and Davis, J. L., 1979. Studies of the electron-nuclear coupling between Fe(III) and ^{14}N in cytochrome P450 and in a series of low spin heme compounds, *J. Biol. Chem.* 254:12,379–12,389.

Pesch, F., Jerina, D. M., Daly, J. W., Lu, A. Y. H., Kuntzman, R., and Conney, A. H., 1972. A reconstituted enzyme fraction that converts naphthalene to *trans*-1,2-dihydroxy-1,2-dihydronaphythalene via naphthalene-1,2-oxide: Presence of epoxide hydrase in cytochrome P450 and P448 fractions, *Arch. Biochem. Biophys.* 153:62–67.

Peterson, J. A., Ishimura, Y., and Griffin, B. W., 1972. *Pseudomonas putida* cytochrome P450: Characterization of an oxygenated form of hemo-protein, *Arch. Biochem. Biophys.* 149:197–208.

Pudzianowski, A. T., and Loew, G. H., 1980. Quantum mechanical studies of model cytochrome P450 hydrocarbon mechanism: A MINDO/3 study of hydroxylation and epoxidation pathways for methane and ethylene, *J. Am. Chem. Soc.* 102:5443–5449.

Rahimtula, A. D., O'Brien, P. J., Seifried, H. E., and Jerina, D. M., 1978. The mechanism of action of cytochrome P450: Occurrence of the NIH shift during hydroxide-dependent aromatic hydroxylations, *Eur. J. Biochem.* 89:133–141.

Selkirk, J. K., Huberman, E., and Heidelberger, C., 1971. An epoxide intermediate in the microsomal metabolism of the chemical carcinogen, dibenz-(a,h)-anthracene, *Biochem. Biophys. Res. Commun.* 43:1010–1016.

Shannon, P., and Bruice, T. C., 1981. A novel P450 model system for the *N*-dealkylation reaction, *J. Am. Chem. Soc.* 103:4580–4582.

Sharrock, M., Munck, E., Debrunner, P. G., Marshall, V., Lipscomb, J. D., and Gunsalus, I. C., 1973. Mossbauer studies of cytochrome P450, *Biochemistry* 12:258–263.

Sharrock, M., Debrunner, P. G., Schulz, C., Lipscomb, J. D., Marshall, V., and Gunsalus, I. C., 1976. Cytochrome P450 and its complexes: Mossbauer parameters of the heme iron, *Biochim. Biophys. Acta* 420:8–26.

Simmoneaux, G., Scholtz, W. F., Reed, C. A., and Lang, G., 1982. Mossbauer spectra of unstable iron porphyrins: Models for compound II of peroxidase, *Biochim. Biophys. Acta* 716:1–7.

Sligar, S. G., 1976. Coupling of spin, substrate, and redox equilibria in cytochrome P450, *Biochemistry* 15:5399–5406.

Sligar, S. G., and Gunsalus, I. C., 1979. Proton coupling in the cytochrome P450 spin and redox equilibrium, *Biochemistry* 18:2290–2295.

Sligar, S. G., Debrunner, P. G., Lipscomb, J. D., Namtvedt, M. J., and Gunsalus, I. C., 1974a. A role of the putidaredoxin COOH-terminus in P450 (cytochrome M) hydroxylations, *Proc. Natl. Acad. Sci. U.S.A.* 71:3906–3910.

Sligar, S. G., Lipscomb, J. D., Debrunner, P. G., and Gunsalus, I. C., 1974b. Superoxide anion from the autoxidation of cytochrome P450, *Biochem. Biophys. Res. Commun.* 61:290–296.

Sligar, S. G., Kennedy, K. A., and Pearson, D. C., 1980. Chemical mechanisms of cytochrome P450 hydroxylation: Evidence for acylation of heme-bound dioxygen, *Proc. Natl. Acad. Sci. U.S.A.* 77:1240–1244.

Stern, J. O., and Peisach, J., 1974. A model compound for study of the CO-adduct of cytochrome P450, *J. Biol. Chem.* 249:7495–7498.

Takamuku, S., Matsumoto, H., Hori, A., and Sakurai, H., 1980. Aromatic hydroxylation by O (^{3}P) atoms, *J. Am. Chem. Soc.* 102:1441–1443.

Tsai, R., Yu, C. A., Gunsalus, I. C., Peisach, J., Blumberg, W., Orme-Johnson, W. H., and Beinhert, H., 1970. Spin state changes in cytochrome P450 on binding of specific substrates, *Proc. Natl. Acad. Sci. U.S.A.* 66:1157–1163.

Tyson, C. A., Lipscomb, J. P., and Gunsalus, I. C., 1972. The role of putidaredoxin and P450 in methylene hydroxylations, *J. Biol. Chem.* 247:5777–5784.

Udenfriend, S., Zaltzman-Nirenberg, P., Daly, J., Guroff, G., Chedsey, C., and Witkop, B., 1967. Intramolecular migration of deuterium and tritium during enzymatic hydroxylation of *p*-deuteroacetanilide and *p*-tritioacetanilide, *Arch. Biochem. Biophys.* 120:413–419.

Ullrich, V., 1979. Cytochrome P450 and biological hydroxylation reactions, *Top. Curr. Chem.* 83:67–104.

Welborn, C. H., Dolphin, D., and James, B. R., 1981. One-electron electrochemical reduction of a ferrous porphyrin dioxygen complex, *J. Am. Chem. Soc.* 103:2869–2871.

Whysner, J. A., Ramseyer, J. A., and Harding, B. W., 1970. Substrate-induced changes in visible absorption and electron spin resonance properties of adrenal cortex mitochondrial P450, *J. Biol. Chem.* 245:5441–5449.

Wilson, L. E., and Harding, B. W., 1970. Studies on adrenal cortical cytochrome P450. III. Effects of carbon monoxide and light on steroid 11B hydroxylation, *Biochemistry* 9:1615–1621.

Other Iron Monooxygenases

16.1 Heme Oxidation

One of the more novel biological reactions in which dioxygen participates is the oxidation of heme to bilirubin. It has long been known that breakdown of heme *in vivo* leads to evolution of carbon monoxide and biliverdin (Sjöstrand, 1952; Ludwig *et al.*, 1957). Bilirubin is then produced by enzymatic reduction of biliverdin. Other facets of the reaction have come to light much more recently. Tenhunen *et al.* (1969) discovered the enzyme that catalyzes this transformation and found that 3 moles of dioxygen are consumed per mole of bilirubin produced and monoxide given off. The reaction also occurs nonenzymatically under a variety of conditions. Substituents on the porphyrin ring are not critical, but metal ion and oxidation state are. Thus, oxophlorins, the first intermediate in the degradation pathway, can be formed by hydrogen peroxide oxygenation of porphyrin chelates of any of several metals with easily accessible 1-higher oxidation states [Fe(III), Co(III), Mn(II), Mn(III) low yield] (Bonnett and Dimsdale, 1972). Chelates of metals without this property are inert. The product has resonance forms (Figure 16-1) A and B, of which B is predominant.

Nonmetallated porphyrins can be oxygenated indirectly via their benzoate esters with benzoylperoxide at 90–100°C (Bonnett *et al.*, 1969). All these metallo oxophlorins are relatively stable, except the ferric compound, which autooxidizes in the presence of air to verdohemin (verdohemins are compounds that hydrolyze to biliverdins). Oxygenation of the ferric oxophlorin can also take place with excess hydrogen peroxide instead of air (Brown, 1976).

If dioxygen is used as oxidant for the entire reaction, then a reductant (e.g., ascorbic acid) must be included to provide the ferrous oxidation state of the

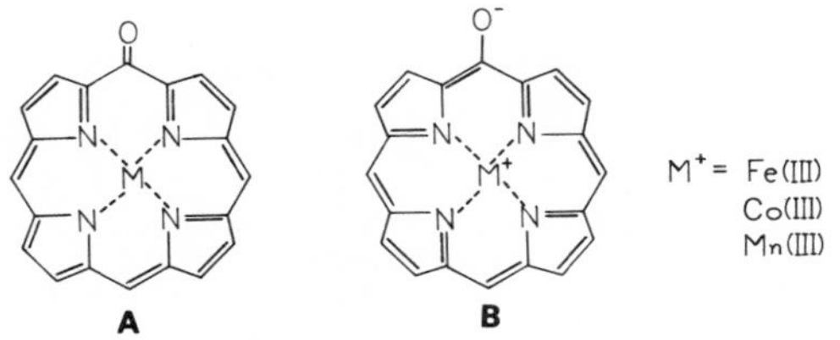

M+ = Fe(II)
Co(III)
Mn(III)

Figure 16-1. Resonance forms of an oxophlorin.

substrate. This fact has been taken to indicate that the initial reaction is essentially a cytochrome-P-450-type hydroxylation in which the "cofactor" is also the aromatic substrate. That the reaction is intramolecular both in solution and on an enzyme surface has also been concluded from a variety of experiments (Brown, 1976). Gotoh and Shikama (1976) reported cursory trapping experiments indicating that superoxide is involved in the reaction, supporting this contention. Hydrogen peroxide reacting with the ferric complex would also be explained by a cytochrome-P-450-type mechanism.

The enzymatic reaction appears to be very similar. Protoporphyrin-IX coordinated to metals that cannot "activate" dioxygen are enzyme inhibitors, as is protoporphyrin-IX containing no metal ion. Cobalt–protoporphyrin-IX is also an inactive substrate (Yoshida and Kikuchi, 1978). Solution experiments (Bonnett and Dimsdale, 1972) suggest that this compound could be hydroxylated to the oxyphlorin, but not cleaved. The enzyme's inability to perform the initial oxygenation has been attributed to inability of the *in vivo* systems to reduce the cobalt.

The initial enzymatic hydroxylation is stereospecific, producing entirely the α-biliverdin product. Comparison of hemoglobin and myoglobin autoxidation and active-site structures suggests that the enzyme imposes this stereo-specificity simply by blocking the other methine positions to attack by iron-bound dioxygen (Brown, 1976). Yoshida et al. (1981) have isolated and characterized a heme–dioxygen complex of heme oxygenase. This was accomplished by passing the reduced form of the enzyme substrate complex through a column to remove all excess reducing agent before exposing it to dioxygen. The reaction cannot proceed until a single electron source is provided, so the intermediate ternary complex can be studied. At slightly acidic pH (6.0), the complex autooxidizes to the ferric form. The rate-determining step of the reaction is not established. Reduction of the ferric heme–heme oxygenase complex to ferrous occurs much faster than heme degradation. Also, the fate of the remaining oxygen atoms is unknown. It was shown that both lactam oxygens in the product biliverdin derive from molecular oxygen rather than water (eliminating the possibility of hydrolysis of a compound such as shown in Figure 16.2). The oxygen lost as carbon monoxide also derives from molecular oxygen.

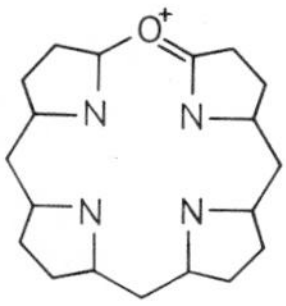

Figure 16-2. A plausible intermediate in heme oxidation that has been ruled out by isotopic labeling experiments.

Finally, it has been shown by $^{16}O_2/^{18}O_2$ labeling experiments (Brown and King, 1978) that the two product lactam oxygen atoms are derived from different dioxygen molecules. Thus, endoperoxidelike intermediates in which the peroxide bridge results from a single dioxygen molecule are also impossible.

It must be noted that this reaction is very complex and is the subject of continued experimentation. Many of the experiments are indirect, and conclusions based on them may be premature. Characterization of the reaction intermediates will ultimately be required for a secure understanding of this reaction.

16.2 Tryptophan Side-Chain Oxidation

A heme protein has been isolated from *Pseudomonas* that catalyzes the side-chain oxidation of tryptophan and other 3-substituted indoles. The product has been isolated only as the quinoxaline formed from *o*-phenylene diamine condensation. The product is believed to be either 3-indolyglyoxal or 3-indolylglyceraldehyde (Takai *et al.*, 1977).

16.3 Indolyl-1,3-alkane α-Hydroxylase

An enzyme similar to the previous one hydroxylates 3-alkyl indoles at the α-position on the side chain. The enzyme is again a heme protein (Roberts and Rosenfeld, 1977).

References

Bonnett, R., and Dimsdale, M. J., 1972. The *meso*-reactivity of porphyrins and related compounds. Part V. The *meso*-oxidation of metalloporphyrins, *J. Chem. Soc. Perkin Trans. 1* 1972:2540–2548.

Bonnett, R., Dimsdale, M. J., and Stephanson, G. F., 1969. The *meso*-reactivity of porphyrins and related compounds. Part IV. Introduction of oxygen functions, *J. Chem. Soc. C* 1969:564–570.

Brown, S. B., 1976. Stereospecific haem cleavage: A model for the formation of bile-pigment isomers *in vivo* and *in vitro, Biochem. J.* 170:23–27.

Brown, S. B., and King, R. F. G. J., 1978. The mechanism of haem catabolism: Bilirubin formation in living rats by [^{18}O] oxygen labeling, *Biochem. J.* 170:297–311.

Gotoh, T., and Shikama, K., 1976. Generation of superoxide radical during autooxidation of oxymyoglobin, *J. Biochem.* 80:397–399.

Ludwig, G. D., Blakemore, W. S., and Drabkin, D. L., 1957. Production of carbon monoxide and bile pigment by haemin oxidation, *Biochem. J.* 88:38P.

Roberts, J., and Rosenfeld, H. J., 1977. Isolation, crystallization and properties of indolyl-1,3-alkaline α-hydroxylase, *J. Biol. Chem.* 252:2640–2647.

Sjöstrand, T., 1952. The formation of carbon monoxide by *in vitro* decomposition of haemoglobin in bile pigments, *Acta Physiol. Scand.* 26:328–333.

Takai, K., Ushiro, H., Noda, Y., Narumiya, S., Tokuyama, T., and Hayaishi, O., 1977. Crystalline hemoprotein from *Pseudomonas* that catalyzes oxidation of side chain of tryptophane and other indole derivatives, *J. Biol. Chem.* 252:2648–2656.

Tenhunen, R., Marver, H. S., and Schmid, R., 1969. Microsomal heme oxygenase: Characterization of the enzyme, *J. Biol. Chem.* 244:6388–6394.

Yoshida, T., and Kikuchi, G., 1978. Reaction of the microsomal heme oxygenase with cobaltic protoporphyrin IX, an extremely poor substrate, *J. Biol. Chem.* 253:8479–8482.

Yoshida, T., Noguchi, M., and Kikuchi, G., 1981. Oxygenated form of heme · heme oxygenase complex and requirement for second electron to initiate heme degradation from the oxygenated complex, *J. Biol. Chem.* 255:4418–4420.

Ribulose Biphosphate Oxygenase

17

17.1 Introduction

Ribulose biphosphate oxygenase/carboxylase is a very interesting enzyme in several respects. It is the enzyme that provides the glycolate associated with photorespiration in C_3 plants by catalyzing the reaction of the substrate with molecular oxygen (Metzler, 1977). If carbon dioxide is encountered instead of dioxygen, the product of the reaction is 3-phosphoglycerate (2 equivalents), which enters the Calvin cycle via reduction to glyceraldehyde-3-phosphate (Figure 17-1). This enzyme thus accounts for all light-dependent carbon fixation as well as the substrate for photorespiration. Since the purpose of photorespiration is not understood, the competing reactions of this enzyme are fascinating.

The enzyme is present in high concentration in chloroplasts. It has a high molecular weight (560,000), consisting of eight larger catalytically active peptide chains and eight smaller chains that are believed to function in control of the enzyme. X-ray data have produced a three-dimensional depiction of a crystalline form of this enzyme.

When $^{18}O_2$ is used for the reaction, one labeled oxygen atom remains in the glycolate carboxyl group and no label is observed in the glycerate molecule (Lorimer et al., 1973). On the other hand, performing the reaction in $H_2^{18}O$ results in ^{18}O incorporation into the carboxylate group of 3-phosphoglycerate. [Incorporation of ^{18}O into the glycolate carboxylate group in this experiment is due to substrate ketone exchange with solvent, which is well known and believed to be catalyzed by an adjacent phosphate group (Sue and Knowles, 1978).]

From the standpoint of the biochemical reactions of dioxygen, this enzyme is of interest because the dioxygen reacts without prior activation. Both oxygenase and carboxylase activities require Mg^{2+} (Andrews et al., 1973), but an earlier report of copper associated with the enzyme has been refuted (Jensen and Bahr, 1977).

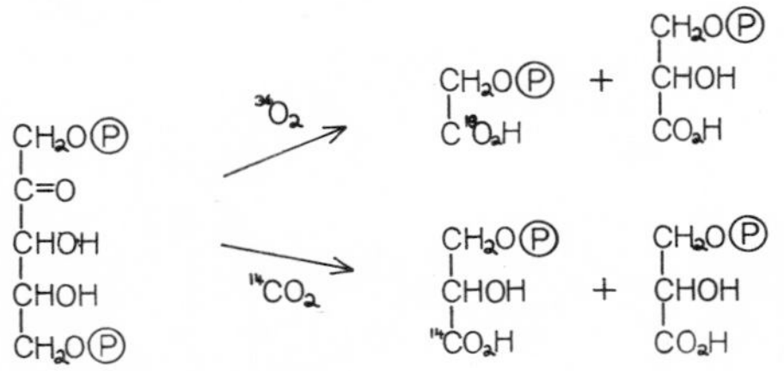

Figure 17-1. Two reactions catalyzed by ribulose bisphosphate oxygenase.

17.2   Chemical Analysis

The aqueous alkaline oxidation of sugars (Figure 17-2) is well known among sugar chemists because it complicates basic reactions of monosaccharides, generating a complex mixture of often inseparable derivatives. The general mechanism for the nonenzymatic reaction (and enzymatic as well) involves α-keto-proton removal to form an enolate anion. These species are known to react readily with dioxygen to form α-ketohydroperoxides, which fragment in water.

The enolates can also isomerize to generate different fragments. However, it has been noted that the solution-phase production distribution cannot be reconciled without the inclusion of a hydride transfer process (Gleason and Barker, 1971a). It was shown that in alkaline solution, removal of hydroxyl protons is also important, resulting in tritium (hydride) transfer (Figure 17-3) (Gleason and

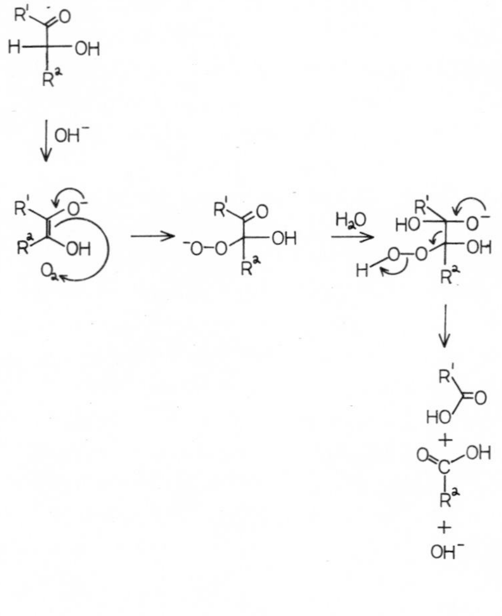

Figure 17-2. Alkaline oxidation of sugars.

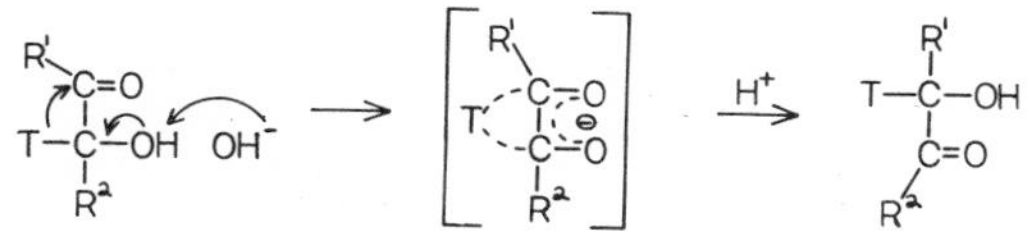

Figure 17-3. Proton removal from a sugar followed by tritium migration.

Barker, 1971b). The solution reaction shows other complications as well. Hence, one of the enzyme's primary functions, in this case, is to prevent side reactions.

17.3 Effectors and Inhibitors

Since the enzyme is the starting point for two major metabolic pathways, one would expect effectors and inhibition. This has been suggested, in fact, as the role of the small subunits associated with the larger catalytic peptide chains (McFadden, 1980). The enzyme is complex in its response to substrate and phosphorylated sugar derivatives. Differences in activity due to order of addition of reactants, preincubation with magnesium and bicarbonate, and the K_m for $CO_{2(sol)}$, as opposed to total carbonate, have caused much confusion in the literature. Taking into account the high concentration of the protein and the low concentrations of inhibitors relative to substrate, it was eventually resolved that the enzyme does exhibit sufficient activity to account for photosynthetic carbon dioxide reduction and for photorespiration at physiological concentrations of CO_2 and O_2. It was also resolved that organic (sugar-derived) inhibitors affect oxygenase and carboxylase activities in the same manner, though pH, temperature, and metal ions can have different effects, indicating that the two activities probably proceed from the same active site. This view is supported by the well-known fact that the two gases competitively inhibit each other's activities (Jensen and Bahr, 1977).

17.4 Reaction Mechanism

Since the enzyme must behave similarly in the two reactions, the carboxylase activity is also of interest (Siegel and Lane, 1973). Addition of carbon dioxide to the enolate (Figure 17-4) is the reverse of decarboxylation of a β-keto acid, but in this case the enolate intermediate isomerizes (in fact, the decarboxylation occurs spontaneously when the intermediate 2-carboxy-3-ketoribitol-1,5-diphosphate is chemically generated). The unusual step in this reaction is addition of water across the carbon–carbon single bond, resulting in fragmentation of the sugar backbone. This fragmentation is not commonly observed in nonenzymatic systems (though it was observed when the unstable 2-carboxy-3-ketoribitol-1,5-

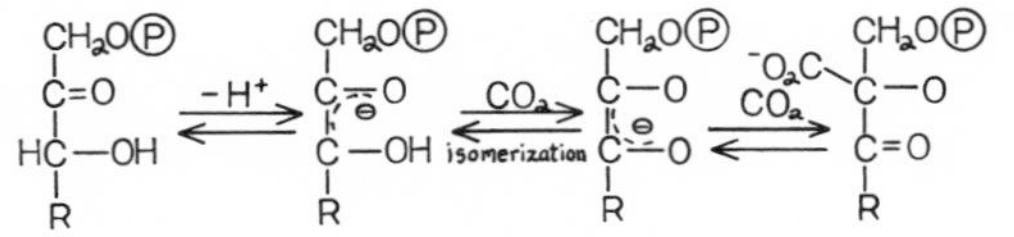

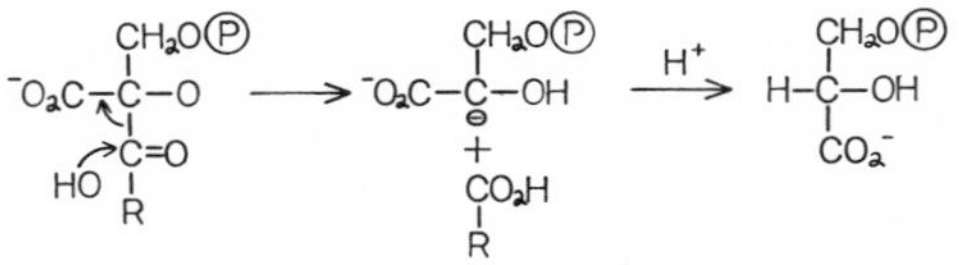

Figure 17-4. Carboxylase reaction of ribulose bisphosphate oxygenase.

diphosphate was chemically synthesized), but the analogous fragmentation of the dioxygen adduct (hydroperoxide) is well known.

Pierce *et al.* (1980) pointed out that the transition-state analogue potent inhibitor that had been studied by other workers is actually a 2-carboxyarabinitol derivative rather than a 2-carboxyribitol derivative. This fact indicates that inversion must occur during the reaction to avoid formation of the L isomer of 3-phosphoglycerate from the C-1–C-3 fragment of the substrate (the enzyme does not exhibit epimerase activity toward the products). This inversion is most easily explained by an enolate form of the intermediate carbanion being stereospecifically attacked by a proton. The divalent metal cation may assist the reaction by stabilizing this enolate (Figure 17-5). Assuming the same reaction geometry for the dioxygen reaction, this assistance would result in the formation of magnesium oxide (Figure 17-6), which is highly insoluble. Studies with bacterial enzyme (Robison *et al.*, 1979) have shown, in fact, that the two reactions behave differently on substitution of Mn^{2+} for Mg^{2+}. The Mn^{2+} enzyme retains about

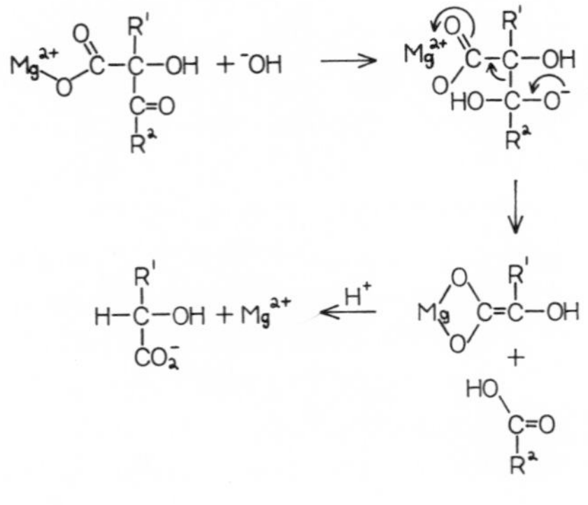

Figure 17-5. Mechanism of the ribulose bisphosphate carboxylase reaction.

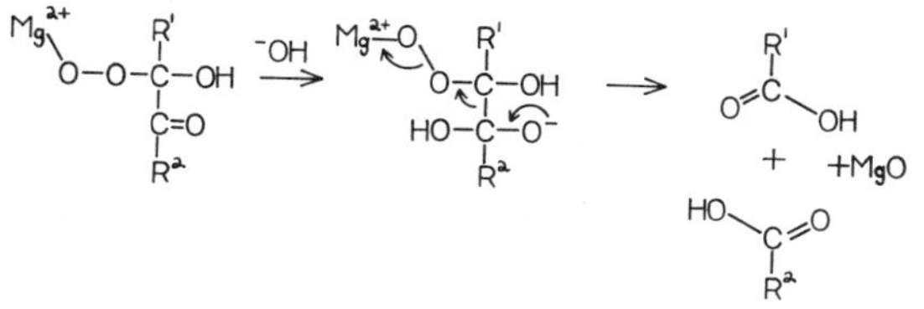

Figure 17-6. Role of magnesium in the ribulose bisphosphate oxygenase reaction.

25% of the carboxylase activity, but doubles the oxygenase activity over Mg^{2+} enzyme; Co^{2+} decreases the oxygenase level to 40% of the Mg^{2+} level while eliminating the carboxylase activity altogether. However, several features suggest a difference between the enzyme found in higher plants and that found in bacteria.

Miziorko (1979) also studied the stable transition-state analogue carboxy-ribitolbisphosphate and found that it forms a stable complex with the enzyme.

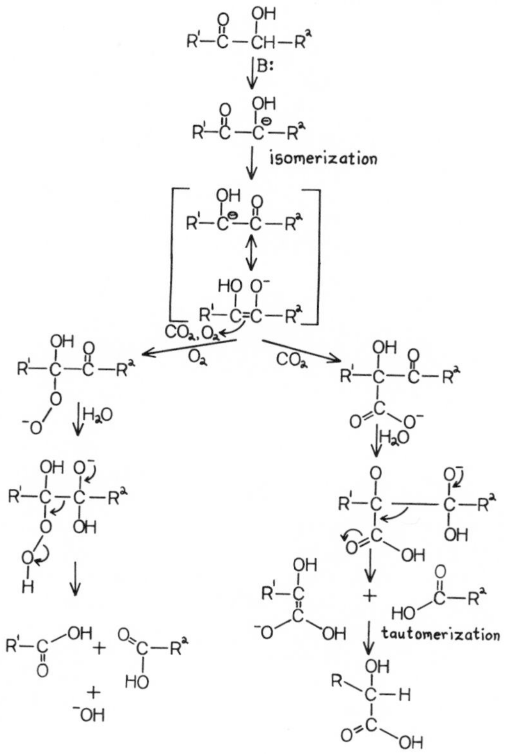

Figure 17-7. A general mechanism for the ribulose bisphosphate oxygenase reaction.

Incubation of this complex with divalent metal ion and bicarbonate led to the detection of another complex containing these components as well as the sugar derivative. Since the transition-state analogue contains a carboxyl group, binding of carbon dioxide to the first complex indicates that the enzyme has two binding sites for carbon dioxide. It was suggested that the activitation that occurs on preincubation of the enzyme with bicarbonate and Mg^{2+} is due to CO_2 binding at the nonreactive site. The general mechanism for this oxygenase reaction, then, is entirely analogous to the carboxylase reaction (Figure 17-7).

By way of explanation, it should be mentioned that carbanions are in general reactive enough to overcome the kinetic spin inversion barrier to reaction with triplet dioxygen (see Chapter 2). This reaction can take the form of anion oxidation to form a carbon radical and superoxide, or it can result in addition, as in the present example, to form a hydroperoxide. Since reaction with carbanions is one of the few ways of overcoming this kinetic barrier (another being formation of a metal–dioxygen complex), it is somewhat surprising that this type of reaction is not more common in biochemical systems. Conversely, the fact that dioxygen carbanion additions are otherwise unknown in biological systems makes the ribulose-1,5-bisphosphate carboxylase/oxygenase reaction all the more intriguing.

References

Andrews, T. J., Lorimer, G. H., and Tolbert, N. E., 1973. Ribulose diphosphate oxygenase. I. Synthesis of phosphoglycolate by fraction-1 protein of leaves, *Biochemistry* 12:11–18.

Gleason, W. B., and Barker, R., 1971a. Oxidation of pentoses in alkaline solution, *Can. J. Chem.* 49:1423–1432.

Gleason, W. B., and Barker, R., 1971b. Evidence for a hydride shift in the alkaline rearrangement of D-ribose, *Can. J. Chem.* 49:1433–1440.

Jensen, R. G., and Bahr, J. T., 1977. Ribulose 1,5-bisphosphate carboxylase oxygenase, *Annu. Rev. Plant Physiol.* 28:379–400.

Lorimer, G. H., Andrews, T. J., and Tolbert, N. E., 1973. Ribulose diphosphate oxygenase. II. Further proof of reaction products and mechanism of action, *Biochemistry* 12:18–23.

McFadden, B. A., 1980. A perspective of ribulose bisphosphate carboxylase/oxygenase, the key catalyst in photosynthesis and photorespiration, *Acc. Chem. Res.* 13:394–399.

Metzler, D. E. (ed.), 1977. *Biochemistry: The Chemical Reactions of Living Cells,* Academic Press, New York, pp. 419–420.

Miziorko, H. M., 1979. Ribulose-1,5-bisphosphate carboxylase: Evidence in support of the existence of distinct CO_2 activator and CO_2 substrate sites, *J. Biol. Chem.* 254:270–272.

Pierce, J., Tolbert, N. E., and Barker, R., 1980. Interaction of ribulosebisphosphate carboxylase/oxygenase with transition state analogues, *Biochemistry* 19:934–942.

Robison, P. D., Martin, M. N., Tabita, F. R., 1979. Differential effects of metal ions on *Rhodospirillum rubrum* ribulosebisphosphate carboxylase/oxygenase and stoichiometric incorporation of HCO_3^- into a Co(III)–enzyme complex, *Biochemistry* 18:4453–4458.

Siegel, M. I., and Lane, M. D., 1973. Chemical and enzymatic evidence for the participation of a 2-carboxy-3-ketoribitol-1,5-diphosphate intermediate in the carboxylation of ribulose-1,5-diphosphate, *J. Biol. Chem.* 248:5486–5498.

Sue, J. M., and Knowles, J. R., 1978. Retention of the oxygens at C2 and C3 of D-ribulose-1,5-bisphosphate in the reaction catalyzed by ribulose-1,5-bisphosphate carboxylase, *Biochemistry* 17:4041–4044.

Dioxygenases

18.1 Lipoxygenase

Lipoxygenase catalyzes several types of oxidations of fatty acids, as shown in Figure 18-1. The best known of these is the dioxygenase function that catalyzes the aerobic oxidation of fatty acids, particularly linoleic acid, to the hydroperoxides. More specifically, lipoxygenase catalyzes the oxidation of *cis,cis*-1,4-pentadienes to the hydroperoxides of *cis,trans*-1,3-dienes (Figure 18-2).

The various lipoxygenases also generally catalyze several anaerobic reactions including reactions of the product (fatty acid hydroperoxide) with the substrate (fatty acid) and cooxidation reactions with apparently nonspecific substrates. The multiple activities and forms of the enzyme complicate its study, since a physiological role has not been established. Changes in enzyme activity during plant development led to the implication that the enzyme is critical in seed germination. Identification of some unusual allylic hydroxylated fatty acids led to the suggestion that the lipoxygenase-formed acid peroxides were reduced to form the hydroxy acids (Morris and Marshall, 1966). Though perhaps valid, neither of these possibilities is either general enough or crucial enough to explain the ubiquitous activity of the enzyme. Another suggestion is that the enzyme is present to scavenge O_2, especially in seeds (Veldink *et al.*, 1977). This seems unlikely, because products, particularly of the anaerobic reaction, are more reactive toward biological molecules than dioxygen is.

The most credible suggestion is that lipoxygenases are involved in the plant response to wounds (Galliard, 1975). Activity is generally high in wounded tissues, and evidence suggests that the active component of the "wound hormone" traumatin is 12-oxo-10-*trans*-dodecenoic acid, which is a metabolite of the product of the lipoxygenase reaction. However, the metabolic significance of lipoxygenase in human platelets is not explained by any of these possibilities.

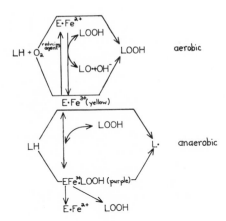

Figure 18-1. Reactions catalyzed by lipoxygenase. (LH) lipid; (LOOH) lipid hydroperoxide.

Ignorance of the physiologically important function of the enzyme has led to confusion about which reaction to study, which iron oxidation state is involved, and what substrate–inhibitor conditions to use for *in vitro* studies. Also, micelle formation by both substrate and product has confused kinetic studies, since the enzyme reacts only with monomeric substrates (Vliegenthart *et al.*, 1979).

In addition, several isozymes often occur together in the same tissue. The isozyme most studied is the soybean lipoxygenase, which provoked interest by crystallizing quite early (Theorell *et al.*, 1947). Soybean has several other isozymes as well. The crystallizable enzyme (isozyme I) has a pH optimum of about 9, whereas the other soybean isozymes (II–IV) have pH optima at 5–7, as do the enzymes isolated from other plants. It is generally assumed that the difference in pH optimum does not reflect a difference in mechanism. However, isozyme I catalyzes the formation of the 13-hydroperoxy acid from linoleic acid and isozyme II catalyzes the formation of 9- and 10-hydroperoxides (Grosch and Laskawy, 1979; Cornelius *et al.*, 1979).

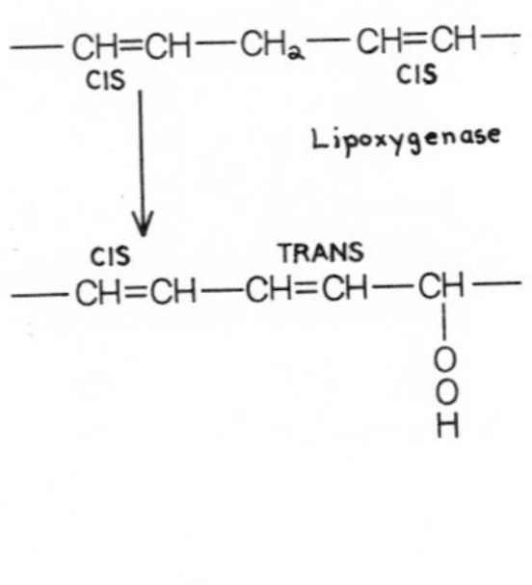

$$— CH=CH—CH_2—CH=CH —$$
$$\quad\ \text{CIS}\qquad\qquad\quad \text{CIS}$$

Lipoxygenase

$$\qquad\ \ \text{CIS}\qquad\quad \text{TRANS}$$
$$— CH=CH—CH=CH—CH —$$
$$\qquad\qquad\qquad\qquad\quad |$$
$$\qquad\qquad\qquad\qquad\quad O$$
$$\qquad\qquad\qquad\qquad\quad O$$
$$\qquad\qquad\qquad\qquad\quad H$$

Figure 18-2. Most important reaction catalyzed by lipoxygenase.

18.2 Lipoxygenase Mechanism

From the standpoint of the dioxygen reaction with which we are primarily concerned, the following pieces of mechanistic evidence are important: (1) Two forms of the enzyme catalyze the aerobic reaction (Pistorius *et al.*, 1976). (2) One reactive form of the enzyme (Egmond *et al.*, 1977) is electron spin resonance (ESR)-silent and colorless, implying that the iron atom is in the ferrous form (Pistorius *et al.*, 1976). (3) The other form, which catalyzes the same reaction, is yellow and shows a broad ESR signal at $g = 6$, indicating that both ferrous and ferric forms are active catalysts. (A value of $g = 6$ is taken to be an indication of unpaired electron density in an orbital with high angular dependence, i.e., a d-orbital of a metal.) (4) Hydrogen is removed from the bisallylic position (C-11 in linoleate) during or before the reaction's rate-limiting step (Egmond *et al.*, 1973; Hamberg and Samuelsson, 1967b). (5) Substituting deuterium for hydrogen at the C-11 position decreases the V_{max} for the reaction by a factor of 9 (Egmond *et al.*, 1973). Such large isotope effects are unusual in biological systems and, though possible for proton abstraction, often indicate abstraction of atoms or hydride ions instead. (6) The hydrogen (proton or atom) is removed stereospecifically and dioxygen is added stereospecifically (Egmond *et al.*, 1972), so that the resulting peroxide is *trans* to the removed hydrogen (Gibian and Gallaway, 1977) (assuming that any intermediate does not twist more than 90°). (7) The activation energy for the reaction is only about 4.0 kcal/mol (Egmond *et al.*, 1977).

The aerobic reaction is superficially similar to the hydroperoxide reaction of singlet dioxygen with an olefin, as discussed in Chapter 3. However, the reaction products are not always identical. For example, while the singlet-dioxygen oxidation of chloresterol yields primarily the 5-hydroperoxide with a small yield of 6-hydroperoxides, the lipoxygenase anaerobic cooxygenation reaction yields both isomers of the 7-hydroperoxide (Teng and Smith, 1973). All these products are allylic hydroperoxides.

The idea that singlet dioxygen *per se* is the reactive species in the lipoxygenase reaction has been dropped for a number of reasons. However, some of the issues that have been mentioned are in fact invalid objections to a singlet-dioxygen mechanism and may confuse mechanistic studies of other reactions. First, since an electrophilic metal–dioxygen complex could react like singlet dioxygen, the 22 kcal of electronic excitation energy required to form singlet dioxygen from ground-state dioxygen may not be required in biological reactions that involve metal ion (Galliard and Chan, 1980; Gibian and Gallaway, 1977). The metal–dioxygen complex would need to have an electrophilic "outer" oxygen atom and to have any unpaired electron density reside primarily on the metal ion.

Second, as discussed in Chapter 3, a concerted mechanism has been largely discarded for the reaction of singlet dioxygen with olefins to form allylic hydroperoxides. Whatever intermediate is involved could be stabilized by an enzyme. Furthermore, proton abstraction would be the result of attack by a basic residue on the protein, not by the "outer" oxygen atom, so that the stereochemistry of proton removal relative to peroxide addition would be determined by the protein. Therefore, the observation that proton removal is the enzymatic case is *trans* to the resulting peroxide whereas it is *cis* in free solution is not a compelling argument against a singlet-dioxygen-like mechanism.

Third, proton removal appears in the kinetics of the nonenzymatic singlet-oxygen reaction. Indeed, deuterium–hydrogen rate-competition studies have been the basis of several experiments detailing the mechanism of that reaction. The fact that the enzymatic reaction shows a deuterium isotope effect does not at all exclude a singlet-dioxygen mechanism.

On the other hand, the kinetic isotope effect observed is quite large (Egmond *et al.*, 1973), especially for an enzymatic reaction. Isotope effects of this magnitude are more commonly observed for hydrogen-atom abstraction than proton abstraction. Hence, this evidence supports a free-radical mechanism over a singlet-oxygen mechanism. Also, formation of the 7-OOH cholesterol derivatives (Teng and Smith, 1973) requires oxidation of cholesterol without migration of the double bond. While this is known to occur for free-radical reactions, it is inexplicable by any proposed singlet-dioxygen mechanism.

Single electron transfers are the crux of the best-supported mechanisms so far suggested. Egmond *et al.* (1977) have developed mechanisms for the aerobic and anaerobic reactions of both ferric and ferrous enzyme forms, while Gibian and Gallaway (1977) and Galliard and Chan (1980) have suggested similar mechanisms for the ferric and ferrous forms, respectively. The essential reactions are shown in Figure 18-3. The evidence for each of these steps is quite strong, but the mechanism leads to conclusions that are chemically suspicious.

For instance, reaction IIa (or IVa) [suggested by Gibian and Gallaway (1977) to be hydrogen-atom abstraction followed by proton loss] requires the enzyme ferric ion to be a much stronger oxidizing agent than any known biological or chemical ferric species. One can calculate the approximate necessary potential as follows:

$$\Delta G \text{ (kcal/mole)}$$

$$2R\cdot + 2H\cdot \rightarrow 2RH \qquad \approx -180$$

$$H_2 \rightarrow 2H\cdot \qquad +104$$

$$\underline{2H^+ + 2e^- \rightarrow H_2} \qquad \underline{0}$$

$$2R\cdot + 2H^+ + 2e^- \rightarrow 2RH \qquad -76$$

$$R \cdot + H^+ + e^- \rightarrow RH \qquad\qquad -38$$

$$38 \text{ kcal/mole} \times \left(+4.3 \times 10^{-2}\, \frac{\text{eV/molecule)}}{\text{kcal/mole}}\right) \quad = +1.63 \text{ V}$$

$$Fe^{+3} + e^- \rightarrow Fe^{+2} \qquad\qquad +0.77 \text{ V}$$

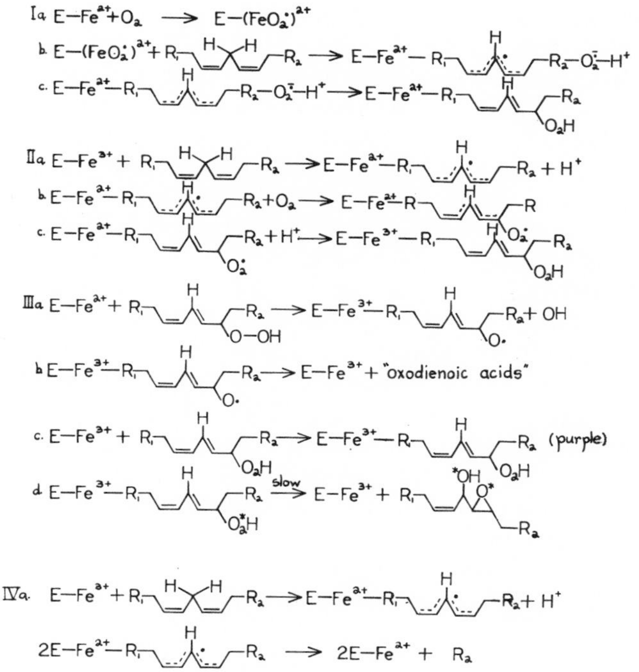

Figure 18-3. Aerobic and anaerobic reactions of ferric and ferrous lipoxygenase forms. (1) Reactions I and II are aerobic; reactions III are either aerobic or anaerobic; reactions IV are strictly anaerobic. (2) Ferric (yellow), ferrous (colorless), and ferric–hydroperoxide complex (purple) all catalyze the aerobic conversion of fatty acid to fatty acid hydroperoxide. (3) The strictly anaerobic reaction is noncatalytic (i.e., a reaction between enzyme and fatty acid) unles some other oxidizing agent, namely, hydroperoxide, is present to convert ferrous enzyme back to ferric (reactions IIIa,b). (4) Reactions III are fast enough relative to reactions I that yellow (ferric) and purple (complex) enzyme forms are always observed in the aerobic reaction. (5) Reactions IV are faster than reactions IIIa,b, so that in the anaerobic reaction of hydroperoxide with fatty acid, the enzyme remains in the ferrous form unless hydroperoxide is present in excess, in which case the enzyme is converted to ferric after the fatty acid is consumed.

Thus, the enzyme must more than double the normal oxidation potential for the ferric-to-ferrous reaction to create a thermodynamic driving force for this reaction. Enzymatic stabilization of the ferrous ion and destabilization of the ferric is contrary to most biological examples. Indeed, if this is the way the enzyme functions (great support for which could be gained by measuring the oxidation potential of the enzyme), then ascertaining the enzymatic structure that allows this remarkable oxidation potential would be very enlightening.

Ions that do have sufficient oxidation potential to oxidize olefinic hydrocarbons to cation radicals are Co^{3+} and Mn^{3+}, and the reaction with Co^{3+} has been studied by Bawn and Sharp (1957). It was found that the reaction required not only the highly reactive metal oxidation state, but also a somewhat unstable sulfate coordinate form of the ion. The rate of reaction with the more stable acetate complex was independent of olefin concentration, indicating that the rate-limiting step was breakdown of the complex. Furthermore, some evidence suggested that dimers of the reactive metal coordination compounds might be involved. The activation energy for the reaction (presumably the radical cation represents the transition state) was 27–29 kcal/mole. These considerations make this mechanism seem quite unlikely for the biochemical reaction. Nonetheless, reduction of the ferric enzyme by linoleic acid has been directly observed (Egmond et al., 1977) under anaerobic conditions (reaction IVa,b in Figure 18-3).

Also, the detailed chemical process by which olefin oxidation occurs is puzzling. The most likely suggestion is, as suggested by Gibian and Gallaway, oxidation followed by proton removal (Figure 18-4). However, Bawn and Sharp (1957) found dienes to be a minor component of the product mixture in their study. Their presence was explained by the sequence in Figure 18-5.

If proton removal were rate-limiting, the diene surely would not be produced. In view of the presence of the solvent nucleophile (H_2O), proton elimination must be very rapid to compete at all with solvent quenching of the cations.

Chemically, the aerobic reaction of ferrous enzyme is much more justifiable. Formation of a complex ferrous ion and dioxygen is well known, and hydrogen

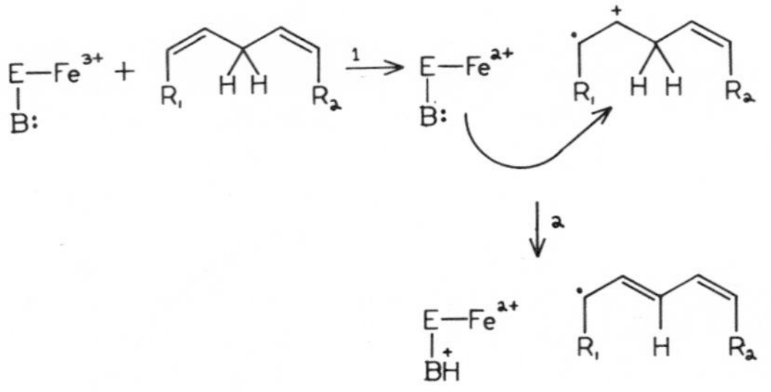

Figure 18-4. Mechanism of olefin oxidation by lipoxygenase suggested by Gibian and Galloway.

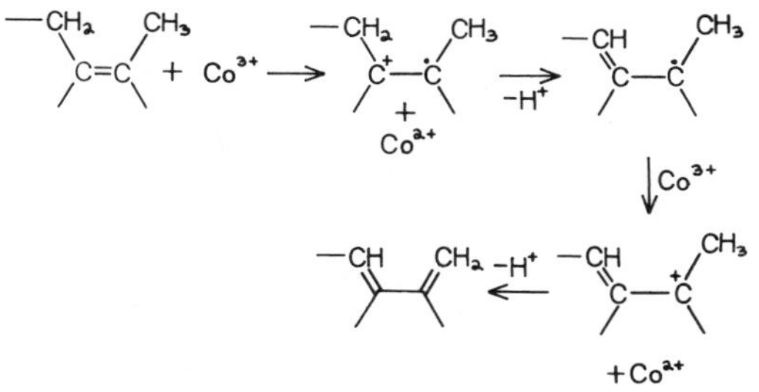

Figure 18-5. An explanation of how dienes are formed in olefin oxidations.

abstraction by this complex (Figure 18-6) is, at least, not too surprising. Furthermore, this abstraction should show a large kinetic isotope effect.

Unfortunately, the stereochemistry is very difficult to rationalize. As pointed out by Egmond et al. (1972), the hydrogen is removed trans to the resulting peroxide. A stereochemical representation of the reaction is shown in Figure 18-7. Clearly, this represents only one of a multitude of possible geometries, but the obvious tendency would be for the oxygen to attack cis to the abstracted atom. Even imagining the two oxygen atoms bridged across the olefin, trans reaction seems to be impossible. Rotation around one of the single bonds between the olefin and the bisallylic carbon results in a geometry in which the H removed and the resulting hydroperoxide are cis as necessary for this reaction mechanism. However, this configuration results in a cis–cis orientation of the double bonds, unlike that found in the actual product.

From this analysis, it would appear that dioxygen cannot be responsible for removal of hydrogen from the bisallylic position unless two molecules are involved in each reactive cycle, yet no other hydrogen acceptor has been proposed (Egmond et al., 1972). Furthermore, studies of a ferrous enzyme–nitric oxide complex (Vliegenthart et al., 1979) concluded that the active aerobic ferrous enzyme forms an initial iron-dioxygen complex.

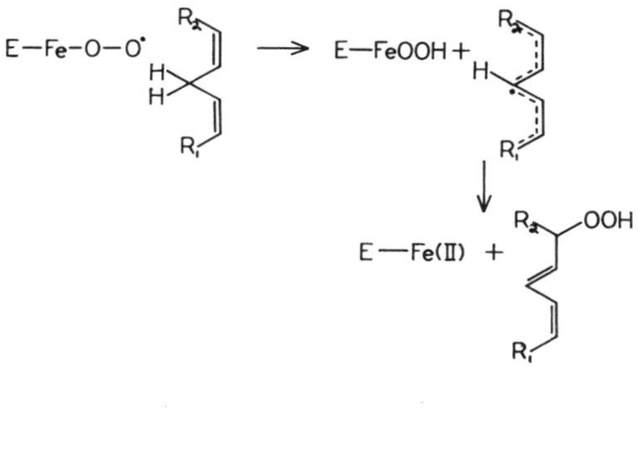

Figure 18-6. A plausible mechanism for lipoxygenase oxidation of an unconjugated diene.

Figure 18-7. Stereochemistry of the lipoxygenase reaction.

Hence, in our view, the lipoxygenase reaction has not been adequately explained. Part of the reason for the mystery is that a physiological role for the catalysis has not been ascertained. We mention the singlet-dioxygen question not because of new evidence in favor of singlet dioxygen, but because oversimplified arguments against it should not be carried to other biochemical reactions and because free-radical mechanisms appear, still, to fall short of entirely explaining the reaction.

18.3 Other Lipoxygenases

A fungal lipoxygenase has been reported that contains heme as an essential cofactor (for a review, see Nozaki, 1979). It shows a different range of specificity than the soybean enzyme and is inhibited by cyanide. The fungal enzyme is stabilized by Co^{2+}. Also, its pH optimum is much higher than even the crystalline soybean lipoxygenase, at pH 12.

Another lipoxygenase has been detected in human platelets (Hamberg and Samuelsson, 1974). This enzyme competes with prostaglandin cyclooxygenase for labeled arachidonic acid. Of the products of these two reactions, 43% was allylic hydroperoxide (Figure 18-8).

Optically pure 10L-^{3}H,3-^{14}C-arachidonic acid was tested as a substrate. The product retained only 14% of the ^{3}H label at C-10 (Hamberg and Hamberg, 1980), indicating stereospecific hydrogen removal, as in soybean lipoxygenase and prostaglandin cyclooxygenase. Since the configuration at C-12 of the product is known to be L, it can be concluded that the two mechanistic events (H abstraction and dioxygen addition) have an antarafacial relationship, as observed in soybean lipoxygenase and prostaglandin cyclooxygenase. Also, measurement of ^{3}H label in unreacted fatty acid as a function of percentage conversion showed

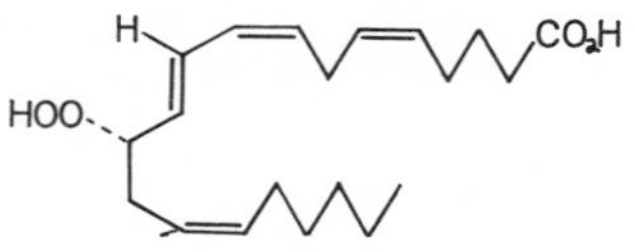

Figure 18-8. One of the products of lipoxygenase oxidation of arachidonic acid.

that there is a rate effect reasonably close to those measured for soybean lipoxygenase and prostaglandin cyclooxygenase.

The authors present evidence that products of the prostaglandin synthetase pathway (endoproxides derived from the same substrate) exert a potent activating effect on the lipoxygenase reaction. This effect was tested *in vivo*. Platelets removed from patients after oral administration of aspirin showed a lag phase in conversion of labeled fatty acid to product that was absent in platelets taken from patients before aspirin administration. Inhibition of cyclooxygenase activity was essentially complete.

18.4   Prostaglandin Synthetase

Prostaglandin biosynthesis provides another example of a reaction that is very similar to a singlet-dioxygen reaction, but proceeds by a mechanism (Samuelsson, 1972) unlike the familiar pathway. All three noncarboxyl oxygen atoms in the product derive from dioxygen. The stoichiometry of the reaction is:

$$C_{20}H_{34}O_2 + 2O_2 + 2GSH \rightarrow C_{20}H_{34}O_5 + H_2O + GSSG$$

One intermediate in the conversion of 8,11,14-all-*cis*-eicosatrienoic acid to prostaglandins is the endoperoxide (Figure 18-9). This product could very easily result from the well known Diels–Alder-type reaction of singlet dioxygen with dienes (Figure 18-10) (see Chapter 3). However, it was shown by a few elegant experiments (Hamberg and Samuelsson, 1967b) that the actual pathway is very similar to the lipoxygenase reaction in that an allylic hydroperoxide is formed before the endoperoxide. As in the lipoxygenase reaction (Hamberg and Samuelsson, 1967b; Hamberg and Hamberg, 1980), hydrogen is removed stereospecifically during the reaction, and substitution of ^3H for Hu in the C-10 position leads to enrichment of the radioactive label in recovered reactant, indicating that abstraction occurs before or during the rate-limiting step. If the endoperoxide were to form before ^3H removal, enrichment would appear in some oxygenated species; hence, hydroperoxide (B in Figure 18-11) is apparently formed before the endoperoxide.

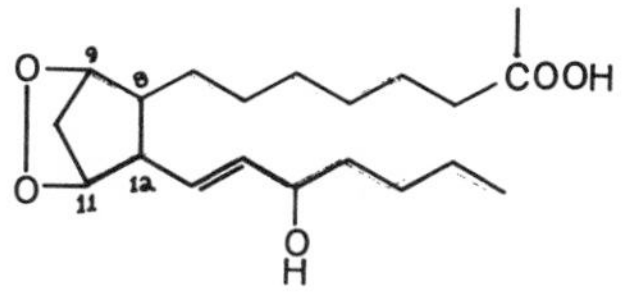

Figure 18-9. An intermediate endoperoxide in the biosynthesis of prostaglandins.

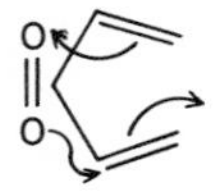

Figure 18-10. Diels–Alder mechanism for endoperoxide formation.

Nugteren *et al.* (1967) prepared the three isomeric all *cis* eicosadienoic acid derivatives of 8,11,14-all-*cis*-eicosatrienoic acid. Only one of these isomers reacted appreciably with the prostaglandin synthetase system. *Cis,cis*-11,14-eicosadienoic acid gave a good yield of 11-hydroxy-12-*trans*,-14-*cis*-eicosadienoic acid, again indicating a lipoxygenaselike reaction and confirming that dioxygen attach at C-11 definitely occurs before attack at C-9.

It is intriguing that the 8,11-eicosadienoic acid showed very little reactivity with these enzymes. Like all the known lipoxygenase-type reactions, H removal apparently requires two adjacent double bonds, one not being sufficient. This fact may well be of mechanistic importance. The possibility exists that more is involved than the difference in difficulty of H abstractions or the difference in stability of a bisallylic radical over an allylic radical. The endoperoxide is then formed in a cyclization reaction (Figure 18-12), which also results in the incorporation of a second molecule of dioxygen.

Free radicals have been implicated in this reaction by liquid nitrogen ESR spectra, by inhibition by the antioxidants "alpha"-tocopherol and propyl gallate, and by sulfite oxidation (Samuelsson *et al.*, 1967). The ESR signal was observed only on mixture of the enzyme and reactants (neither separately was ESR-active) and was observed at $g = 2$. These pieces of evidence do not indicate whether a radical actually participates in the dioxygen addition, the cyclization, or both.

The stereochemistry of the autoxidation reaction by which triolefinic hydroperoxy fatty acids are converted to prostaglandinlike bicyclic endoperoxides has been studied (O'Connor *et al.*, 1981). Remarkably, the uncatalyzed reaction gives structurally similar products but shows a strong preference for a *cis* relationship between the alkyl ring substituents, rather than a *trans* relationship. This points to a strong mechanistic effect that the enzyme must exert on this reaction.

As expected by analogy with the lipoxygenase reaction, the cyclooxygenase system, which was first purified in 1975, contains nonheme iron. In addition, the active enzyme was found (Miyamoto *et al.*, 1976; Hemler *et al.*, 1976) to

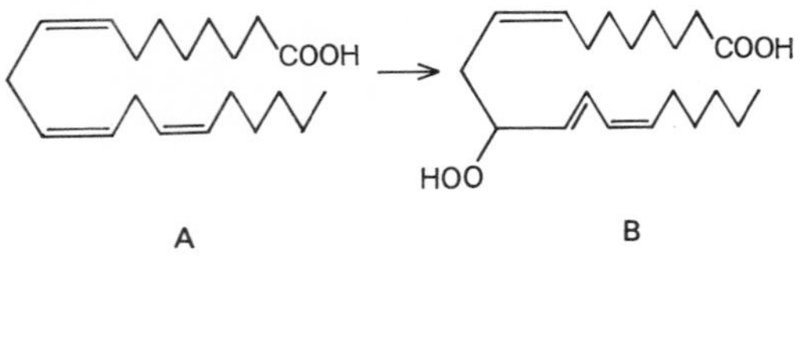

Figure 18-11. Formation of a hydroperoxide (B) in a step prior to endoperoxide formation in prostaglandin biosynthesis.

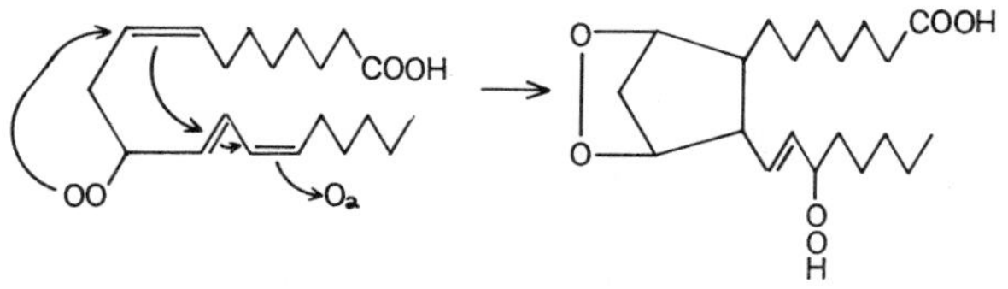

Figure 18-12. Formation of endoperoxide from hydroperoxide in prostaglandin biosynthesis.

be associated with iron-containing hemin. Later work (Ohki *et al.*, 1979) showed that heme was required for both cyclooxygenase and peroxidase activity. The separate catalyses have proved inseparable despite all efforts at purification, leading to the possibility that the peroxidase activity is artifactual (a common property of heme-containing enzymes) and that *in vivo* the reduction occurs via another enzyme, such as glutathione peroxidase. The prostaglandin hydroperoxidase activity, however, does not involve glutathione, but can accept hydrogen from hydroquinones such as epinephrine and guaiacol. The reduction is also stimulated, mysteriously, by quinones and indoles, but the stimulators are not altered during the reaction.

Early studies (Rahimtula and O'Brien, 1976; Marnett *et al.*, 1974, 1975) with trapping agents suggested that singlet dioxygen is produced either as a by-product (Marnett *et al.*, 1975) of the peroxidase activity or as an oxygenating intermediate in the cyclooxygenase activity of prostaglandin synthetase. The studies were inconclusive due to several problems with purity and methodology. Later work (Marnett *et al.*, 1979) showed that singlet dioxygen is not involved in the peroxidase activity, but that the observed cooxygenation of diphenyl isobenzofuran is a general and interesting phenomenon. The chief substrate requirement for cooxygenation is the C-15 hydroperoxy substituent of the fatty acid derivative. However, oxygen atoms from the hydroperoxide function were not incorporated into the product dibenzoylbenzene, and many equivalents of product were produced relative to hydroperoxide used. One of the oxygen atoms found in the cooxygenation product derives from atmospheric dioxygen, the other from the reacting diphenylisobenzofuran (DPBF) molecule. Two free-radical chain mechanisms were suggested to account for the observations based on known DPBF chemistry. The initiation of either free-radical chain can be ascribed to an interaction of the peroxidase with the substrate (or nonsubstrate) fatty acid hydroperoxide. This information, combined with the low-temperature ESR detection of a free radical on addition of prostaglandin G or arachidonic acid to prostaglandin synthetase, indicates that the peroxidase center contains an entity capable of one-electron transfer to or from an aliphatic peroxide, as is the case with other peroxidases.

Isolation of the endoperoxide (Nugteren and Hazelhof, 1973) led to the

finding that it is not merely an intermediate in the synthesis of the stable prostaglandins, but that it has potent biological activity (Nicolaou and Gasic, 1978)
and serves a physiological function. Furthermore, it is transformed into a greater
number of derivatives than originally realized (see Chapter 5).

18.5 Pyrrolases

Activated aromatic rings are not cleaved by most organic oxidizing agents,
but the reduced forms of dioxygen cause this cleavage, as do the most vigorous
oxidizing reagents. Orthoquinones can be oxidatively cleaved by alkaline hydrogen peroxide (Grinstead, 1964) or by superoxide. The pyridine complex of
methoxy cupric chloride will oxidize o-benzoquinone or catechol to methyl
muconic acid in the absence of dioxygen (Figure 18-13) (Rogic and Demmin,
1978). In this reaction, the copper acts as an oxidant and not merely as a dioxygen
carrier. The pyridine complex of cupric chloride has no effect on carbon–carbon
bond cleavage of o-quinones. One might speculate that the methoxy cupric
chloride is a source of methoxy radicals for o-quinone oxidation. Molecular
oxidation will oxidize catechols to orthoquinone with or without metal-ion catalysis. The dioxygen is apparently reduced to water during the reaction, indicating that hydrogen peroxide can be involved without significant ring cleavage
(Grinstead, 1964). However, the best metal catalysts for the oxidation are Mn(II)
and Co(II), the latter of which has been shown to be an inhibitor of the enzyme
protocatechuate 3,4-dioxygenase.

Orthoquinones have been dismissed as intermediates in the biological reaction because of the lack of color change during catalysis and because carbonyl
trapping reagents such as aniline have been unreactive with any intermediate in
the reaction (Mason, 1957). Furthermore, hydrogen peroxide has been essentially
ruled out as an intermediate oxidizing agent. Several distinctions can be drawn
among the enzymes that catalyze the oxidative cleavage of aromatic rings. It is

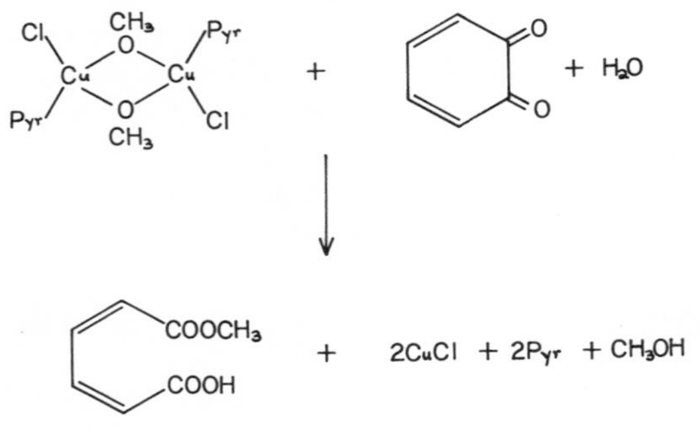

Figure 18-13. A model carbon–carbon cleavage by a copper catalyst.

not yet clear whether the various types utilize a similar mechanism or not, but some clear differences have been firmly established. All the substrates for this reaction are electron-rich aromatic systems and, except for those involving heme, contain one phenolic functionality and one other electron-donating functional group that can exchange protons (either $-OH$ or $-NH_2$).

18.6   Tryptophan Dioxygenase

The heme-containing dioxygenases include L-tryptophan oxygenase, which oxidatively cleaves the indole portion of L-tryptophan to form N-formyl kynurenine (Figure 18-14). Both atoms of oxygen from the dioxygen are found in the N-formyl kynurenine (Hayaishi *et al.*, 1957a), so that the enzyme is a dioxygenase.

The enzyme contains two heme groups (Schutz and Feigelson, 1972) and four peptide chains (Poillon *et al.*, 1969) held together by noncovalent forces. There has been controversy over whether copper is an essential part of the enzyme. Claims for an essential copper (Maeno and Feigelson, 1965; 1968; Brady *et al.*, 1972; Feigelson and Brady, 1974) have been negated by obtaining an active enzyme containing much less than one equivalent of copper per heme (Makino and Ishimura, 1976).

18.7   Indoleamine 2,3-Dioxygenase

The other known heme-containing dioxygenase is indoleamine 2,3-dioxygenase. This enzyme acts on either D- or L-tryptophan as well as a number of related derivatives: serotonin, melatonin, 5-hydroxytryptophan, and tryptamine. Its mechanism is, as yet, only poorly understood. Only the ferrous form of the enzyme is catalytically active with dioxygen, but the ferric form reacts identically with superoxide ion (Hayaishi *et al.*, 1977; Ohnishi *et al.*, 1977). In fact, an oxidative side reaction leading to the ferric form (and hence to inactivation unless a superoxide source is present) can be reversed by reduction of the inactive form with superoxide, which then also acts as substrate (Taniguchi *et al.*, 1979). With

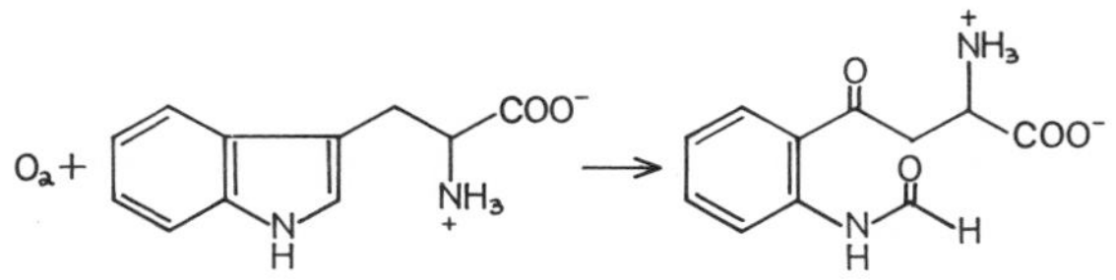

Figure 18-14. Reaction catalyzed by L-tryptophan oxygenase.

the best substrate, D- or L-tryptophan, the enzyme oxidation is two orders of magnitude slower than catalysis, so that any adequate reducing agents could maintain the aerobic reaction. But with poorer substrates, the inactivation and catalysis reactions are of about the same rate, so that superoxide would be necessary and presumably limiting to the catalysis. In these cases, superoxide, rather than dioxygen, can be considered the substrate.

18.8   Intradiol and Extradiol Pyrrolases

The nonheme iron aromatic ring cleavage enzymes are of two apparently distinct types. Enzymes that cleave between two phenolic groups (i.e., "intradiol" pyrrolases) to form muconic acid or its derivatives have generally been found to utilize Fe(III) throughout the catalytic cycle. Other enzymes cleave adjacent to one or the other phenolic group (i.e., "extradiol" pyrrolases) to form hydroxy muconic semialdehyde derivatives using Fe(II). [The early literature generally assumed that the active intradiol pyrrolases contained Fe(II), since dioxygen–ferrous complexes are well known and dioxygen–ferric are not. The ferric in the active form of these enzymes is one of the curious and revealing anomalies of this reaction.] However, at least one example has been found of an intradiolpyrrolase cleaving in an extradiol fashion with a substitute substrate (Fujiwara *et al.*, 1975).

18.8.1   Intradiol Pyrrolases

Of the intradiol pyrrolases, the best studied are pyrocatechase, which converts catechol to muconic acid, and protocatechuate 3,4-dioxygenase, which cleaves 3,4-dihydroxybenzoic acid to form 2-carboxymuconic acid. Both oxygens of dioxygen appear in the muconic acid (Hayaishi *et al.*, 1957b). Protocatechuate 3,4-dioxygenase will also cleave pyrogallol, but the product lactonizes to give 2-pyrone-6-carboxylic acid (Saeki *et al.*, 1980). Macroscopically, these enzymes are quite different, but the behavior on addition of substrate or dioxygen or both, in terms of ESR, color changes, and kinetics, is quite similar (Nozaki, 1979). Protocatechuate 3,4-dioxygenase has proved more useful for mechanistic studies than pyrocatechase because (1) the two hydroxyl groups are distinguishable in the substrate and (2) the carboxyl group, which is involved in binding but not catalysis (Que *et al.*, 1977), can be modified to generate competitive inhibitors that are more easily studied than the actual substrate. 4-Nitrocatechol (Tyson, 1975) and protocatechualdehyde have been used extensively in this manner. Protocatechualdehyde has tentively been shown to react with an amino group in an active site of protocatechuate 3,4-dioxygenase by NaBT$_4$ reduction of the enzyme–substrate complex (Que *et al., 1977*).

The hydroxyl groups of protocatechuic acid, on the other hand, are catalytically active. They are also distinguishable by ESR spectra of flourinated substrate analogues and by inhibition constants (Que *et al., 1977*; May *et al., 1978*). Evidence indicates that substituents at the 4-position (benzoate numbering system) affect the ligand field of the ferric ion by displacing a water ligand (Que *et al., 1977*) from the site, but that the oxidation state of the iron is unaffected. The resulting spectrum shows a different absorption if the displacing group is itself a ligand (i.e., phenoxide anion—substrate) than if it is not a ligand (i.e., flouride—potent inhibitor). Substrate binding (Peisach *et al., 1972*) also changes the rhombic field of the native enzyme into a tetragonal field. Early work concluded that the rhombic field of the native enzyme consisted of sulfur atoms arranged in a manner similar to rubredoxin. More recently, Mossbauer and Raman spectroscopy have shown strong evidence for tyrosine being at least one of the ligands in the native enzyme (Tatsumo *et al., 1978*; Keyes *et al., 1978*; Que and Heistand, 1979) that is not displaced by substrate binding (Felton *et al., 1978*). Though Raman spectra could not detect any iron–sulfur ligation, they did not exclude the possibility, but EPR and Mossbauer data make it appear unlikely. Binding substrate did change these spectra in a manner compatible with a change in the ligand field from rhombic to tetragonal.

A "ternary complex" taken to be a reactive intermediate containing substrate, dioxygen, and enzyme was detected spectrally in 1972 (Fujisawa *et al., 1972*). A similar spectrum was measured for substrate analogues, which were then further studied because of their longer lifetimes. The Mossbauer and ESR spectra, in particular, were studied for the ternary complex (Que *et al., 1976, 1977*). It was shown that although the ligand geometry is similar in the enzyme–substrate complex and the "ternary complex," the field (as reflected in the zero field-splitting parameters) changes markedly. It was also shown that neither the oxidation state nor the high-spin state changes on binding of dioxygen and that the iron–substrate interaction is highly ionic.

A chemical, spectral, and chromatographic study of the enzyme–substrate–dioxygen "ternary complex" revealed that by this stage the reaction with dioxygen has already occurred. Hence, the fate of the dioxygen molecule on binding was not ascertained by the studies of the "ternary complex." Given the conditions under which the "ternary complex" was first detected and those under which end product was isolated from it, the assimilation–cleavage reaction must occur very quickly. [It should be mentioned that the electron paramagnetic resonance spectrum for the enzyme–product complex is reported to be different than for the "ternary complex" (Que *et al., 1976*).]

Hamilton (1974) proposed a mechanism for the reaction before many of

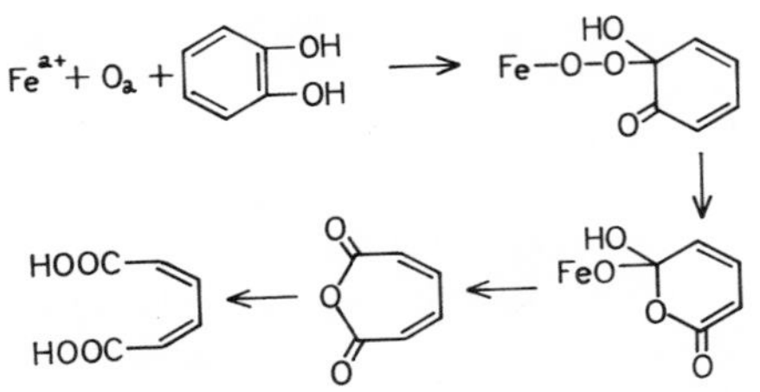

Figure 18-15. A proposed mechanism for the intradiol pyrrolase reaction.

the inhibition, ESR, and Mossbauer studies had been published, on the basis of the thermodynamic calculations. A major consideration in this proposal was that dioxetane ring opening is an unlikely pathway due to the endothermicity of formation and especially the exothermicity of ring cleavage. Instead, it suggested that dioxygen forms a complex with iron that then reacts with the substrate (Figure 18-15). Several problems with this mechanism are readily apparent: (1) Dioxygen is not known to complex or bind with Fe(III), which is the iron species unanimously agreed on. (2) Later experiments have shown a definite substrate iron-binding that was not predicted by this mechanism. (3) The hydroperoxide rearrangement seems unreasonable. Assuming that the iron–oxygen complex is formed and behaves analogously (Figure 18-16) to the peracid in oxidation of an α-diketone (Figure 18-17), one could easily imagine the identical anhydride product. The problem is that FeO^+ would then have to nucleophilically attack the anhydride to get the second atom of the dioxygen molecule back into the product. Such a reaction seems unlikely; at the very least, it would be a kinetically slow (hence detectable) step.

A similar mechanism (Figure 18-18) was proposed by Que et al. (1977). In it, the iron is specifically bound to the 4-hydroxyl and the carboxyl group forms a salt with a lysine residue. Noting that the substrate is a highly electron-rich system and citing evidence that hydroquinone anions react with oxygen to form semiquinones and superoxide, they propose that these two radicals form and then combine to give a hydroperoxide. Inhibition of intradiol pyrrolase activity by a cupric salt has also been taken as evidence that a discrete superoxide intermediate is formed during the reaction (Mayer et al., 1979). The iron atom simultaneously migrates from the hydroxyl position to the peroxy anion to form

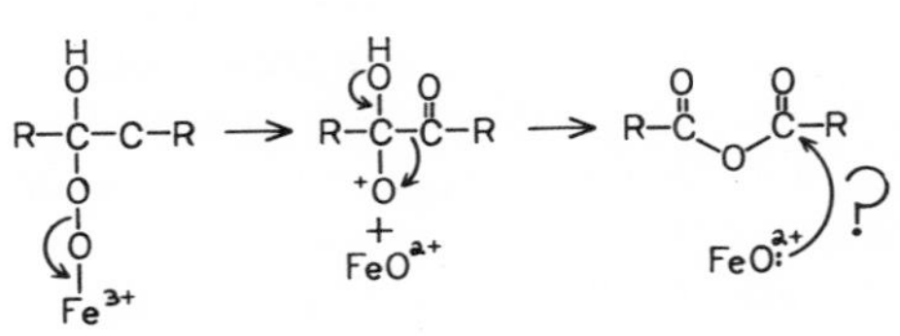

Figure 18-16. Hypothetical reaction for α-diketone oxidation by an iron–dioxygen complex analogous to the peracid oxidation of a diketone.

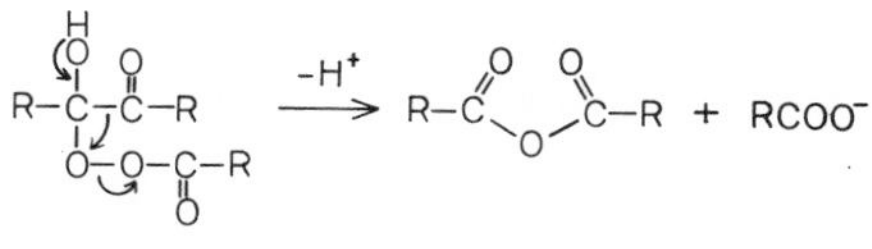

Figure 18-17. Peracid oxidation of an α-di-ketone.

$$R-\overset{H}{\underset{O}{\overset{O}{C}}}-\overset{O}{\underset{O-C-R}{C}}-R \xrightarrow{-H^+} R-\overset{O}{C}\underset{O}{\overset{O}{C}}-R + RCOO^-$$

what, according to this mechanism, should be the "ternary complex." The final step of the proposal is rearrangement to the anhydride as suggested by Hamilton.

In support of this mechanism, Lauffer *et al.* (1981) have shown that when bound to ferric ion in aprotic solvents, 3,5-di-*t*-butyl-catechol is very easily oxidized to its semiquinone form by dioxygen. The species formed has unpaired electron density primarily on the catechol moiety, and the iron remains ferric. Interestingly, if the ferric–catechol complex is deprotonated, it does not react with dioxygen. Apparently, the anion forms a bidentate complex (6-coordinated iron) that is protected from oxidation by chelation. Deprotonation does not affect the iron oxidation state. Also, the bidentate complex may be formed by reduction of the semiquinone complex with superoxide. The ring-cleavage portion of the enzymatic reaction did not occur in this model study, which is surprising since all the supposedly essential components are present.

However, the reaction still seems somewhat mysterious. The activation–protection of the catechol ligand by ferric ion is not expected, although it is possible that the initial complex is highly ionic and that the bidentate complex is much more covalent. Also, if the suggested "ternary complex" were, in fact, the structure shown, we believe that the study of Nakata *et al.* (1978) would have detected some compounds other than product. Another possible pathway

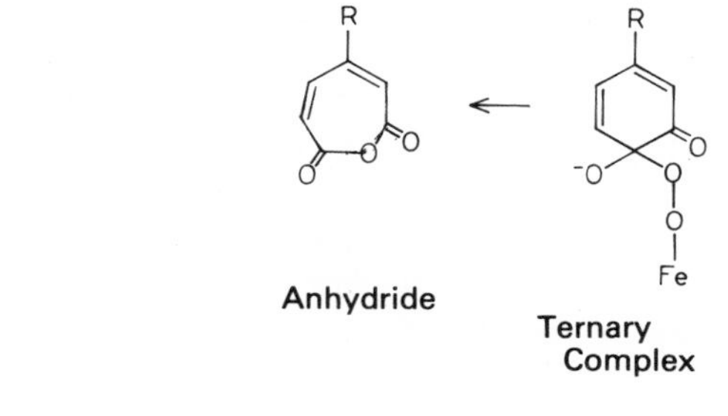

Figure 18-18. A mechanism for the intradiol-pyrrolase-catalyzed oxidation of catechols.

Anhydride

Ternary Complex

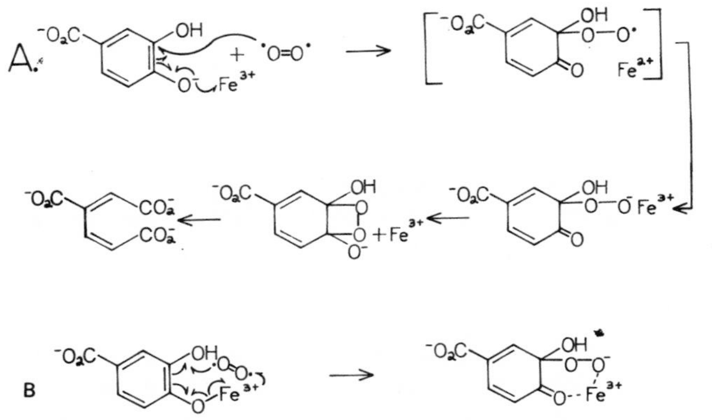

Figure 18-19. A suggested mechanism for the intradiol pyrrolase reaction. Mechanism B is a concerted form of mechanism A.

is suggested in Figure 18-19. In this case, the iron is acting as an oxidant toward the catechol and as a reductant toward the dioxygen moieties in the activated complex. Note that neither ferrous ion nor low-spin iron could perform this catalysis. The electronic changes that occur are exactly those expected of the ferric salt of a semiquinone anion in the presence of oxygen. Since the reaction is concerted (or almost concerted), one would expect to observe no detectable ferrous ion and no enzyme–dioxygen complex or ferric dioxygen bond.

Reactions subsequent to hydroperoxide formation could occur via either Hamilton's rearrangement or a dioxetane. We consider the dioxetane pathway to be the more likely of the two because dioxetane formation would be aided by iron catalysis (Fe^{3+} coordinated to carboxyl oxygen would strongly promote nucleophilic attack at the carbonyl carbon), whereas, as mentioned above, anhydride hydrolysis would be hindered by a strong positive charge. We regard the large exothermicity of the dioxetane opening to be of minor importance in view of the large number of degrees of freedom available for energy dissipation in an aquated high-molecular-weight molecule. However, the intermediacy of the anhydride is probably testable (see below).

Another possible mechanism (Figure 18-20) is a metathesislike reaction (see Chapter 3) in which the metal ion catalyzes (i.e., lowers the activation barrier) for 1,2-dioxetane formation and subsequent rearrangement. Dioxygen would not be a ligand to the metal in a classic sense, but would overlap orbitals of the metal that were already mixed with those of the substrate. Spin angular momentum would be conserved if the triplet dioxygen reacted with iron in the $\frac{3}{2}$

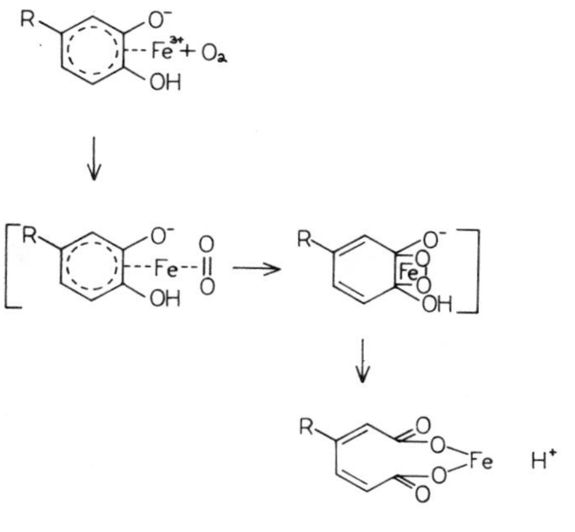

Figure 18-20. A mechanism for the intradiol pyrrolase reaction in which the formation of the dioxetane is allowed by the iron.

spin excited state to yield ground-state $\frac{5}{2}$ spin iron and covalent product. The thermodynamic argument against dioxetane formation and cleavage mentioned by Hamilton is entirely avoided in this mechanism, as is the question of the role of the metal ion in catalysis. It should be emphasized that with olefins, this type of metathesis reaction occurs quickly at room temperature and that the cyclic intermediate is, in fact, an activated complex that does not dissociate from the metal.

The ESR and inhibition studies are more difficult to rationalize by this mechanism. The inequality of the hydroxyl groups must be explained by some factor other than iron coordination, since the coordination must be with a set of π electrons on the aromatic nucleus. Also, it is not known whether the ESR and Raman spectra of the enzyme–substrate complex are compatible with the structure drawn and whether or not the necessary spin state would be populated at room temperature.

Perhaps interesting experiments would be (1) to test the enzyme–product and "ternary complex" relationship by using muconimide (Figure 18-21), since this compound must be very much like the anhydride product proposed previous to hydrolysis by both earlier mechanisms, and (2) to repeat the chemical study of the "ternary complex" (Nakata et al., 1978) by using $^{18}O_2$ during incubation.

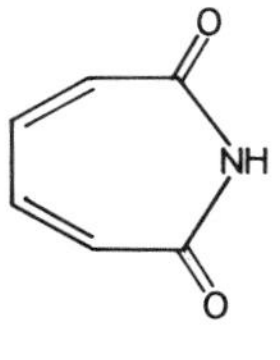

Figure 18-21. A possible substrate or transition-state inhibitor for intradiol pyrrolase.

If the anhydride is the productlike intermediate bound to the enzyme prior to product formation, work-up will cause hydrolysis away from the enzyme, so that only 1 atom of labeled oxygen will be incorporated into the isolated product.

It should be pointed out that none of the proposed mechanisms explains why some of the competitive inhibitors are not reactive, so that this is not a good criterion for evaluating proposed mechanisms. There is no obvious reason that 4-nitrocatechol in particular is not a substrate for any of the mechanisms. In view of the rapid reaction between poor substrate and dioxygen, this fact may very well call into question spectral experiments exploring the "ternary complex" of competitive inhibitors of the reaction.

18.8.2 Extradiol Pyrrolases

Unlike the intradiol pyrrolases, extradiol pyrrolases seem to perform as one would expect from known iron–dioxygen chemistry. The iron atoms are in the ferrous form. That dioxygen does not bind to the enzyme in the absence of substrate is not too surprising in that substrate binds equally well with or without the iron ion. It has been noted, furthermore, that superoxide, the presence of which is expected during catalysis due to the ferrous state of the iron, does nonenzymatically cleave catechol derivatives (Foote and Moro-oka, 1976) to lactonized products of the hydroxy muconic semialdehyde oxidation level. Therefore, it seems likely that the substrate binds in such a way as to make the ferrous

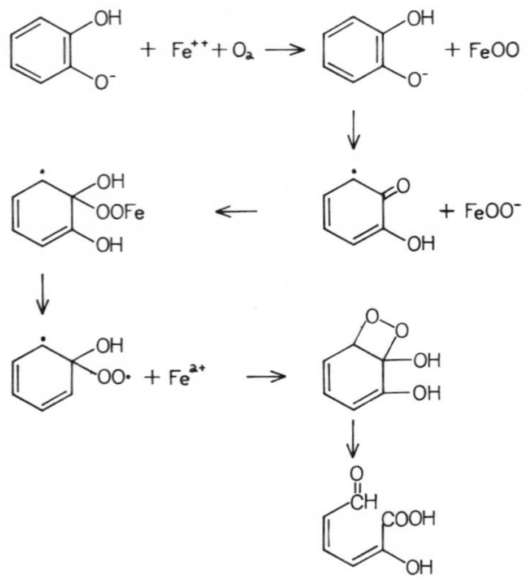

Figure 18-22. Mechanism of the extradiol pyrrolase reaction.

ion accessible to the dioxygen molecule. Superoxide ferric complex would then abstract (Figure 18-22) a hydrogen atom or an electron from the substrate to form the semiquionone. Combination of the resulting species and homolytic cleavage of the iron–oxygen bond would result in a diradical, which could combine to form a dioxetane.

The role of the metal in this mechanism is fundamentally different from its role in the intradiol cleavage reaction. In this case, the metal catalyzes the reaction by reducing the dioxygen molecule to the superoxide oxidation level. The ferric–superoxide complex also promotes the oxidation of the hydroquinone over deprotonation, which is the common reaction of uncomplexed superoxide. Homolytic cleavage of the iron–dioxygen bond is unusual, but not unimaginable, especially if the ligand field stabilizes the low-spin ferrous species.

18.9 α-Keto-Acid-Dependent Hydroxylases

Another group of dioxygenases are the α-keto-acid-dependent hydroxylases. This class of enzymes oxidizes a wide range of substrates. The usual cosubstrate is α-ketoglutaric acid, and in the well-studied case of p-hydroxyphenylpyruvate, only a single substrate is required. Two mechanisms (Figures 18-23 and 18-24) have been suggested that for the p-hydroxyphenylpyruvate reaction (Lindblad et al., 1970). These mechanisms differ only in the order in which steps occur. The chemistry of the two mechanisms is quite similar. Both require an electrophilic attack on the aromatic nucleus and a nucleophilic attack on the keto carbon. Both also involve steps without precedent in organic chemistry.

In mechanism 1, a dioxygen–iron complex is required in which the "outer" oxygen atom is electrophilic and has little unpaired electron density (similar to

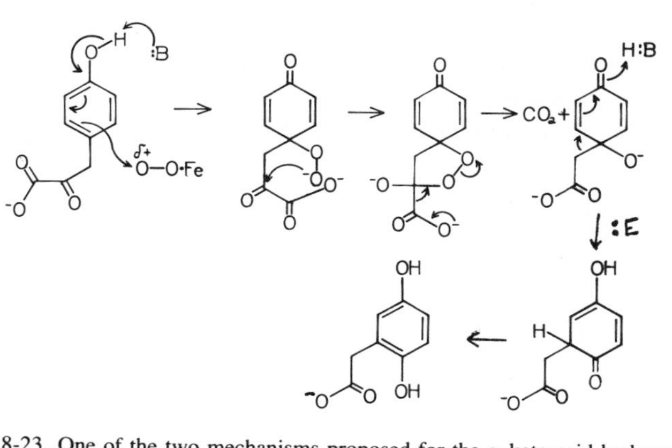

Figure 18-23. One of the two mechanisms proposed for the α-keto acid hydroxylases.

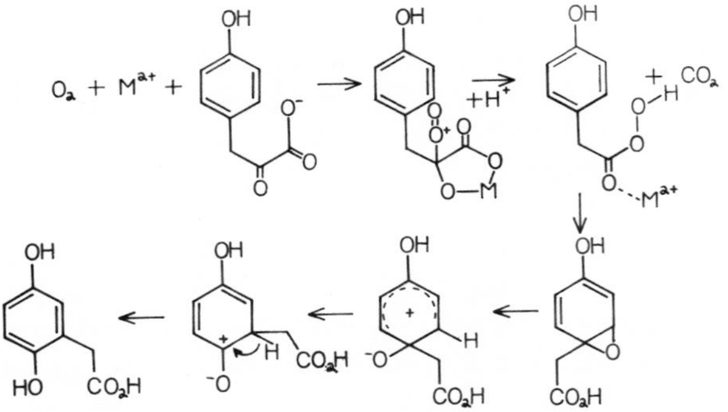

Figure 18-24. Second of the two mechanisms proposed for the α-keto acid hydroxylases.

the species required in the hypothetical singlet-dioxygen-like mechanism in the lipoxygenase reaction). Hamilton (1971) objected to this mechanism because of the unprecedented iron–dioxygen complex and because the reaction also occurs without the *p*-hydroxyl group that stabilizes the phenyl cation after the initial electrophilic attack. However, the unstabilized cation (A in Figure 18–25) required in the mechanism of Figure 18-23 could be very little more stable than the corresponding cation required in the mechanism of Figure 18-24. The mechanisms comply with, and are indistinguishable by, kinetic analysis (Rundgron, 1977). Also, it is known that superoxide causes oxidative decarboxylation of α-keto acids (see Chapter 4), so this part of the reaction is easily rationalized. However, the means of coupling decarboxylation to hydroxylation of organic substrates is much more intriguing.

The mechanism of the α-ketoglutarate-dependent hydroxylation of substrates less reactive than aromatic rings remains a mystery. These substrates include proline, lysine, γ-butyrobetaine, and thymidine, all of which are hydroxylated in positions not "activated" or involving π-bonds. Ferrous ion and ascorbate are required in all cases. Also, other factors, including guanosine or adenosine triphosphate (Takeda *et al.*, 1976; Wondrack *et al.*, 1979), catalase

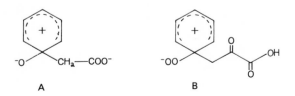

Figure 18-25. Cations required in the mechanisms of Figures 18-23 (A) and 18-24 (B).

(Takeda *et al.*, 1976; Wondrack *et al.*, 1978), albumin, and sulfhydryl reagents (Takeda *et al.*, 1976), can increase activity severalfold. These effects are interrelated and not entirely explained by maintenance of the reduced iron ion.

The requirement for α-ketoglutarate is remarkable and very specific. In view of the thermodynamic driving force for decarboxylation, an α-keto acid is a very good reductant for dioxygen. However, whether the iron–dioxygen complex reacts first with substrate or first with keto acid in these cases is not clear. A mechanism has been proposed (Figure 18-26) (Lindstedt and Lindstedt, 1970) that suggests that an anion or (nucleophilic) radical derived from the substrate reacts first with the iron–dioxygen complex to form a peroxy anion that nucleophilically attacks the carbonyl group of the cosubstrate. This mechanism shares with Witkop's mechanism for *p*-hydroxyphenylpyruvate dioxygenase a requirement for an electrophilic ferric–superoxide (ferrous–dioxygen) complex.

The major problem, though, is that formation of the initial carbanion must require a base much stronger than any normally available to enzymes. Likewise, no superoxide species is capable of removing inactivated aliphatic hydrogen atoms to form the initial radical.

Siegel (1979) has suggested a mechanism in which an iron–oxygen species acquires a high oxidation state (like that of P-450) by coupling with the exothermic decarboxylation of the α-keto acid (Figure 18-27). The resultant Fe^{4+} species could then oxidize unactivated hydrocarbons in the same manner as

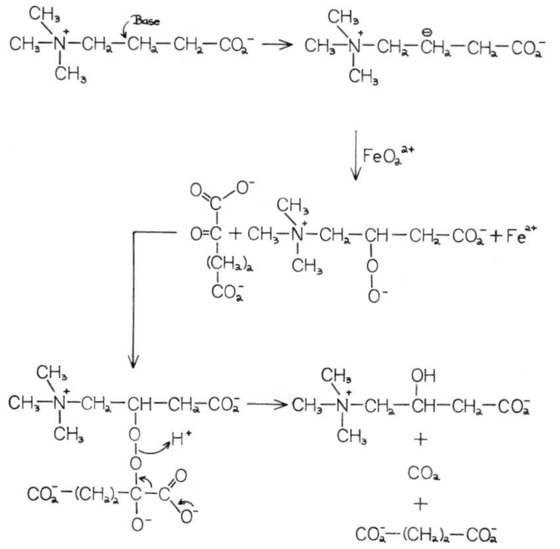

Figure 18-26. Mechanism of α-ketoglutarate-dependent hydroxylations.

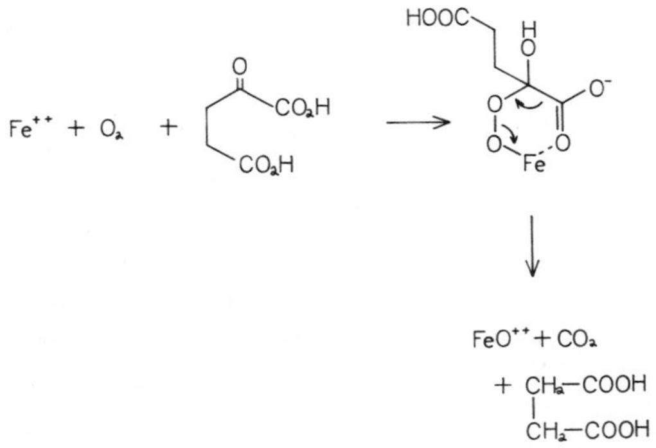

$Fe^{++} + O_2 +$ [structure] $\longrightarrow$ [structure]

$FeO^{++} + CO_2$

$+ \; \underset{CH_2-COOH}{CH_2-COOH}$

Figure 18-27. A proposed mechanism for proline hydroxylase.

cytochrome P-450. This mechanism explains the need for the oxidizing agent. It is also consistent with the observation that decarboxylation of the ketoacid can occur in the absence of the substrate, but oxidation cannot occur without the keto acid.

Ferric–superoxide complexes react as electrophiles in catalase, peroxidase, and cytochrome P-450. In those cases, the mechanism presumably proceeds through a species with a $4+$ charge: $FeO^{3+} + $ porphyrin$^+$. Since the porphyrin is not available in the α-ketoglutarate-dependent dioxygenases, the reactivity of these ferric–superoxide complexes cannot be stated with certainty.

18.10   2-Methyl-3-hydroxy-5-carboxy-pyridine dioxygenase

Investigation into the degradative pathway of pyridoxine in *Pseudemonas spp.* led to the discovery of two novel flavin dioxygenases that catalyze the cleavage of an activated aromatic ring (Sparrow *et al.*, 1969). All other enzymes known to catalyze this type of reaction require the active participation of iron (see Section 18.8).

The two reactions are closely related, but the enzymes are clearly distinguishable, exhibiting high substrate specificity (Figure 18-28). The stoichiometry and kinetics of one of the two (2-methyl-3-hydroxypyridine-5-carboxylic acid oxygenase) have been clearly established (Kishore and Snell, 1981a). Incorporation studies showed that some of the product derives two oxygen atoms from dioxygen, but some of the product derives one new oxygen in the carboxylate group from water. This could be explained either by an intermediate that could exchange oxygen with solvent (presumably before the carboxylate group is formed) or by a split reaction pathway after the initial reaction as occurred. The enzyme is a tetramer with 2 moles of FAD and two catalytic sites per tetramer that function identically and independently. Apoenzyme, prepared by acid–ammonium

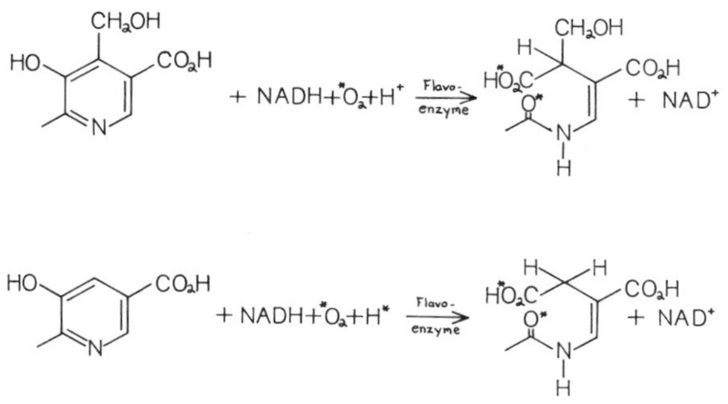

Figure 18-28. Reactions catalyzed by 2-methyl-3-hydroxy-5-carboxypyridine dioxygenase.

sulfate precipitation, is not reconstituted by FMN, FMN + adenine, or 5-deaza-FAD, but regains most of its activity with FAD. 1-Deaza-FAD complexes with the enzyme, but is catalytically inert (Kishore and Snell, 1981b). Fluorescence measurements clearly show energy transfer from substrate to flavin, indicating that the two components are fairly close (<80 Å) to each other.

The kinetic mechanism of the enzyme has been determined (Kishore and Snell, 1981a) and is shown in Figure 18-29. Also, it was shown that the 4R hydrogen is transferred to FAD during the reaction, while the 4S hydrogen is retained in NAD^+. The proton at C-4 of the product derives from solvent, not from NADH. This could be due either to direct participation of a solvent molecule in the reaction pathway or to hydride transfer from flavin after the reduced flavin has been exposed to solvent. (Exchange of the N^5 proton of reduced flavin is very rapid and leakage of electrons from reduced enzyme to flavins in solution has been observed, indicating that the flavin N^5 position may be quite exposed. However, it was also concluded that N^5 is required for flavin binding from the fact that 5-deaza-FAD neither binds to the enzyme nor inhibits FAD binding.)

This enzyme is extremely interesting because of the comparison it offers to the flavin monooxygenases. Bruice (Muto and Bruice, 1981a,b) has shown that some intermediate derived from the anion of flavin-4a-hydroperoxide can transfer two oxygens to phenolate anions to generate peroxide anions. The peroxides decompose according to their particular structure, giving ring-opened

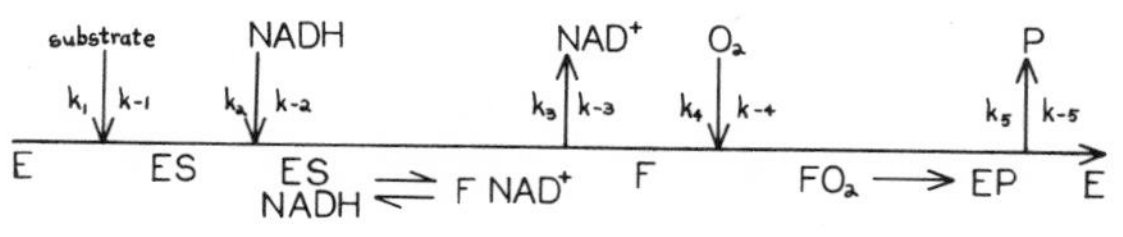

Figure 18-29. Kinetic mechanism of 2-methyl-3-hydroxy-5-carboxypyridine dioxygenase.

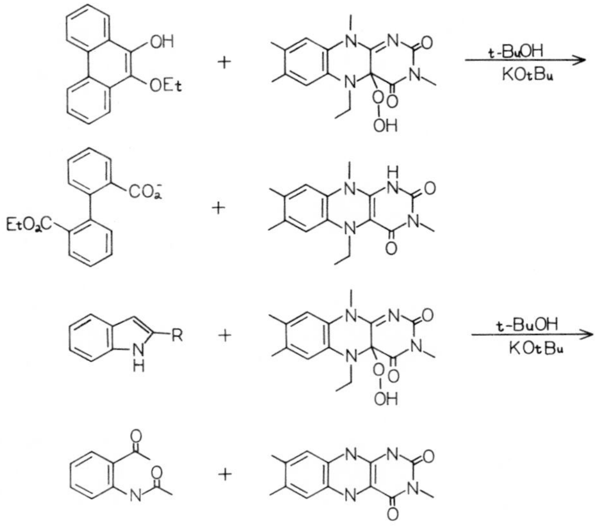

Figure 18-30. A model reaction for a flavin dioxygenase.

products in some cases, including 10-ethoxy-9-phenanthrol anion and indole derivatives (Figure 18-30). Kinetics have shown that an intermediate is involved before oxygen transfer, but its identity is not clear. Some possibilities have been eliminated, and among those remaining are the flavin-4a,10a-dioxetane (Figure 18-31) and an oxygen-flavin complex of some sort.

The flavin-derived product is reduced flavin, which could presumably react with those substrates that give quinones, to form oxidized flavin and hydroquinone (Figure 18-32). This, then, represents a possible mechanism for the flavin monooxygenase reaction, as well as for the dioxygenase reaction. Hence, it would be very helpful to have a clear understanding of the enzymatic dioxygenase reaction. Projection of the mechanism to the monooxygenase substrates could strongly support or reject the peroxide transfer mechanism for those enzymes.

From the enzyme kinetics, it can be concluded that 4a-hydroxperoxyflavin

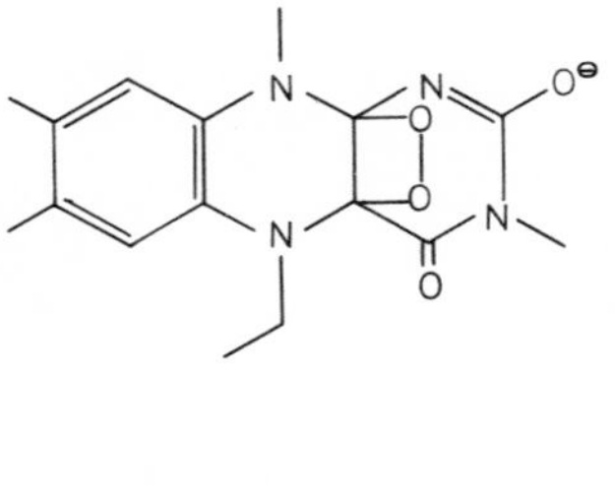

Figure 18-31. A possible intermediate in the flavin dioxygenase reaction.

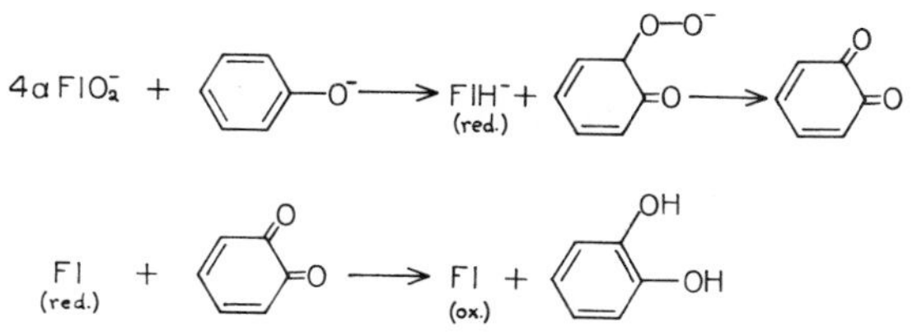

Figure 18-32. A possible mechanism of flavin monooxygenase reaction related to the dioxygenase.

is the initial "active oxygen" species. Whether it transfers oxygen or rearranges to a more reactive form is not known. The identity of the initial product after oxygen transfer is also of interest. The fact that only a fraction of the new carboxylate oxygen derives from molecular oxygen, and that some of it comes from solvent, limits the number of possible mechanisms. The authors (Kishore and Snell, 1981c) suggest that after oxygen transfer, two routes of decomposition are possible, one requiring hydration, the other a dioxetane, the relative importance of which determines the incorporation from solvent or dioxygen (Figure 18-33). Another possibility is that some intermediate, such as in Figure 18-33, has a limited opportunity to exchange with solvent before the carboxylate is formed. Unfortunately, the incorporation measured from $H_2^{18}O$ and $^{18}O_2$ experiments amounts to only about 1.5 oxygen atoms/molecule, so that the final resolution of this question is not available.

The most important feature of the reaction for comparison with monooxygenase enzymes is the nature of the initial reaction between substrate and oxygenated flavin intermediate. The authors suggest, as a precedent, the alkaline reaction between a pyridinium derivative and hydrogen peroxide (Figure 18-34).

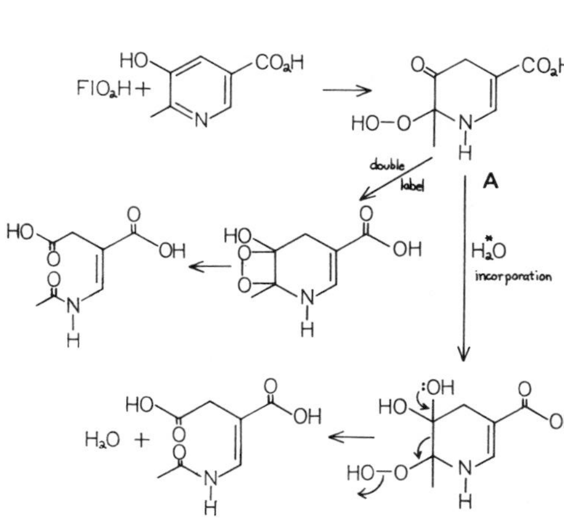

Figure 18-33. Possible mechanisms of the 2-methyl-3-carboxypyridine hydroxylase reaction.

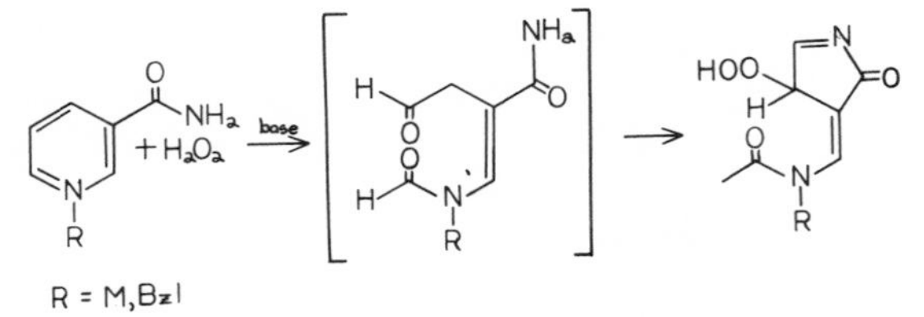

R = M,Bzl

Figure 18-34. A chemical reaction analogous to the 2-methyl-3-hydroxy-5-carboxypyridine hydroxylase reaction.

Such a mechanism could easily account for all observations of this reaction (Figure 18-35). If this mechanism were known to be true, one could conclude that the monooxygenases must follow a different pathway, since the nitrogen would be expected to exert little effect on breakdown of A (Figure 18-36); hence, the monooxygenases would be dioxygenases. However, another possible mechanism has been suggested (Muto and Bruice, 1981b) that could also explain the reaction (Figure 18-37). But in this mechanism, the hetero-ring atom would be expected to affect the outcome of the reaction, so that a clear distinction from the monooxygenases is not possible.

The enzyme shares many similarities with known monooxygenases (Kishore and Snell, 1981c). In fact, the major difference between them is in the behavior of the 1- and 5-deazaflavins with the enzyme. In monooxygenases, 1-deaza-FAD binds to the enzyme and functions as an NADH oxidase; i.e., it reacts with dioxygen and then eliminates hydrogen peroxide before being re-reduced by NADH or NADPH. In the dioxygenase, binding occurs, but oxidase activity does not. With 5-deaza-FAD, the monooxygenases generally form an inert complex. However, the dioxygenase does not bind 5-deaza-FAD, and binding of

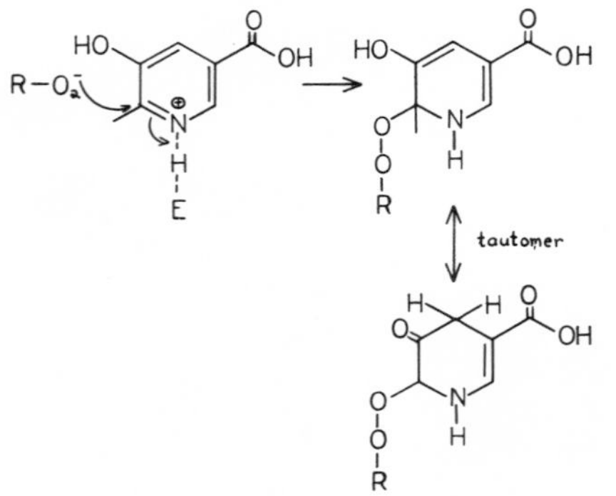

Figure 18-35. Mechanism of the attack of a peroxide on a 2-methyl-3-hydroxy-5-carboxypyridine. (R) H or flavin. The oxidized flavin could be eliminated from the intermediate at any time after attack of peroxide on substrates.

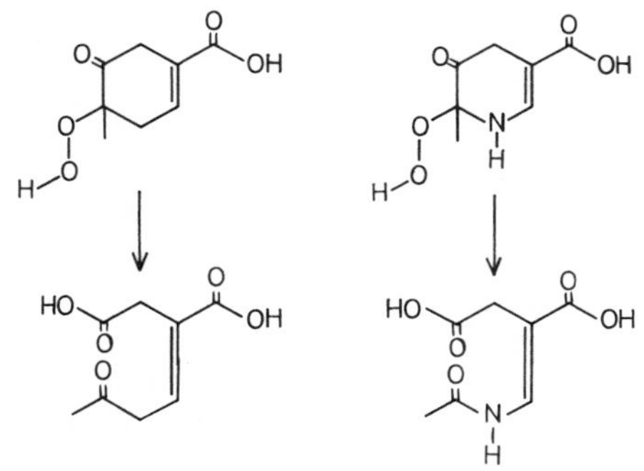

Figure 18-36. Breakdown of the hydroperoxide intermediates of a flavin dioxygenase with or without a nitrogen in the ring.

FAD is not inhibited by 5-deaza-FAD. This has been taken as an indication that N^5 plays an important binding role in the dioxygenase, but not in the monoxygenases. It is interesting to speculate, conversely, that in the monooxygenases, the N^5 nitrogen is free to play a role in the catalytic mechanism, while in the dioxygenase it is not.

The other known flavin dioxygenase is 2-nitropropane dioxygenase. This enzyme contains ferric ion in addition to FAD in mole-equivalent ratios of 1 : 1 : 1 enzyme. Incubation of enzyme and substrate in the absence of dioxygen leads to reduction of the flavin, but not of the ferric ion. The products of the reaction are acetone and nitrite (Figure 18-38), the two oxygen atoms being incorporated into two acetone molecules.

18.11   Quercetinase

Quercetinase is a copper-containing enzyme. It utilizes dioxygen to cleave quercetin to carbon monoxide and 2-protocatechuoyl phloroglucinol (Figure 18-

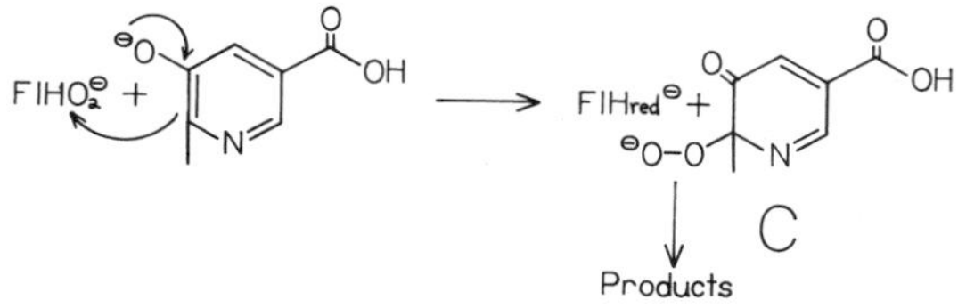

Figure 18-37. Another possible mechanism for the flavin dioxygenase reaction. Compound B is the unidentified intermediate formed from flavin 4a-hydroperoxide anion. Some intermediate after C would be reduced by FlHred⁻.

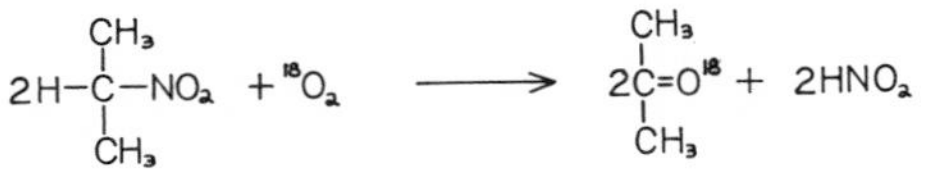

$$2H-\underset{\underset{CH_3}{|}}{\overset{\overset{CH_3}{|}}{C}}-NO_2 + {}^{18}O_2 \longrightarrow 2\underset{\underset{CH_3}{|}}{\overset{\overset{CH_3}{|}}{C}}{=}O^{18} + 2HNO_2$$

Figure 18-38. Reaction catalyzed by 2-nitro-propane dioxygenase.

39). Both atoms of dioxygen appear in the 2-protocatechuoyl phloroglucinol carboxylic acid and none in the carbon monoxide (Krishnamurty and Simpson, 1970). The proposed mechanism proceeds through a cyclic peroxide [A in Figure 18-39 (Oka *et al.*, 1972)].

The enzyme is inhibited by diethyldithiocarbamate and other copper inhibitors, but not by iron inhibitors. The emission spectrum shows the presence of copper in the cupric form in the resting state (Oka and Simpson, 1971). There are 2 g-atoms copper/mole enzyme, but probably only one copper atom per active center because the enzyme binds 2 moles of substrate.

18.12   Benzene Dioxygenase

Dioxygenases are also known that essentially add hydrogen peroxide across one of the double bonds of benzene derivatives (Figure 18-40). *In vivo*, the

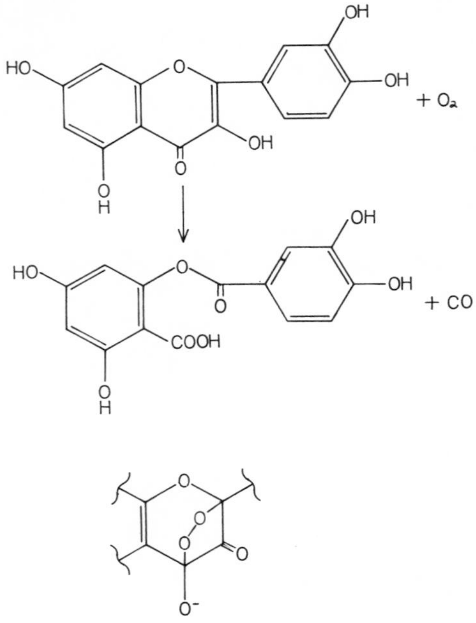

Figure 18-39. Reaction catalyzed by quercitinase showing the proposed cyclic peroxide intermediate (A).

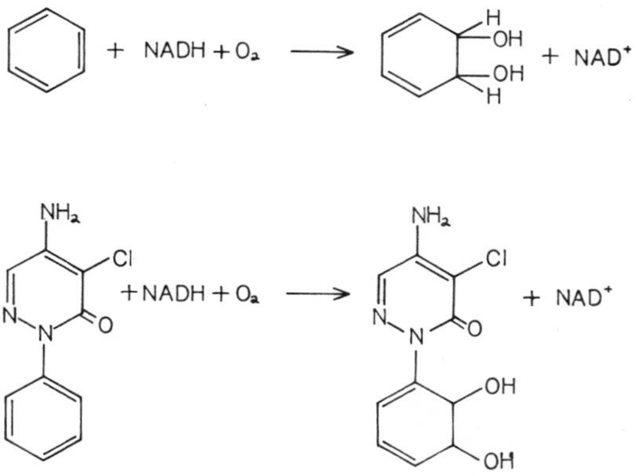

Figure 18-40. Reactions catalyzed by benzene dioxygenases.

resultant *cis*-1,2-dihydroxy-3,5-cyclohexadiene derivatives are oxidized (dehydrogenated) to catechols by other enzymes. Since dioxygen is the oxidizing agent for the dioxygenase reaction, a reduction must also occur.

These enzymes are responsible for the ability of microbes to degrade aromatics. The benzene dioxygenase and pyrazon dioxygenase activities are both three-protein systems involving two nonheme iron proteins and one flavoprotein, which constitutes a mini-electron-transport chain. NADH supplies the reducing power. In both cases, one of the ferredoxin-type proteins acts as the hydroxylase and the flavoprotein is the species initially reduced by the NADH.

A benzoate dioxygenase two-enzyme microbial system is also known that utilizes ferrous ion and NADH.

Two other enzymes of this type are known. Both hydroxylate anthranilate, and both involve ferrous ion and NADH (or NADPH) reducing power (Figure 18-41). In these cases, however, substituents are eliminated from the ring to regenerate the aromatic system without the aid of a subsequent dehydrogenase.

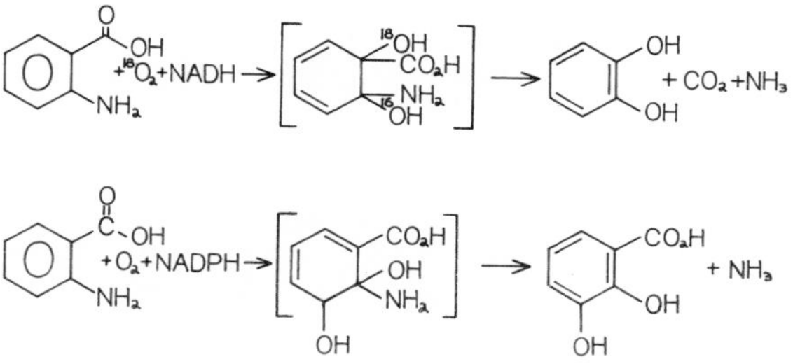

Figure 18-41. Reactions catalyzed by anthranilate hydroxylases.

The order and detailed mechanism of these reactions are entirely unknown. A complete list of known dioxygenases is available (Nozaki, 1979).

References

Bawn, C. E. H., and Sharp, J. A., 1957. Reaction of the cobaltic ion. IV. Oxidation of olefins by cobaltic salts, *J. Chem. Soc. London* 1957:1854–1865.

Brady, F., Monaco, M. E., Forman, H. J., Shutz, G., and Feigelson, P., 1972. On the role of copper in activation of and catalysis by tryptophan 2,3-dioxygenase, *J. Biol. Chem.* 247:7915–7922.

Cornelius, P. A., Vanos, G., Ryke-Schilder, R., and Vliegenthart, J. F. G., 1979. 9-L-Linoleyl hydroperoxide, a novel product from the oxygenation of linoleic acid by type-2-lipoxygenases from soybeans and peas, *Biochim. Biophys. Acta* 575:479–484.

Egmond, M. R., Vliegenthart, J. F. G., and Boldingh, J., 1972. Stereospecificity of the hydrogen abstraction at carbon atom ω-8 in the oxygenation of linoleic acid by lipoxygenases from corn germ and soya beans, *Biochem. Biophys. Res. Commun.* 48:1055–1060.

Egmond, M. R., Veldink, G. A., Vliegenthart, J. F. G., and Boldingh, J., 1973. C-11 H-abstraction from linoleic acid, the rate-determining step in lipoxidase catalysis, *Biochem. Biophys. Res. Commun.* 54:1178–1184.

Egmond, M. R., Fasella, P. M., Veldink, G. A., Vliegenthart, J. F. G., and Boldingh, J., 1977. On the mechanism of action of soybean lipoxygenase-1: A stopped flow kinetic study of the formation and conversion of yellow and purple enzyme species, *Eur. J. Biochem.* 76:469–479.

Feigelson, P., and Brady, F. O., 1974. Heme-containing oxygenases, in *Molecular Mechanisms of Oxygen Activation,* O. Hayaishi and T. Hayaishi (eds.), Academic Press, New York, pp. 87–134.

Felton, R. H., Cheung, L. D., Phillips, R. S., and May, S. W., 1978. A resonance Raman study of substrate and inhibitor binding to protocatechuate 2,3-dioxygenase, *Biochem. Biophys. Res. Commun.* 85:844–850.

Foote, C. S., and Moro-oka Y., 1976. Chemistry of superoxide ion. I. Oxidation of 3,5-di-*t*-butyl catechol with KO_2, *J. Am. Chem. Soc.* 98:1510–1514.

Fujisawa, H., Keitaro, H., Uyeda, M., Okumo, S., Nozaki, M., and Hayaishi, O., 1972. Protocatechuate 3,4-dioxygenase. III. An oxygenated form of enzyme as reaction intermediate, *J. Biol. Chem.* 247:4422–4428.

Fujiwara, H., Golovleva, L. A., Saeki, Y., Nozaki, M., and Hayaishi, O., 1975. Extradiol cleavage of 3-substituted catechols by an intradiol dioxygenase, pyrocatechase, from pseudomonad, *J. Biol. Chem.* 250:4848–4855.

Galliard, T., 1975. Degradation of plant lipids by hydrolytic and oxidative enzymes, in *Recent Advances in the Chemistry and Biochemistry of Plant Lipids,* T. Galliard and E. I. Mercer (eds.), Academic Press, New York, pp. 319–357.

Galliard, T., and Chan, H. W. S., 1980. Lipoxygenases, in *Biochemistry of Plants: A Comprehensive Treatise,* Vol. 4, *Lipids: Structure and Function,* P. K. Stumpf and E. E. Conn (eds.), Academic Press, New York, pp. 131–161.

Gibian, M. J., and Gallaway, R. A., 1977. Chemical aspects of lipoxygenase reactions, in *Bioorganic Chemistry I,* E. E. van Tamelen (ed.), Academic Press, New York, pp. 111–113.

Grinstead, R. R., 1964. Metal catalyzed mechanism of pyrocatechase action, *Biochemistry* 3:1308–1314.

Grosch, W., and Laskawy, G., 1979. Co-oxidation of carotenes requires one soybean lipoxygenase isozyme, *Biochim. Biophys. Acta* 575:439–445.

Hamberg, M., and Hamberg, G., 1980. On the mechanism of the oxygenation of arachidonic acid by human platelet lipoxygenase, *Biochem. Biophys. Res. Commun.* 95:1090–1097.

Hamberg, M., and Samuelsson, B., 1967a. Oxygenation of unsaturated fatty acids by vesicular gland of sheep, *J. Biol. Chem.* 242:5344–5354.

Hamberg, M., and Samuelsson, B., 1967b. On the specificity of the oxygenation of unsaturated fatty acids catalyzed by soybean lipoxygenase, *J. Biol. Chem.* 242:5329–5335.

Hamberg, M., and Samuelsson, B., 1973. Detection and isolation of the endoperoxide intermediate in prostaglandin biosynthesis, *Proc. Natl. Acad. Sci. U.S.A.* 70:899–903.

Hamberg, M., and Samuelsson, B., 1974. Prostaglandin endoperoxides: Novel transformations of arachidonic acid in human platelets, *Proc. Natl. Acad. Sci. U.S.A.* 71:3400–3404.

Hamilton, G. A., 1971. The proton in biological redox reactions, in *Progress in Bioorganic Chemistry*, Vol. 1, E. T. Kaiser and F. J. Keady (eds.), Wiley Interscience, New York, pp. 142–148.

Hamilton, G. A., 1974. Chemical models and mechanisms for oxygenases, in *Molecular Mechanisms of Oxygen Activation*, O. Hayaishi (ed.), Academic Press, New York, pp. 443–445.

Hayaishi, O., 1969. Electronic aspects of the catalysis of the oxygenases, *Ann. N. Y. Acad. Sci.* 158:318–335.

Hayaishi, O., Rothberg, S., Mehler, A. H., and Saito, Y., 1957a. Studies on oxygenase: Enzymatic formation of kynurenine from tryptophane, *J. Biol. Chem.* 299:889–896.

Hayaishi, O., Katagiri, M., and Rothberg, S., 1957b. Studies on oxygenases: Pyrocatechase, *J. Biol. Chem.* 229:905–920.

Hayaishi, O., Hirata, F., Ohnishi, T., Henry, J.-P., Rosenthal, I., and Katoh, A., 1977. Indoleamine 2,3-dioxygenase: Incorporation of $^{18}O_2^-$ and $^{18}O_2$ into the reaction products, *J. Biol. Chem.* 252:3548–3550.

Hemler, M., Lands, W. E. M., and Smith, W. L., 1976. Purification of the cyclooxygenase that forms prostaglandins, *J. Biol. Chem.* 251:5575–5579.

Hou, C. T., Lillard, M. O., and Schwartz, R. D., 1976. Protocatechuate 3,4-dioxygenase from *Acinetobacter calcoaceticus*, *Biochemistry* 15:582–588.

Ishimura, Y., Nozaki, M., and Hayaishi, O., 1967, Evidence for an oxygenated intermediate in the tryptophan pyrrolase reaction, *J. Biol. Chem.* 242:2574–2576.

Ishimura, Y., Nozaki, M., Hayaishi, O., Nakamura, T., Tamura, M., and Yamazaki, I., 1970. The oxygenated form of L-tryptophan 2,3-dioxygenase as reaction intermediate, *J. Biol. Chem.* 245:3593–3602.

Keyes, W. E., Loehr, T. H., and Taylor, M. L., 1978. Raman spectral evidence for tyrosine coordination of iron in protocatechuate 3,4-dioxygenase, *Biochem. Biophys. Res. Commun.* 83:941–945.

Kishore, G. M., and Snell, E. E., 1981. Kinetic investigations on a flavoprotein oxygenase: 2-Methyl-3-hydroxypyridine-5-carboxylic acid oxygenase, *J. Biol. Chem.* 256:4228–4233.

Kishore, G. M., and Snell, E. E., 1981b. Interaction of 2-methyl-3-hydroxypyridine-5-carboxylic acid oxygenase with FAD, substrates and analogues, *J. Biol. Chem.* 256:4234–4240.

Kishore, G. M., and Snell, E. E., 1981c. 2-Methyl-3-hydroxypyridine-5-carboxylic acid oxygenase: Molecular and catalytic characteristics, in *Oxygen and Oxy-radicals in Chemistry and Biology*, M. A. Rodger and E. L. Powers (eds.), Academic Press, New York, pp. 521–534.

Krishnamurty, H. G., and Simpson, F. J., 1970. Degradation of rutin by *Aspergillus flavus:* Studies with oxygen 18 on the action of a dioxygenase on quercetin, *J. Biol. Chem.* 245:1467–1471.

Lauffer, R. B., Heistand, R. H., II, and Que, J., Jr., 1981. Dioxygenase model studies: Reaction of oxygen with iron catecholates, *J. Am. Chem. Soc.* 103:3947–3949.

Lindblad, B., Lindstedt, B. G., and Lindstedt, S., 1970. The mechanism of enzymic formation of homogentisate from *p*-hydroxyphenyl-pyruvate, *J. Am. Chem. Soc.* 92:7446–7449.

Lindstedt, G., and Lindstedt, S., 1970. Cofactor requirements of γ-butyrobetaine hydroxylase from rat liver, *J. Biol. Chem.* 245:4178–4186.

Maeno, H., and Feigelson, P., 1965. The participation of copper in tryptophan pyrrolase action, *Biochem. Biophys. Res. Commun.* 21:297–302.

Maeno, H., and Feigelson, P., 1968. Studies on the interaction of carbon monoxide with tryptophan oxygenase of pseudomonads, *J. Biol. Chem.* 243:301–305.

Makino, R., and Ishimura, Y., 1976. Negligible amount of copper in hepatic L-tryptophan 2,3-dioxygenase, *J. Biol. Chem.* 251:7722–7725.

Marnett, L. J., Wlodawer, P., and Samuelsson, B., 1974. Light emission during the action of prostaglandin synthetase, *Biochem. Biophys. Res. Commun.* 60:1286–1294.

Marnett, L. J., Wlodawer, P., and Samuelsson, B., 1975. Co-oxygenation of organic substrates by the prostaglandin synthetase of sheep vesicular gland, *J. Biol. Chem.* 250:8510–8517.

Marnett, L. J., Bienkowski, M. J., and Pagels, W. R., 1979. Oxygen 18 investigation of the prostaglandin synthetase-dependent co-oxidation of diphenylisobenzofuran, *J. Biol. Chem.* 254:5077–5082.

Mason, H. S., 1957. Mechanisms of oxygen metabolism, *Adv. Enzymol.* 19:74–322.

May, S. W., Phillips, R. S., and Oldham, C. D., 1978, Interaction of protocatechuate 3,4-dioxygenase with fluoro substituted hydroxybenzoic acid and related compounds, *Biochemistry* 17:1853–1860.

Mayer, R., Widon, J., and Que, L. J., Jr., 1979. Involvement of superoxide in the reactions of the catechol dioxygenases, *Biochem. Biophys. Res. Commun.* 92:285–291.

Miyamoto, T., Ogino, N., Yamamoto, S., and Hayaishi, O., 1976. Purification of prostaglandin endoperoxide synthetase from bovine vesicular gland microsomes, *J. Biol. Chem.* 251:2629–2636.

Morris, L. J., and Marshall, M. O., 1966. Occurrence of *cis,trans*-linoleic acid in seed oils, *Chem. Ind. (London)* 1966:1493–1494.

Muto, S., and Bruice, T. C., 1981a. Dioxygen transfer from 4a-hydroperoxy flavin anion. 2. Oxygen transfer to the 10 position of 9-hydroxyphenanthrene anions and to 3,5-di-*t*-buty-catechol anion, *J. Am. Chem. Soc.* 102:4472–4480.

Muto, S., and Bruice, T. C., 1981b. Dioxygen transfer from 4a-hydroperoxy flavin anion. 3. Oxygen transfer to the 3-position of substituted indoles, *J. Am. Chem. Soc.* 102:7559–7564.

Nakata, H., Yamauchi, T., and Fujisawa, H., 1978. Studies on the reaction intermediate of protocatechuate 3,4-dioxygenase: Formation of enzyme–product complex, *Biochim. Biophys. Acta* 527:171–181.

Nicolaou, K. C., and Gasic, G. P., 1978. Synthesis and biological properties of prostaglandin endoperoxides, *Angew. Chem. Int. Ed. Engl.* 17:293–312.

Nozaki, M., 1979. Oxygenases and dioxygenases, in *Topics in Current Chemistry; Biochemistry 78*, M. J. S. Dewar, (eds.), Springer-Verlag, pp. 145–186.

Nugteren, D. H., and Hazelhof, E., 1973. Isolation and properties of intermediates in prostaglandin biosynthesis, *Biochim. Biophys. Acta* 326:448–461.

Nugteren, D. H., Beerthius, R. L., and Van Dorp, D. A., 1967. Biosynthesis of prostaglandins, *Prostaglandins* (Proceedings of the 2nd Nobel Symposium, Stockholm) 1966:45–50.

O'Connor, D. E., Micheliech, E. D., and Coleman, M. C., 1981. Isolation and characterization of bicycloendoperoxides derived from methyl linolenate, *J. Am. Chem. Soc.* 103:223–224.

Ohki, S., Ogino, N., Yamamoto, S., and Hayaishi, O., 1979. Prostaglandin hydroperoxidase, an integral part of prostaglandin endoperoxide synthetase from bovine vesicular and gland microsomes, *J. Biol. Chem.* 254:839.

Ohnishi, T., Hirata, F., and Hayaishi, O., 1977. Indoleamine 2,3-dioxygenase: Potassium superoxide as a substrate, *J. Biol. Chem.* 252:4643–4647.

Oka, T., and Simpson, F. J., 1971. Quercetinase, a dioxygenase containing copper, *Biochem. Biophys. Res. Commun.* 43:1–5.

Oka, T., Simpson, F. J., and Krishnamurty, H. G., 1972. Degradation of rutin by *Aspergillus flavus:* Studies on specificity, inhibition and possible reaction mechanism of quercitinase, *Can. J. Microbiol.* 18:493–508.

Peisach, J., Fugisawa, H., Blumberg, W. E., and Hayaishi, O., 1972. The role of substrate in the binding of O_2 to the non-heme iron containing oxygenase, protocatechuate 3,4-dioxygenase, *Fed. Proc. Fed. Am. Soc. Exp. Biol.* 31 (Abstr. No. 1304).

Pistorius, E. K., Alexrod, B., and Palmer, G., 1976. Evidence for the participation of iron in lipoxygenase reaction from optical and electron spin resonance studies, *J. Biol. Chem.* 251:7144–7148.

Poillon, W. M., Maeno, H., Koike, K., and Feigelson, P., 1969. Tryptophan oxygenase of *Pseudomonas acidovorous:* Purification, composition and subunit structure, *J. Biol. Chem.* 244:3447–3456.

Que, L. J., Jr., and Heistand, R. H. II, 1979. Resonance Raman studies on pyrocatechase, *J. Am. Chem. Soc.* 101:2219–2221.

Que, L. J., Jr., Lipscomb, J. D., Zimmermann, R., Munck, E., Orme-Johnson, N. R., and Orme-Johnson, W. H., 1976. Mossbauer and EPR spectroscopy on protocatechuate 3,4-dioxygenase from *Pseudomonas aeruginosa, Biochim. Biophys. Acta* 452:320–334.

Que, L. J., Jr., Lipscomb, J. D., Minck, E., and Wood, J. M., 1977. Protocatechuate 3,4-dioxygenase inhibitor studies and mechanistic implications, *Biochim. Biophys. Acta* 485:60–74.

Rahimtula, A., and O'Brien, P. J., 1976. The possible involvement of singlet oxygen in prostaglandin biosynthesis. *Biochem. Biophys. Res. Commun.* 70:893–899.

Rogic, M., and Demmin, T. R., 1978. Cleavage of carbon–carbon bonds—Copper(II) induced oxygenolysis of *o*-benzoquinones, catechols and phenols: On the question of nonenzymatic oxidation of aromatics and activation of molecular oxygen, *J. Am. Chem. Soc.* 100:5472–5487.

Rundgren, M., 1977. Steady state kinetics of 4-hydroxy-phenylpyruvate dioxygenase from human liver (III), *J. Biol. Chem.* 252:5094–5099.

Saeki, Y., Nozaki, M., and Senoh, S., 1980. Cleavage of pyrogallol by non-heme iron-containing dioxygenase, *J. Biol. Chem.* 255:8465–8471.

Samuelsson, B., 1972. Biosynthesis of prostaglandins, *Fed. Proc. Fed. Am. Soc. Exp. Biol.* 31:1442–1450.

Samuelsson, B., Granstrom, E., and Hamberg, M., 1967. The mechanism of the biosynthesis of prostaglandins, *Prostaglandins* (Proceedings of the 2nd Nobel Symposium, Stockholm) 1966:31–44.

Schutz, G., and Feigelson, P., 1972. Purification and properties of rat liver tryptophan oxygenase, *J. Biol. Chem.* 247:5327–5332.

Siegel, B., 1979. α-Ketoglutarate dependent dioxygenase: A mechanism for prolyl hydroxylase action, *Bioorg. Chem.* 8:219–226.

Sparrow, L. G., Ho, P. P. P., Sandaram, T. K., Zach, D., Nyns, E. J., and Snell, E. E., 1969. The bacterial oxidation of vitamin B_6. VII. Purification, properties and mechanism of action of an oxygenase which cleaves the 3-hydroxypyridine ring, *J. Biol. Chem.* 244:2590–2600.

Takeda, K., Kawai, S., Tetsuka, T., and Konno, K., 1976. Stimulation of rotocollagen-proline hydroxylase, activity by nucleoside triphosphate, *Biochem. Biophys. Res. Commun.* 69:957–961.

Taniguchi, T., Sono, M., Hirata, F., Hayaishi, O., Tamura, M., Hayaishi, K., Iizuki, K., and Ishimura, Y., 1979. Indoleamine 2,3-dioxygenase: Kinetic studies on the binding of superoxide anion and molecular oxygen to enzyme, *J. Biol. Chem.* 254:3288–3294.

Tatsumo, Y., Saeki, Y., Iwaki, M., Yagi, T., and Nozaki, M., 1978. Resonance Raman spectra of protocatechuate 3,4-dioxygenase: Evidence for coordination of tyrosine residue to ferric ion, *J. Am. Chem. Soc.* 100:4614–4615.

Teng, J. I., and Smith, L. L., 1973. Steroid metabolism. XXIV. On the unlikely participation of singlet molecular oxygen in several enzyme oxygenations, *J. Am. Chem. Soc.* 95:4060–4061.

Theorell, H., Bergstrom, S. J., and Akeson, A., 1947. Crystalline lipoxidase, *Acta Chem. Scand.* 1:571–576.

Tyson, C. A., 1975. 4-Nitrocatechol as a colorimetric probe for non-heme dioxygenases, *J. Biol. Chem.* 250:1765–1770.

Veldink, G. A., Vliegenthart, J. F. G., and Boldnigh, J., 1977. Plant lipoxygenases, *Prog. Chem. Fats Other Lipids* 15:131–166.

Vliengenthart, J. F. G., Veldink, G. A., and Boldingh, J., 1979. Recent progress in the study on the mechanism of action of soybean lipoxygenase, *J. Agric. Food Chem.* 27:623–626.

Wondrack, L. M., Hsu, C.-A., and Abbott, M. T., 1978. Thymine 7-hydroxylase and pyrimidine deoxyribonucleoside 2'-hydroxylase activities in *Rhodotorula glutinis*, *J. Biol. Chem.* 253:6511–6515.

Wondrack, L. M., Warn, B. J., Saewert, M. D., and Abbott, M. T., 1979. Substitution of nucleoside triphosphates for ascorbate in the thymine 7-hydroxylase reaction of *Rhodotorula glutinis*, *J. Biol. Chem.* 254:26–29.

Bioluminescence

19.1 General Concepts

Certain exothermic chemical reactions produce light instead of heat. Chemiluminescent reactions also occur in living systems. The term usually used for light production from living systems is *bioluminescence*. Fireflies, bacteria, and other organisms produce bioluminescence.

One explanation for a reaction that produces light instead of heat is that orbital symmetry restrictions prevent the ground-state reactants from producing products in the ground state. This is true for the cleavage of four-membered rings to produce two double bonds (Adam, 1975). The cleavage of cyclobutane to form two ethylenes will not occur thermally because there is simply not enough energy from the reaction to produce the ethylenes in the excited state. However, when a dioxetane such as trimethyl dioxetane (Kopecky and Mumford, 1969) or tetramethyl dioxetane (Turro and Lechtken, 1972) is cleaved to acetaldehyde and acetone or acetone only, respectively, there is enough energy to excite the products electronically. The dioxetane ring is highly strained and carbonyls are quite stable.

The mechanism described above implies a concerted cleavage of dioxetanes to produce two carbonyl groups, with electronic excitation of the products. Another mechanism for producing light is called the chemically initiated electron-exchange luminescence (CIEEL) mechanism (Schuster, 1979). In the CIEEL mechanism, an electron is transferred from an activator (an aromatic compound that will fluoresce) to the dioxetrane or other light-producing substrate. The radical anion of the substrate then cleaves to give another radical anion, which in turn reduces the radical cation of the activator. If the last electron is placed in an empty orbital, the activator is in an excited state, able to fluoresce.

19.2 Firefly Bioluminescence

The substrate for the light-producing enzyme luciferase in the firefly is called *luciferin*. The luciferin structure has been proved by synthesis (White *et*

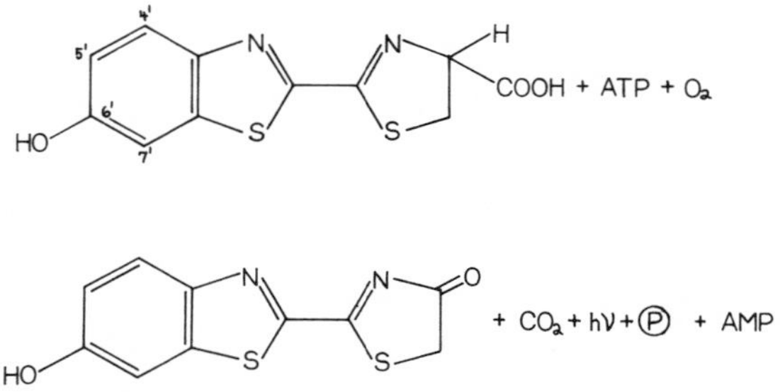

Figure 19-1. Luciferase reaction of fireflies.

al., 1961). Luciferase catalyzes the reaction of luciferin with dioxygen and ATP to produce light, CO_2, AMP, phosphate, and the oxidatively decarboxylated luciferin (Figure 19-1).

It is interesting to note that if the hydroxyl group is at the 4'-, 5'-, or 7'-position instead of the 6'-position, the luciferin is inactive; however, the 4'-,6'-dihydroxy compound is active (White and Worther, 1966). The mechanism (Figure 19-2) of this reaction has been studied by means of model compounds and isotopic labeling with ^{18}O dioxygen (White *et al.*, 1975, 1980).

One oxygen atom of the CO_2 formed is derived from dioxygen as required by this mechanism (Shimomura *et al.*, 1977). The carbanion is probably partially stabilized by the ring sulfur. In the CIEEL mechanism, the other half of the

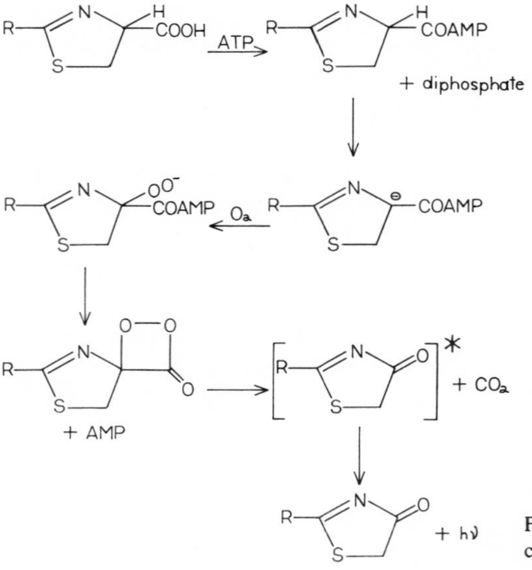

Figure 19-2. Mechanism of the firefly luciferase reaction.

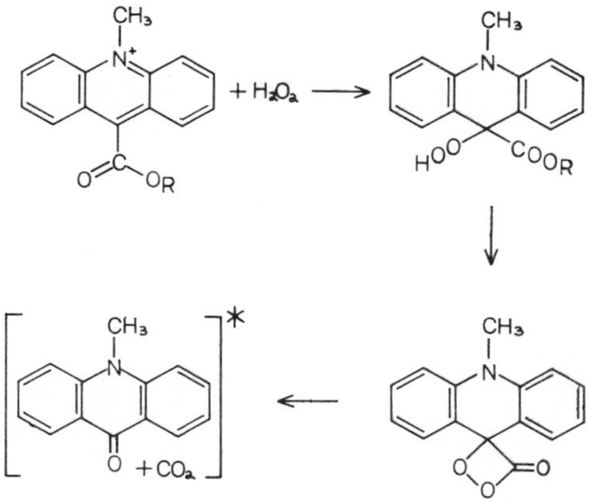

Figure 19-3. Mechanism of a chemiluminescent reaction similar to the firefly luciferase reaction.

molecule designated R could act as the activator. The compound formed by an amino group for the hydroxyl group in luciferin will act as a substrate, but the light produced has the wavelength shifted (White *et al.*, 1966). Compounds with acetylamino or trifluoracetylamino groups substituted for the hydroxyl group are inactive. This may be support for the CIEEL mechanism or may be merely another example of enzyme stereospecificity.

The luciferin reaction is very similar to the chemiluminescent oxidation of an ester by hydrogen peroxide to produce CO_2 and a ketone (McCapra, 1970) (Figure 19-3).

19.3 *Cypridina* Bioluminescence

The structure of luciferin in *Cypridina*, a crustacean, is known from synthesis (Kishi *et al.*, 1966). The mechanism probably also utilizes a dioxetane to produce light (Figure 19-4) (Shimomura and Johnson, 1971).

One oxygen of the CO_2 is derived from dioxygen as required by this mechanism (Hart *et al.*, 1978). Hart and co-workers point out that this result is also consistent with a hydroperoxide intermediate decomposing directly to the product without proceeding through the dioxetane (Figure 19-5).

19.4 Bacterial Bioluminescence

The bioluminescence of bacteria is catalyzed by flavins. These reactions proceed through intermediates similar to the hydroxylation reactions catalyzed

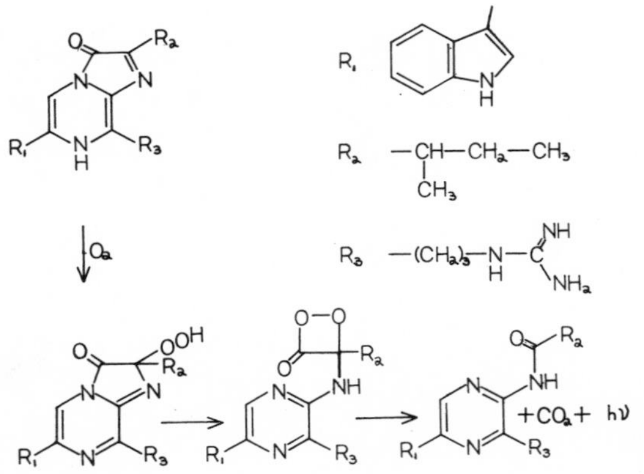

Figure 19-4. Mechanism of *Cypridina* bioluminescence.

by flavins. This type of luminescence occurs in bacteria that grow symbiotically with fish at great depths in the ocean. The enzyme contains 1 mole of $FMNH_2$ per mole of enzyme (Merghen and Hastings, 1971).

The overall reaction is the oxidation of a long-chain aldehyde, typically decanal, and a two-electron reductant by dioxygen to produce the corresponding fatty acid (McCapra and Hysert, 1973; Dunn *et al.*, 1973), light, water, and oxidized reductant.

The overall sequence of the reaction (Figure 19-6) has been studied by Hastings and Balny (1975). Reduced FMN, on the enzyme luciferase (complex I), combines with dioxygen to form complex II (Hastings and Gibson, 1963), absorbing around 372 nm with a fluorescent band at 485 nm (Hastings *et al.*, 1973). Complex II is presumably the 4a-hydroperoxide of FMN. The complex has been isolated at low temperatures (Hastings *et al.*, 1973). Complex II will decompose to give FMN and hydrogen peroxide and does not require the aldehyde for its formation (Hastings *et al.*, 1973). However, complex II can be stabilized by the addition of long-chain alcohols (Tu, 1979).

Complex II reacts with a long-chain aldehyde to produce complex IIA, (Figure 19-7) which has been considered merely as an enzyme–aldehyde–hydroperoxy flavin complex. Complex IIA produces a light emitter

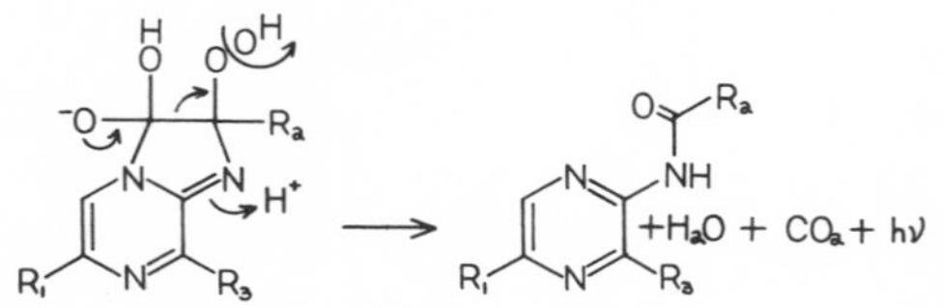

Figure 19-5. An alternate mechanism of *Cypridina* bioluminescence.

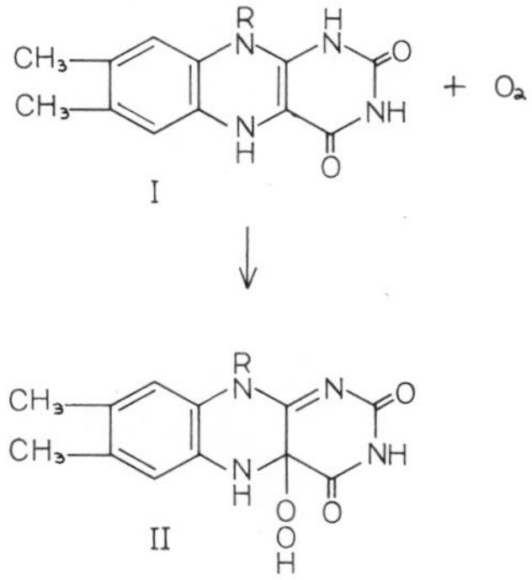

Figure 19-6. First step in bacterial bioluminescence.

that decomposes to FMN and the fatty acid with concurrent emission of light. The 4a-hydroperoxide of 5-methylflavin will react with aldehydes to produce luminescence in very low yield in the absence of an enzyme (Kemal and Bruice, 1976). The details of these reactions are unclear. A dioxetane (McCapra and Hysert, 1973) and an oxazetidine have both been proposed as intermediates (Lowe et al., 1976).

Bruice and co-workers (Kemal et al., 1977; Shepherd and Bruice, 1980) proposed that a mixed-flavin aldehyde peroxide is formed that decomposes to a 4a-hydroxyflavin and acid with the production of light (Figure 19-7). The 4a-hydroxyflavin eliminates water to form the oxidized flavin.

After light emission, the FMN requires a two-electron reductant to form the reduced form of the flavin to start the cycle again. With the exception of the light-emission step, many of these steps are reversible. The rate-determining step of these reversible steps depends on the length of the aldehyde chain (Baumstark et al., 1979).

A light emitter has been isolated from *Photobacterium phosphoreum*—6,7-dimethyl-8-ribityllumazine (Figure 19-8). This compound serves as both a light emitter and a flavin precursor (Koka and Lee, 1979).

19.5   Limpet Bioluminescence

Limpets also produce luminescence. Little is known of the actual mechanism, but the substrate and the overall reaction are known. The reaction (Figure 19-9) requires dioxygen, water, a luciferase, and a purple protein. The substrate is a formyl ester of an enol that is oxidized and hydrolyzed during the light-emitting step to a ketone and two molecules of formic acid (Shimomura et al., 1972).

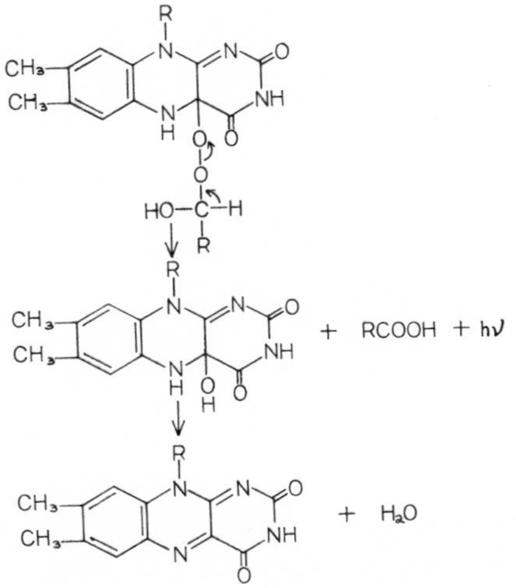

Figure 19-7. A mechanism proposed for the bacterial bioluminescence reaction.

19.6 Centipede Bioluminescence

Certain centipedes exhibit bioluminescence, but again, little is known about the process. The maximum emissions are at 510 and 480 nm. The process requires dioxygen, a substrate, and an enzyme (Anderson, 1980).

References

Adam, W., 1975. Biological light: α-Peroxylates as bioluminescent intermediates, *J. Chem. Ed.* 52:138–145.

Anderson, J. A., 1980. Biochemistry of centipede bioluminescence, *Photochem. Photobiol.* 31:179–181.

Baumstark, A. L., Cline, T. W., and Hastings, J. W., 1979. Reversible steps in the reaction of aldehydes with bacterial luciferase intermediates, *Arch. Biochem. Biophys.* 193:449–455.

Dunn, D. K., Michaliszyn, G. A., Bogacki, I. G., and Merghen, E. A., 1973. Conversion of aldehyde to acid in the bacterial bioluminescent reaction, *Biochemistry* 12:4911–4918.

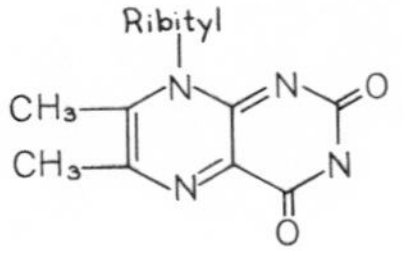

Figure 19-8. A light emitter found in *Photobacterium phosphoreum*.

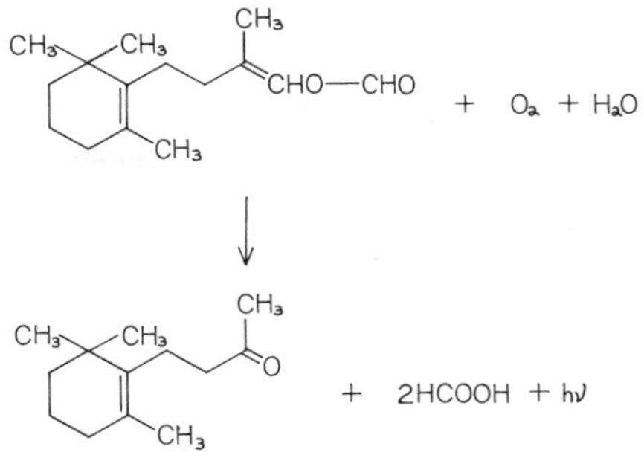

Figure 19-9. Reaction that produces light in limpets.

Hart, R. C., Stempel, K. E., Boyer, P. D., and Cormier, M. J., 1978. Mechanism of the enzyme-catalyzed bioluminescent oxidation of coelenterate-type luciferin, *Biochem. Biophys. Res. Commun.* 81:980–986.

Hastings, J. W., and Balny, C., 1975. The oxygenated bacterial luciferase intermediate, *J. Biol. Chem.* 250:7288–7293.

Hastings, J. W., and Gibson, Q. H., 1963. Intermediates in the bioluminescent oxidation of reduced flavin mononucleotide, *J. Biol. Chem.* 238:2537–2554.

Hastings, J. W., Balny, C., LePeuch, C., and Douzou, P., 1973. Spectral properties of an oxygenated luciferase–flavin intermediate isolated by low temperature chromatography, *Proc. Natl. Acad. Sci. U.S.A.* 70:3468–3472.

Kemal, C., and Bruice, T. C., 1976. Simple synthesis of a 4a-hydroperoxy adduct of a 1,5-dihydroflavin: Preliminary studies of a model for bacterial luciferase, *Proc. Natl. Acad. Sci. U.S.A.* 73:995–999.

Kemal, C., Chan, T. W., and Bruice, T. C., 1977. Chemiluminescent reactions and electrophilic oxygen donating ability of 4a-hydroperoxyflavins: General synthetic method for the preparation of N^5-alkyl-1,5-dihydroflavins, *Proc. Natl. Acad. Sci. U.S.A.* 74:405.

Kishi, T., Goto, T., Inoue, S., Suguira, S., and Kishimoto, H., 1966. *Cypridina* bioluminescence. III. Total synthesis of *Cypridina* luciferin, *Tetrahedron Lett.* 29:3445–3450.

Koka, P., and Lee, J., 1979. Separation and structure of the prosthetic group of the blue fluorescence protein from the bioluminescent bacterium *Photobacterium phosphoreum, J. Biol. Chem.* 76:3068–3072.

Kopecky, K. R., and Mumford, C., 1969. Luminescence in the thermal decomposition of 3,3,4-trimethyl-1,2-dioxetane, *Can. J. Chem.* 47:709–711.

Lowe, J. N., Ingraham, L. L., Alspach, J., and Rasmussen, R., 1976. A proposed symmetry forbidden oxidation mechanism for the bacterial luciferase catalyzed reaction, *Biochem. Biophys. Res. Commun.* 73:465–469.

McCapra, F., 1970. The chemiluminescence of organic compounds, *Pure Appl. Chem.* 24:611–629.

McCapra, F., and Hysert, D. W., 1973. Bacterial bioluminescence—identification of fatty acid as product, its quantum yield and a suggested mechanism, *Biochem. Biophys. Res. Commun.* 52:298–304.

Merghen, E. A., and Hastings, J. W., 1971. Binding site determination from kinetic data: Reduced flavin mononucleotide binding to bacterial luciferase, *J. Biol. Chem.* 246:7666–7674.

Schuster, G. B., 1979. Chemiluminescence of organic peroxides: Conversion of ground state reactants to excited state products by the chemically initiated electron exchange luminescence mechanism, *Acc. Chem. Res.* 12:366–373.

Shepherd, P. T., and Bruice, T. C., 1980. Formation of a nonchemiluminescent excited state species in the decomposition of 4a(alkylperoxy) flavins, *J. Am. Chem. Soc.* 102:7774–7776.

Shimomura, O., and Johnson, F. H., 1971. Mechanism of the luminescent oxidation of *Cypridina* luciferin, *Biochem. Biophys. Res. Commun.* 44:340–346.

Shimomura, O., Johnson, F. H., and Kohama, Y., 1972. Reactions involved in bioluminescence systems of limpet (*Latia neritoides*) and luminous bacteria, *Proc. Natl. Acad. Sci. U.S.A.* 69:2086–2089.

Shimomura, O., Goto, T., and Johnson, F. H., 1977. Source of oxygen in the CO_2 produced in the bioluminescent oxidation of firefly luciferin, *Proc. Natl. Acad. Sci. U.S.A.* 74:2799–2802.

Tu, S.-C., 1979. Isolation and properties of bacterial luciferase-oxygenated flavin intermediate complexed with long-chain alcohols, *Biochemistry* 18:5940–5945.

Turro, N. J., and Lechtken, P., 1972. Thermal decomposition of tetramethyl-1,2-dioxetane: Selective and efficient chemelectronic generation of triplet acetone, *J. Am. Chem. Soc.* 94:2886–2888.

White, E. H., and Worther, H., 1966. Analogs of firefly luciferin III, *J. Org. Chem.* 31:1484–1488.

White, E. H., McCapra, F., Field, G. F., and McElroy, W. D., 1961. The structure and synthesis of firefly luciferin, *J. Am. Chem. Soc.* 83:2402–2403.

White, E. H., Worther, H., Seliger, H. H., and McElroy, W. D., 1966. Amino analogs of firefly luciferin and biological activity thereof, *J. Am. Chem. Soc.* 88:2015–2019.

White, E. H., Miano, J. D., and Umbreit, M., 1975. On the mechanism of firefly luciferin luminescence, *J. Am. Chem. Soc.* 97:198–200.

White, E. H., Steinmetz, M. G., Miano, J. D., Wildes, P. D., and Morland, R., 1980. Chemi and bioluminescence of firefly luciferin, *J. Am. Chem. Soc.* 102:3199–3208.

Dioxygen Toxicity

20.1 Introduction

We are so accustomed to an aerobic life that we tend to forget that dioxygen and some of the partial reduction products of dioxygen can be very dangerous chemicals. Organisms that require dioxygen commonly require it within relatively narrow limits of concentration. Too high a concentration of dioxygen can be just as lethal to the organism as too low a concentration. Many obligate anaerobes can be killed by dioxygen levels slightly above very low threshold concentrations.

The lethal action of dioxygen has been compared to the lethal action of ionizing radiation with the conclusion that the actual causes are the same (Gerschman *et al.*, 1954). Both phenomena are considered to be the result of free radicals caused by the irradiation or by reactions of dioxygen with organic molecules.

Dioxygen oxidizes a small amount of hemoglobin to methemoglobin during the normal dioxygen-carrying function of hemoglobin. However, because ground-state dioxygen does have such strong kinetic barriers to oxidations, the toxicity of dioxygen is mostly the result of either the odd electron reduction products, superoxide ion and hydroxyl radical, or to singlet dioxygen instead of ground-state dioxygen *per se*. All these agents are strong oxidizing agents and hence potentially dangerous.

Hydrogen peroxide is produced in the cell. One source is from reduced flavins, which react extremely rapidly with dioxygen to produce hydrogen peroxide (Massey *et al.*, 1971; Smith *et al.*, 1974). The intermediate in the reaction is probably the 4a-hydroperoxide (cf. Figure 20-1). Hydrogen peroxide can damage hemoglobin. This damage is repaired by an NADH reductase. Hydrogen peroxide can also oxidize the double bonds of unsaturated lipids in cell membranes, forming hydroperoxides that subsequently break down to smaller molecules. The enzymes catalase and peroxidase help keep the concentrations of hydrogen peroxide at a low level in the cell and in this manner help protect

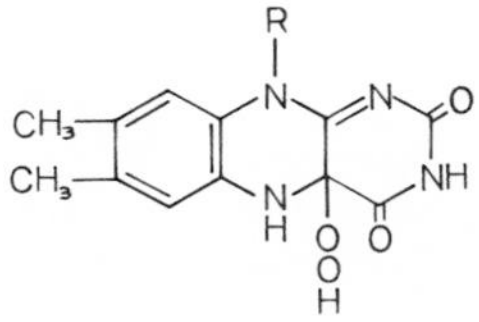

Figure 20-1. A proposed reaction intermediate in the reduction of dioxygen by reduced flavins.

organisms from the damaging effects of oxygen toxicity. Acatalasemia is a disease in which the patient has a low level of catalase (Aebi and Wyss, 1978). This disease is ubiquitous but particularly prevalent in Korea.

Glutathione peroxidase (see Chapter 6) (Tappel, 1981; Rotruck *et al.*, 1973) reduces the level of hydrogen peroxide in the cell by utilizing the hydrogen peroxide to oxidize glutathione. The oxidized glutathione can be reduced by NADPH. Glutathione peroxidase contains selenium but does not contain a heme group.

Hydroxyl radicals are damaging to biological systems and in the presence of hydrogen peroxide are more damaging to erythrocyte ghosts as measured by permeability than can be accounted for by independent reactions of hydrogen peroxide and hydroxyl radical (Kong and Davison, 1980). One possibility is the reaction of hydroxyl radical with hydrogen peroxide:

$$\cdot OH + H_2O_2 \rightarrow HO_2\cdot + H_2O$$

However, the concentration of hydroxyl radical is probably too low for this reaction to be important. The hydroxyl radical probably initiates the hydrogen peroxide oxidation of the substrate by a free-radical mechanism (Kong and Davison, 1980):

$$RH_2 + \cdot OH \rightarrow RH\cdot + H_2O$$

$$RH\cdot + H_2O_2 \rightarrow OH\cdot + OH^- + H^+$$

Tien *et al.* (1981) have studied the effect of buffer on microsomal peroxidation of lipids by following malondialdehyde formation. NADPH and iron were required. Tris and ADP chelation of iron greatly accelerated oxidation. Tris buffer (as opposed to NaCl) had little effect on the reaction, indicating that hydroxyl radical is probably not involved, since Tris is a very effective hydroxyl radical trap.

The authors pointed out that if peroxides are present, then a peroxidation reaction can be facilitated by ferrous ion and a system to maintain its reduced state (e.g., superoxide), but if peroxides are not present, ferrous ion will not stimulate peroxidation. Therefore, a separate process must be involved (which apparently does not depend on superoxide or hydroxyl radical) for initial lipid hydroperoxide formation. A system was developed for lipid peroxidation dependent on hydroxyl radical (Fenton's reagent). Addition of ADP caused the system to lose its sensitivity to hydroxyl radical traps such as mannitol, indicating that in the presence of biologically ubiquitous iron-chelating agents, hydroxyl radical is insignificant. These results led to the conclusion that initiation of lipid peroxidation is caused by perferryl ion or ferryl ion chelated to ADP: $ADP-Fe^{2+}O_2$ or $ADP-FeO^{2+}$.

20.2 Superoxide Toxicity

Biological reactions tend to go by two-electron reductions of dioxygen, so that one-electron reduction intermediates are not major products. However, biological systems will also generate substantial quantities of superoxide (Fridovich, 1972, 1974). Superoxide ion can be formed by many types of oxidation of organic compounds by dioxygen. Examples are the autoxidation of hydroquinones, reduced flavins (Misra and Fridovich, 1972a), epinephrine (Misra and Fridovich, 1972b), ferredoxins (Misra and Fridovich, 1971), and probably even hemoglobin and myoglobin (Misra and Fridovich, 1972c; Gotoh and Shikama, 1976). There is a difference in opinion as to the cytotoxicity of superoxide ion. Superoxide is potentially very dangerous in that it is a powerful oxidant. It is believed to be a factor in the production of cancer (Totter, 1980). It is believed to oxidize many biological materials (Fridovich, 1979), including virus (Lavelle et al., 1973), lipids (Kellogg and Fridovich, 1975), and membranes (Goldberg and Stern, 1976; Kellogg and Fridovich, 1977).

On the other hand, superoxide ion appears to be a rather innocuous agent in many chemical reactions (Fee and Hildebrand, 1974; Fee, 1980; Bors et al., 1980). Note in Chapter 4 that superoxide ion often reacts as a base rather than as an oxidizing agent. Superoxide ion becomes a rather poor nucleophile in aqueous solutions compared to its nucleophilicity in aprotic solvents; it adds only to highly activated double bonds and does not abstract hydrogen atoms or electrons. This viewpoint (Fee, 1981) states that such an unreactive species could not account for DNA degradation, lipid peroxidation, depolymerization of polysaccharides, the hydroxylation of aromatic compounds, and the death of the cell as is observed in doxygen toxicity. However, there could be a danger in equating cell toxicity of a compound to its reactivity. The drastic effects observed in vitro in the presence of superoxide ion have been explained (Fee, 1981) by a metal-catalyzed Haber–Weiss reaction, the reduction of a metal ion by superoxide ion followed by the reaction of the metal ion with hydrogen peroxide to form the highly reactive hydroxyl radicals.

Claims have been made that superoxide ion produces hydroxy radicals in biological systems by the reaction with hydrogen peroxide in the uncatalyzed Haber and Weiss (1934) reaction:

$$O_2^- + H_2O_2 \rightarrow OH^- + OH\cdot + O_2$$

However, this reaction is slow compared with other reactions of superoxide ion and so may be of little significance in biological systems (see Section 4.3 in Chapter 4). The reaction requires metal catalysis to be of importance (Halliwell, 1976) because the uncatalyzed reaction is probably too slow to compete with disproportionation reactions (Melhuish and Sutton, 1978).

Superoxide dismutase (SOD), discussed in Chapter 4, is claimed to be an important factor in determining dioxygen toxicity (Fridovich, 1975) because it keeps the concentration of superoxide low. Generally, aerobic bacteria contain SOD, whereas anaerobic bacteria do not (McCord et al., 1971), although there is at least one exception. *Mycoplasma pnuemoniae* generates superoxide ion but appears to lack SOD (Lynch and Cole, 1980). *Escherichia coli* contains both the Mn SOD and Fe SOD in aerobic culture, but mostly Fe SOD in anaerobic culture (Hassan and Fridovich, 1977a). Inhibition of the synthesis of the Mn SOD resulted in dioxygen toxicity. The synthesis of the Mn SOD depends on the rate of respiration rather than the actual amount of dioxygen. SOD in *E. coli* is at a low activity under fermentation conditions, but increases under high respiration conditions (Hassan and Fridovich, 1977b).

The authors would like to take a middle ground in this controversy over the toxicity of superoxide ion. There is no evidence that superoxide ion is a normal metabolite needed for synthesis. Thus, to the cell it is a xenobiotic that needs to be eliminated regardless of its reactivity. The presence of SOD can be rationalized on this basis, if on no other.

However, despite the low reactivity of superoxide ion, it is a dangerous cell component because it has the potential to form some very reactive reagents. The most likely mechanism is the metal-catalyzed Haber–Weiss reaction (Halliwell et al., 1980), in which hydrogen peroxide is formed by the dismutation of superoxide ion. Metal ions are ubiquitous in biological systems, so it is conceivable that superoxide ion, hydrogen peroxide, and metal ions could all be at the same spot. The real oxidant may well be the hydroxyl radical or alkoxy radicals derived from the hydroxyl radical (Bors et al., 1980).

A fascinating aspect of dioxygen toxicity is the killing of bacteria by phagocytic cells in mammals. The best studied of these cells are the neutrophiles (Babior, 1981), which use NADPH to reduce dioxygen to superoxide ion. The superoxide ion dismutes to hydrogen peroxide and dioxygen. The hydrogen peroxide reacts with chloride ion by an enzyme-catalyzed process to form the cytotoxic reagent, hypochlorite ion. Hypochlorite ion is toxic itself, or it could react with another chloride ion (Hill and Okolow-Zubkowska, 1981) to produce hydroxyl radicals in a process that would be a chloride-ion-catalyzed Haber–Weiss reaction. Another mechanism may be the production of hydroxyl radicals by a metal-catalyzed Haber–Weiss reaction.

20.3 Measurement of Lipid Oxidation

Measurement of *in vivo* lipid hydroperoxide content is required in the study of dioxygen toxicity and general free-radical damage. Chemiluminescence of the very low levels of singlet dioxygen and excited-state carbonyls produced in the hydroperoxidation process has been used as one type of assay for this process (Boveris *et al.*, 1981). Another method has been to assay for lipid peroxides by means of thiobarbituric acid (Porter *et al.*, 1978).

Tappel and co-workers (Tappel and Dillard, 1981) have developed an assay to measure exhaled saturated hydrocarbons that result from cleavage of carbon free-radical intermediates. Pentane is formed to the extent of approximately 0.002 mole/mole lipid hydroperoxide present, by metal-ion reduction of R_1R_2HCOOH to form $R_1R_2HCO\cdot$ and $\cdot OH$, followed by β-scission to form $R_1\cdot + R_2HCO$ and H abstraction to form R_1H (pentane) and another radical. Even low levels of pentane can be measured accurately by gas chromatography.

That pentane is a product of lipid peroxidation has been shown by (1) *in vivo* reaction of metal ions with lipid hydroperoxides; (2) an inverse relationship between amounts of the dietary antioxidants vitamin K and selenium and pentane production in live rats; (3) a direct relationship between conjugated diene (another product of lipid peroxidation) in the liver and exhaled pentane, in rats injected with halogenated hydrocarbons.

Using the pentane assay, conclusions have been drawn regarding many agents that possibly participate in lipid peroxidation. It was found that vitamin-C-treated rats were more susceptible to oxidation caused by a strong oxidant than rats not treated with vitamin C. Also, the participation of iron was established by increase of exhaled pentane following injection of large amounts of iron dextrose. Exercise also increases pentane exhalation.

20.4 Cytotoxic Compounds

Certain cytotoxic compounds probably act on the cell by producing active forms of reduced oxygen. For example, 6-hydroxydopamine, 6-aminodopamine, 6,7-dihydroxytryptamine, and dialuric acid are all cytotoxic agents and form hydrogen peroxide, superoxide ion, and hydroxyl radicals on reaction with dioxygen (Cohen and Heikkila, 1974). Similarly, the diabetes-producing toxicity of alloxan is probably the result of the production of reactive forms of oxygen by

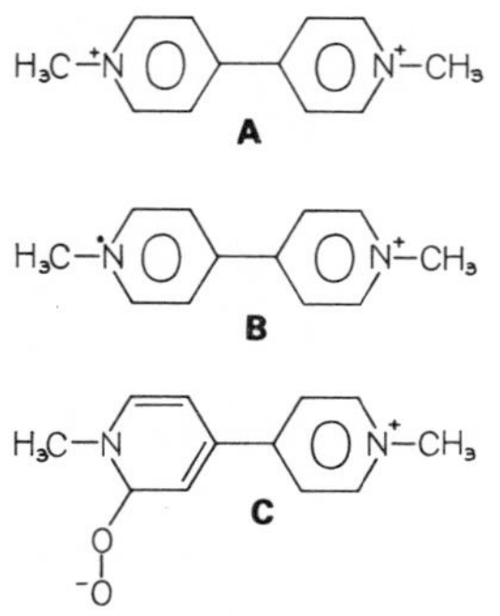

Figure 20-2. Methyl viologen [paraquat (A)] and the radical cation (B) and peroxy anion (C) derived from methyl viologen.

the aerobic oxidation of dialuric acid to alloxan and superoxide ion. The alloxan is reduced in the cell to dialuric acid. The cytotoxic effect of alloxan can be reduced by the addition of SOD or catalase (Grankvurst *et al.,* 1979). The cytotoxic reagent probably reacts with dioxygen to produce superoxide ions.

Methyl viologen (paraquat) (A in Figure 20-2) is an herbicide with potent toxicity in many biological systems. The toxicity is not well understood, but one hypothesis maintains that superoxide, which is known to be formed by the reaction of a radical cation (B) with dioxygen, is the general toxic agent. Sawyer's group (Nanni *et al.,* 1981) has shown that cation B in fact reacts rapidly with superoxide, probably forming the peroxy anion (C). The peroxy anion (C) can rearrange or react to form the products observed in the reaction. It is suggested that the toxicity is actually due to the oxygen-transfer ability of C (demonstrated by reaction with dimethylsulfoxide to form dimethylsulfone).

Superoxide has been shown to react very rapidly with polyhalogenated hydrocarbons including carbon tetrachloride, chloroform, and *p,p'*-DDT in aprotic solvents (Roberts and Sawyer, 1981). This is a very unusual reaction, apparently a nucleophilic attack on the halogen. The reaction is complex, producing several reactive intermediates that are clearly detrimental products. In addition, superoxide ion will add to chloroalkenes to form peroxyradical carbanions (Calderwood *et al.,* 1983). It has been suggested that these reactions may be the cause of the oxidative toxicity of some chlorinated hydrocarbons that are otherwise unreactive.

Many dyes are toxic due to oxidations they cause in the presence of light (Spikes, 1977; Spikes and Glad, 1964). These are called photosensitizers, and the effect is often called "photodynamic action." Photodynamic action can occur by several mechanisms, but the most prevalent is the production of singlet dioxygen. The photosensitizer, S, must have a relatively stable and easily available triplet state. The triplet state is formed by crossover from an excited singlet state:

$$S_o \xrightarrow{\ h\nu\ } {}^1S$$

$${}^1S \rightarrow {}^3S$$

The triplet state may abstract an electron from, or donate an electron to, a substrate, or it may form singlet dioxygen:

$${}^3S + A \rightarrow S^{\cdot-} + A^{\cdot+}$$

$${}^3S + A \rightarrow S^{\cdot+} + A^{\cdot-}$$

$${}^3S + {}^3O_2 \rightarrow {}^1O_2 + SO$$

In the second reaction, A may be triplet dioxygen, so that $A^{\cdot-}$ would be superoxide ion. The third mechanism is believed to be the most important. Hundreds of compounds may act as sensitizers, including the acridines, anthraquinones, azines, porphyrins, flavins, thiazines, and xanthines.

Dyes will catalyze the photooxidation of proteins and amino acids (Spikes and McKnight, 1971). Five amino acids are susceptible to dye-sensitized photooxidations: cysteine, histidine, methionine, tryptophan, and tyrosine (Spikes and McKnight, 1971). Proflavin, hematoporphyrin, and Rose Bengal will sensitize the photooxidation of methionine to the sulfoxide. Methylene blue will sensitize the photooxidation of tryptophan, methionine, histidine, and tyrosine residues in proteins. Methylene blue sensitizes the photooxidation of the histidine model, 4-methylimidazole, to form acetyl urea. Cysteine is oxidized to cysteic acid and tryptophan is oxidized to kynurenine. The oxidation of tyrosine causes ring-opening and carbon dioxide production, but the other products are not known.

Since porphyrins are sensitizers, the addition of free porphyrins to erythrocytes will cause hemolysis in the presence of light and dioxygen (Spikes, 1977). The damage to the cell is on the membrane. Amino acids and cholesterol in the cell membrane are destroyed, and certain membrane enzymes are inactivated. Certain patients accumulate free porphryins in the blood, and these patients have photosensitive blood.

The plant genus *Hypericum* is toxic to grazing animals because of a photosensitizer called hypericin (Figure 20-3) in its leaves. Buckwheat leaves also contain a photosensitizing agent.

Terthienyl (Figure 20-4) is a natural nematocide that occurs in marigolds (*Tagetes*). On irradiation, it produces singlet dioxygen (Bakker *et al.*, 1979). It is interesting to speculate that this is its mechanism of toxicity.

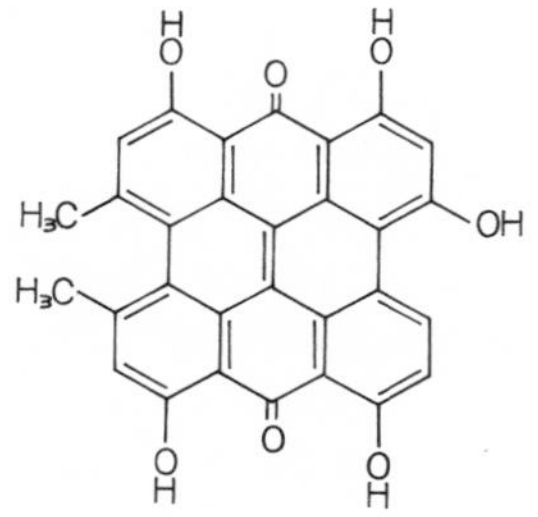

Figure 20-3. Hypericin, a toxic material in *Hypericum*.

Phagocytosis is the act of leukocytes in destroying bacteria. The majority of research on phagocytic activity has been performed with relatively short-lived polymorphonuclear leukocytes. While these leukocytes are engulfing their victim cells, a cyanide-independent 10- to 20-fold increase in the rate of dioxygen consumption occurs. Superoxide and hydrogen peroxide are formed via an NADPH oxidase reaction. A disease, chronic granulomatosis, is associated with a lack of this respiratory burst, a lack of hydrogen peroxide and superoxide production, and frequent repetitive infections. After the victim cell has been engulfed, the resulting vacuole fuses with granules containing microbicidal enzymes, including myeloperoxidase. The components necessary for microbicidal activity are hydrogen peroxide, chloride ion, and myeloperoxidase.

Singlet dioxygen has been suggested as the active agent in the microbicidal action (for a review, see Krinsky, 1979). This was based on the observation of chemiluminescence during the phagocytosis. Since chloride ion and hydrogen peroxide are involved in this reaction, the hypochlorite oxidation of peroxide was suggested as the reaction that generates excited molecules. The chemiluminescence was then ascribed to either singlet-dioxygen emission or excited carbonyl products of singlet-dioxygen reactions. Aqueous disproportionation of superoxide was suggested as another route for the formation of singlet dioxygen, but the results of Foote *et al.* (1980) make this an unlikely possibility. However, Foote (1976) has pointed out that the only evidence for singlet dioxygen participating in phagocytes is based on the weak argument that chemiluminescence is correlated with bactericidal activity. Held and Hurst (1978) have found that the inhibitors to microbicidal action, 2,5-dimethylfuran, β-carotene, and histidine, actually reduce HOCl instead of reacting with singlet dioxygen. Thus, the active agent may actually be HOCl, although HOCl has been proposed as a source of singlet dioxygen (Rosen and Klebanoff, 1977). The microbicidal action of hy-

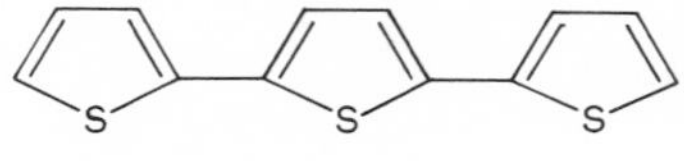

Figure 20-4. Terthienyl, a natural nematocide in marigolds.

drogen peroxide and chloride mixtures was pointed out by Stelmaszynska and Zglicyznski (1974), who found that hydrogen peroxide and chloride ion in the presence of myeloperoxidase will chlorinate aliphatic amino compounds including amino acids to form *N*-chloramines.

Cartenoid-producing (wild-type) *Sarcinea lutea* are immune to human polymorphonuclear leukocytes, but colorless mutants are destroyed rapidly by the leukocytes. This difference is attributed to the singlet-oxygen-quenching effect of carotenoid pigments. Diazabicyclo [2.2.2] octane also prevents the microbicidal action of the myeloperoxidase system at a low concentration. Other trapping, quenching, and D_2O studies have also supported the suggestion that singlet dioxygen is produced at some point in the myeloperoxidase–hydrogen peroxide–halide system. However, hypochlorous acid gave these same results in a manner that precluded singlet-dioxygen production via the familiar hypohalite–hydrogen peroxide route; i.e., excess hydrogen peroxide inhibited and excess chloride ion facilitated the reaction.

Hydrogen peroxide, superoxide, hydroxyl radical, and singlet dioxygen have all been implicated in the phagocytotic process, as have chlorite ion or hypochlorite or both. Which oxidizing agents are produced directly by the enzymes and which are secondary products is still a matter of conjecture.

The toxicity of cupric ion to erythrocytes is inhibited in an atmosphere of nitrogen or in the presence of carbon monoxide (Barnes and Frieden, 1983). This observation suggests that the cupric ion catalyzes the autoxidation of hemoglobin, producing superoxide ion or possibly another intermediate that is responsible for the toxic effect. Cupric ion is known to catalyze the aerobic oxidation of hemaglobin (Rifkind, 1974) to produce superoxide ion and hydrogen peroxide (Winterbourn *et al.,* 1976).

20.5 Beneficial Functions of Dioxygen

Before finishing this discussion, we should remind the reader that dioxygen has many beneficial functions. Dioxygen is essential for the biodegradation of natural organic compounds as well as of xenobiotics. Many of the reactions carried out by the pseudomonad soil bacteria discussed in Chapter 12 lead to the destruction of xenobiotics. Almost every conceivable type of aromatic compound (Mason, 1957) in organisms can be hydroxylated as a first step in degradation. Many of these compounds are drugs or toxins that are naturally found in the organism. Bacteria will oxidize aromatic hydrocarbons such as benzene, biphenyl, and naphthalene, and also paraffinnic hydrocarbons (Dagley, 1975). *Pseudomonas oleovorans* oxidizes normal *n*-octane to *n*-octanol and *Ps. putida* oxidizes benezene to *cis*-benzene glycol. Bacteria can oxidize cyclohexane to cyclohexanone and further to 6-hydroxyhexanoic acid. Pseudomonads will even oxidize polyvinyl alcohol to a polymer containing keto groups (Morita *et al.*, 1979) that can be subsequently hydrolyzed.

References

Aebi, H. E., and Wyss, S. R., 1978. in *The Metabolic Basis of Inherited Disease,* J. B. Standbury, J. B. Wyngaarden, and D. S. Fredrickson (eds.), Fourth Edition, McGraw-Hill, New York, pp. 1792–1808.

Babior, B. M., 1981. Oxygen as a weapon: How neutrophils use oxygen to kill bacteria, in *Oxygen and Life,* Royal Society of Chemistry, Burlington House, London, pp. 107–118.

Bakker, J., Gommers, F. J., Niewenhuis, I., and Wynberg, H., 1979. Photoactivation of the nematocidal compound α-terthienyl from roots of marigolds (*Tagetes* species): A possible singlet oxygen role, *J. Biol. Chem.* 254:1841–1844.

Barnes, G. and Frieden, E., 1983. Oxygen requirement for cupric ion induced hemolysis, *Biochem. Biophys. Res. Commun.* 115:680–684.

Bors, W., Saran, M., and Czapski, G., 1980. The nature of the intermediates during biological oxygen activation, in *Biological and Clinical Aspects of Superoxide and Superoxide Dismutase,* W. H. Bannister and J. V. Bannister (eds.), Proceedings of the Federation of European Biochemical Societies Symposium No. 62, Elsevier, New York, pp. 1–31.

Boveris, A., Cadenas, E., and Chance, B., 1981. Ultraweak chemiluminescence: A sensitive assay for oxidative radical reactions, *Fed. Proc. Fed. Am. Soc. Exp. Ziol.* 40:195–198.

Calderwood, T. S., Neuman, R. C., Jr., and Sawyer, D. T., 1983. Oxygenation of chloroalkenes by superoxide in aprotic media, *J. Am. Chem. Soc.* 105:3337–3339.

Cohen, G., and Heikkila, R. E., 1974. The generation of hydrogen peroxide, superoxide radical, and hydroxyl radical by 6-hydroxydopamine, dialuric acid, and related cytotoxic agents, *J. Biol. Chem.* 249:2447–2452.

Dagley, S., 1975. A biochemical approach to some problems of environmental pollution, *Essays Biochem.* 11:81–138.

Fee, J. A., 1980. Is superoxide toxic?, in *Biochemical and Clinical Aspects of Superoxide and Superoxide Dismutase,* W. H. Bannister and J. V. Bannister (eds.), Proceedings of the Federation of European Studies Symposium No. 62, Elsevier, New York, pp. 41–48.

Fee, J. A., 1981. A comment on the hypothesis that oxygen toxicity is mediated by superoxide, in *Oxygen and Life,* Royal Society of Chemistry, Burlington House, London, pp. 77–97.

Fee, J. A., and Hildebrand, P. G., 1974. On the development of a well-defined source of superoxide ion for studies with biological systems, *FEBS Lett.* 39:79–82.

Foote, C. S., 1976. Photosensitized oxidation and singlet oxygen: Consequence in biological systems (W. A. Pryor, ed.), *Free Read. Biol.* 2:85–133.

Foote, C. S., Shook, F. C., and Abakerli, R. A., 1980. Chemistry of superoxide ion. 4. Singlet oxygen is not a major product of dismutation, *J. Am. Chem. Soc.* 102:2503–2504.

Fridovich, I., 1972. Superoxide radical and superoxide dismutase, *Acc. Chem. Res.* 5:321–326.

Fridovich, I., 1974. Superoxide dismutase, *Adv. Enzymol.* 41:35–97.

Fridovich, I., 1975. Oxygen: Boon and bane, *Am. Sci.* 63:54–59.

Fridovich, I., 1979. Superoxide and superoxide dismutase, in *Advances in Inorganic Biochemistry I,* G. L. Eichorn and L. G. Marzilli, (eds.), Elsevier Press, New York, pp. 67–91.

Gerschman, R., Gilbert, D. L., Nye, S. W., Dwyer, P., and Fenn, W. O., 1954. Oxygen poisoning and X-irradiation: A mechanism in common, *Science* 119:623–626.

Goldberg, B., and Stern, A., 1976. Superoxide anion as a mediator of drug-induced oxidative hemolysis, *J. Biol. Chem.* 251:6468–6470.

Gotoh, T., and Shikama, K., 1976. Generation of superoxide radical during autoxidation of oxymyoglobin, *J. Biol. Chem.* 80:397–399.

Grankvurst, K., Marklund, W., Sehlin, J., and Taljedal, I., 1979. Superoxide dismutase, catalase and scavengers of hydroxyl radicals protect against the toxic action of alloxan on pancreatic islet cells *in vitro, Biochem. J.* 182:17–25.

Haber, F., and Weiss, J., 1934. The catalytic decomposition of hydrogen peroxide by iron salts, *Proc. R. Soc. London* 147:332–351.

Halliwell, B., 1976. An attempt to demonstrate a reaction between superoxide and hydrogen peroxide, *FEBS Lett.* 72:8–10.

Halliwell, B., Richmond, R., Wong, S. F., and Gutteridge, J. M. C., 1980. The biological significance of the Haber–Weiss reaction, in *Biological and Clinical Aspects of Superoxide and Superoxide Dismutase*, W. H. Bannister and J. V. Bannister (eds.) Proceedings of the Federation of European Biochemical Societies Symposium No. 62, Elsevier, New York, pp. 32–40.

Hassan, H. M., and Fridovich, I., 1977a. Physiological function of superoxide dismutase in glucose-limited chemostat cultures of *Escherichia coli, J. Bacteriol.* 130:805–811.

Hassan, H. M., and Fridovich, I., 1977b. Regulation of superoxide dismutase synthesis in *Escherichia coli:* Glucose effect, *J. Bacteriol.* 132:505–510.

Held, A. M., and Hurst, J. K., 1978. Ambiguity associated with the use of singlet trapping agents in myeloperoxidase-catalyzed reactions, *Biochem. Biophys. Res. Commun.* 81:878–885.

Hill, H. A. O., and Okolow-Zubkowska, M. J., 1981. The exploitation of molecular oxygen by human neutrophils: Spin-trapping of radicals produced during the respiration burst, in *Oxygen and Life,* The Royal Society of Chemistry, Burlington House, London, pp. 98–106.

Kellogg, E. W., III, and Fridovich, I., 1975. Superoxide, hydrogen peroxide, and singlet oxygen in lipid peroxidation by a xanthine oxidase system, *J. Biol. Chem.* 250:8812–8817.

Kellogg, E. W., III, and Fridovich, I., 1977. Liposome oxidation and erythrocyte lysis by enzymatically generated superoxide and hydrogen peroxide, *J. Biol. Chem.* 252:6712–6728.

Kong, S., and Davison, A. J., 1980. The role of interactions between O_2, H_2O_2, e^- and O_2^- in free radical damage to biological systems, *Arch. Biochem. Biophys.* 204:18–29.

Krinsky, N. I., 1979. Biological roles of singlet oxygen, *Singlet Oxygen: Org. Chem. A Ser. Monogr.* 40:597–641.

Lavelle, F., Michelson, A. M., and Dimitrjevic, L., 1973. Biological protection by superoxide dismutase, *Biochem. Biophys. Res. Commun.* 55:350–357.

Lynch, R. E., and Cole, B. C., 1980. *Mycoplasma pneumoniae:* A prokaryote which consumes oxygen and generates superoxide but which lacks superoxide dismutase, *Biochem. Biophys. Res. Commun.* 96:98–105.

Mason, H. S., 1957. Mechanisms of oxygen metabolism, *Adv. Enzymol.* 19:79–234.

Massey, W., Palmer, G., and Ballou, D., 1971. On the reactions of reduced flavins and flavoproteins with molecular oxygen, in *Flavins and Flavoproteins*, H. Kamin (ed.), University Park Press, Baltimore, pp. 349–362.

McCord, J. M., Keele, B. B., Jr., and Fridovich, I., 1971. An enzyme-based theory of obligate anaerobiosis: The physiological function of superoxide dismutase, *Proc. Natl. Acad. Sci. U.S.A.* 68:1024–1027.

Melhuish, W. H., and Sutton, H. C., 1978. Study of the Haber–Weiss reaction using a sensitive method for detection of OH radicals, *J. Chem. Soc. Chem. Commun.* 1978:970–971.

Misra, H., and Fridovich, I., 1971. The generation of superoxide radical during the autoxidation of ferredoxins, *J. Biol. Chem.* 246:6886–6890.

Misra, H., and Fridovich, I., 1972a. The univalent reduction of oxygen by reduced flavins and quinones, *J. Biol. Chem.* 247:188–192.

Misra, H., and Fridovich, I., 1972b. The role of superoxide anion in the autoxidation of epinephrine and a simple assay for superoxide dismutase, *J. Biol. Chem.* 247:3170–3175.

Misra, H., and Fridovich, I., 1972c. The generation of superoxide radical during the autoxidation of hemoglobin, *J. Biol. Chem.* 247:6960–6962.

Morita, M., Hamada, N., Sakai, K., and Watanabe, Y., 1979. Purification and properties of secondary alcohol oxidase from a strain of *Pseudomonas, Agric. Biol. Chem.* 43:1225–1235.

Nanni, E. J., Jr., Angelis, C. T., Dickson, J., and Sawyer, D. T., 1981. Oxygen activation by radical coupling between superoxide ion and reduced methyl viologen, *J. Am. Chem. Soc.* 103:4268–4270.

Porter, N. A., Dixon, J., and Randas, I., 1978. Cyclic peroxides and the thiobarbituric assay, *Biochim. Biophys. Acta* 441:506–512.

Rifkind, J. M., 1974. Copper and the autoxidation of hemoglobin, *Biochemistry* 13:2475–2481.

Roberts, J. L., Jr., and Sawyer, D. T., 1981. Facile degradation by superoxide ion of carbon tetrachloride, chloroform, methylene chloride, and *p,p′*-DDT in aprotic media, *J. Am. Chem. Soc.* 103:712–714.

Rosen, H., and Klebanoff, S. J., 1977. Formation of singlet oxygen by the myeloperoxidase-mediated antimicrobial system, *J. Biol. Chem.* 252:4803–4810.

Rotruck, J. T., Pope, A. L., Ganther, H. E., Swanson, A. B., Hofeman, D. G., and Hoekstra, W. G., 1973. Selenium: Biochemical role as a component of glutathione peroxidase, *Science* 179:588–590.

Smith, S. B., Bristlein, M., and Bruice, T. C., 1974. Electrophilicity of the 8 position of the isoalloxazine (flavine) ring system: Comment on the mechanism of oxidation of dihydroisoalloxazine, *J. Am. Chem. Soc.* 96:3696–3697.

Spikes, J. D., 1977. Photosensitization, in *The Science of Photobiology,* K. Smith (ed.), Plenum Press, New York, pp. 87–112.

Spikes, J. D., and Glad, B. W., 1964. Photodynamic action, *Photochem. Photobiol.* 3:471–487.

Spikes, J. D., and McKnight, M. L., 1971. Dye-sensitized photooxidation of proteins, *Ann. N. Y. Acad. Sci.* 171:149–161.

Stelmaszynska, T., and Zgliczynski, J. M., 1974. Myeloperoxidase of human neutrophilic granulocytes as chlorinating enzyme, *Eur. J. Biochem.* 45:305–312.

Tappel, A. L., 1981. Glutathione peroxidase and its selenocysteine active site. *Selenium in Biology and Medicine,* Ed. by Spallholz, d. E., Martin, d. L., and Ganther, H. E., Avi Publishing, Westport, Conn., pp. 44–53.

Tappel, A. L., and Dillard, C. J., 1981. *In vivo* peroxidation: Measurement via exhaled pentane and protection by vitamin E, *Fed. Proc. Fed. Am. Soc. Exp. Biol.* 40:174–178.

Tien, M., Svinger, B. A., and Aust, S. D., 1981. Superoxide dependent lipid peroxidation, *Fed. Proc. Fed. Am. Soc. Exp. Biol.* 40:179–182.

Totter, J. R., 1980. Spontaneous cancer and its possible relationship to oxygen metabolism, *Proc. Natl. Acad. Sci. U.S.A.* 77:1763–1767.

Winterbourn, C. C., McGrath, B. M., and Carrell, R. W., 1976. Reactions involving superoxide and normal and unstable hemoglobins, *Biochem J.* 155:493–502.

Further Chemistry of Singlet Dioxygen

21.1 A Word of Explanation

A considerable amount of research has been reported on the reactions of singlet dioxygen. Much of this material has little or no relevance to biological reactions at our present level of understanding of these reactions. However, a knowledge of these reactions may prove to be useful in our future understanding of biological oxidations; for this reason, this material is summarized in this chapter.

21.2 Reaction of Singlet Dioxygen with Olefins to Form Allylic Hydroperoxides

The mechanism of the allylic hydroperoxide reaction has not been established. Many possibilities have been suggested, and despite years of extensive research, only one has been unanimously rejected. 1,2-Dioxetanes, suggested to explain the carbonyl products, are stable enough to isolate under some conditions (or at least detect unequivocally) and have been shown to give exclusively carbonyl products without rearrangement to allylic hydroperoxides. The possibility of a (2 + 2) electrocyclic reaction to form a dioxetane was explored by Kearns to explain the carbonyl products. It was shown theoretically that such a reaction could occur for electron-rich olefins even though the analogous bisolefin-to-cyclobutane conversion is symmetry-forbidden. However, dioxetanes have been prepared and shown to decompose only to carbonyl compounds and not to allylic hydroperoxides (Kopecky and Mumford, 1969). Dioxetanes are therefore believed to be the intermediates that lead to carbonyl products, but they are not the initial intermediates; that is, (2 + 2) electrocyclic ring formation is not believed to operate in this case.

The remaining mechanisms are all controversial. The earliest postulate was a "concerted" mechanism, which invoked a six-membered cyclic transition state. A "peroxide" mechanism was suggested as an alternative, as were an open zwitterion and an open-chain diradical. One important complicating feature of this reaction is its low activation energy (generally $\leqslant 5$ kcal/mole). The fact that any kinetic barrier to reaction is very small has the effect of minimizing the amount of information one can obtain from solvent effects, isotope effects, and polar substituent effects.

The open-chain zwitterion is believed to lead to the dioxetane, at least in most of those cases that given carbonyl cleavage products. The magnitude of this pathway depends on the stability of the carbonium ion, which explains the close correlation between "electron richness" of the olefin and the ratio of carbonyl products. Enamines (Figure 21-1) give this reaction entirely, and enol ethers give mixtures, the proportions of which depend on the solvent (Saito *et al.*, 1972; Dewar, 1975; Frimer *et al.*, 1977). The kinetics have been studied and are consistent.

The reaction of singlet dioxygen and dimethylindole (Figure 21-2) is an example of the reaction of singlet dioxygen with enamines in general. It proceeds first through a zwitterionic peroxide that then converts to a dioxetane (Saito *et al.*, 1979). This reaction is interesting because of its relationship to tryptophan metabolism. At room temperature, the product with singlet dioxygen is the cleaved pyrrole ring, which must have been formed from the dioxetane. However, in methanol at $-70°C$, a methoxy derivative of the hydroperoxide can be isolated. These results are consistent with the formation of a zwitterionic peroxide followed by dioxetane formation. Gorman *et al.* (1979) have pointed out that since the activation energies of the carbonyl-forming reactions and the allylic-hydroperoxide-forming reactions are all essentially zero, and the activation entropies are all negative (-18 to -34 e.u.), the transition states must be similar. Hence, it is unlikely that the transition state for the indole–singlet oxygen reaction is zwitterionlike, since the zwitterion is very unlikely to be involved in the allylic-hydroperoxide-forming reaction (see below). This is not to say, of course, that

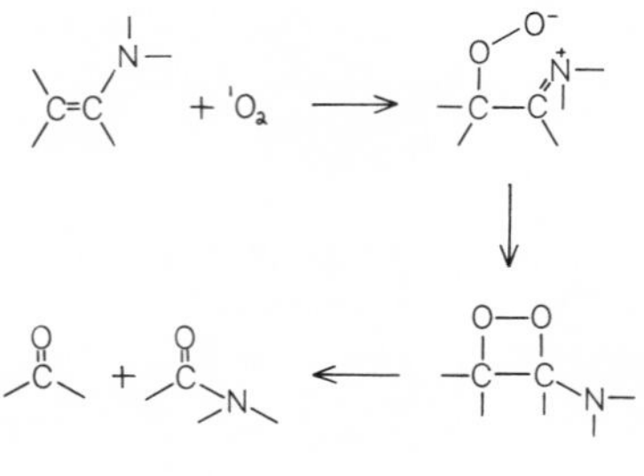

Figure 21-1. Reaction of singlet dioxygen with an enamine.

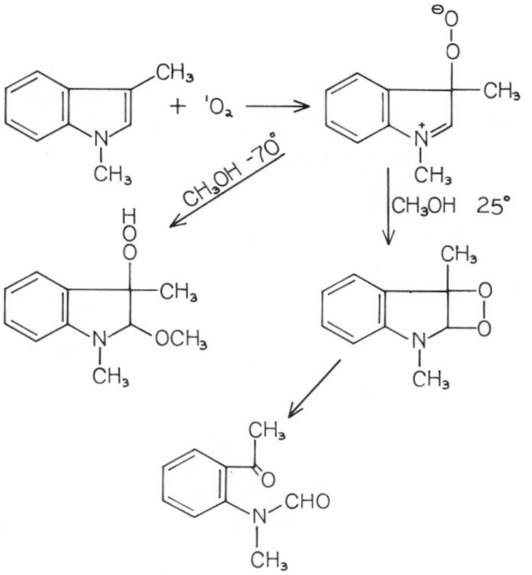

Figure 21-2. Reaction of singlet dioxygen with methylindole.

a zwitterion could not be involved as a later intermediate in the pathway to carbonyl products.

21.3 Concerted Cyclic Transition State Hypothesis in Allylic Hydroperoxides

Early evidence was interpreted to mean that the olefin–singlet dioxygen reaction leading to allylic hydroperoxides passed through a concerted cyclic transition state in which one oxygen is coordinated to a p-orbital of the olefin and the other oxygen accepts a proton from the allylic carbon (Figure 21-3).

There are three arguments in support of this transition state. The first is that the singlet dioxygen is added with retention of configuration, Nickon and Bagli, 1961. Since the oxygen in a concerted reaction must be added on the same side that the hydrogen is removed, retention is expected. Second, the double bond

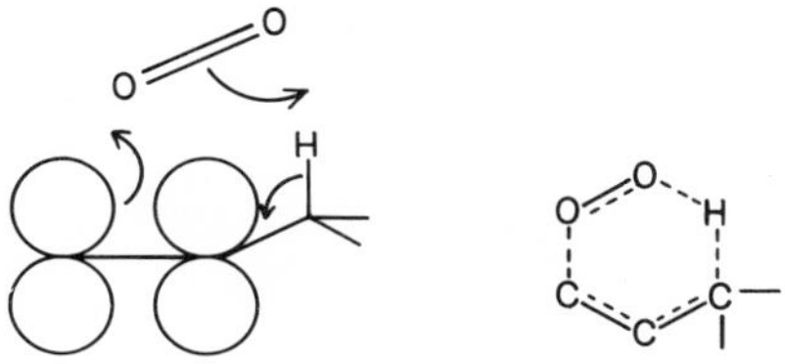

Figure 21-3. A proposed mechanism for the olefin–singlet dioxygen reaction.

always shifts because the proton removal forms a new double bond. Third, the reaction proceeds at the fastest rate when the C–H bond at the allylic position is parallel to the p-orbital of the double bond as shown in the transition state (Schenck *et al.*, 1964; Nickon and Mendelson, 1965; Gollnick and Schenck, 1964). This mechanism does not explain the formation of the observed carbonyl side products or some of the stereospecificity involved. Furthermore, the evidence does not establish the time sequence of hydrogen removal in relation to dioxygen addition.

21.4 Perepoxide Hypothesis

A perepoxide reaction intermediate has been considered as shown in Figure 21-4. This intermediate could lead to either the allylic peroxide product (note the geometry in Figure 21-5) or the dioxetane cleavage products. Perepoxides have been considered as intermediates in the flavin monooxygenase reactions (Dmitrienko *et al.*, 1977).

The perepoxide explains many of the features of the "ene" reaction that the concerted mechanism does not. For instance, the stereochemistry is retrained and some further stereospecificity is explained, the preference for reaction with electron-rich olefins is clear, and the formation of the three-numbered ring is symmetry-allowed. However, other facets of the reaction are difficult to understand in terms of a perepoxide.

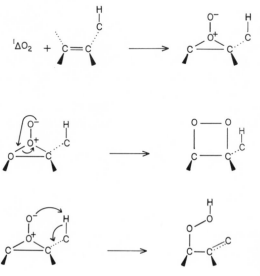

Figure 21-4. Proposed perepoxide intermediate in the singlet dioxygen–olefin reaction.

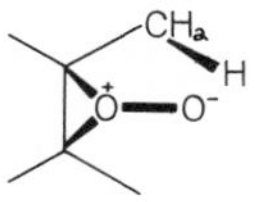

Figure 21-5. Geometric representation of perepoxide.

The pathway leading to the dioxetane via the perepoxide appears to offer no energetic advantage over the much simpler (and therefore probably quicker) peroxy zwitterion (Figure 21-6). In fact, it has been shown that for the case of enol ethers, the perepoxide does not rearrange to the dioxetane (Frimer, 1979). However, it is possible that the different products arise from entirely separate pathways.

The N-oxide and sulfoxide three-membered ring analogues of the perepoxide are known compounds [Kopecky and van de Sande (1972) and references therein]. Perepoxides have, in fact, been studied as intermediates in the reaction of per-oxyhalohydrins in methanolic base (Kopecky and van de Sande, 1972). By studying a series of substituted peroxyhalohydrins, the authors were able to show that (1) the hydroperoxide migrated (Figure 21-7); (2) the peroxy anion was required for reaction; (3) cleavage to singlet dioxygen and olefin, and recombination, did not occur; (4) an equilibrating zwitterion was not involved; and (5) the nature of the halogen does not affect the product ratio (Kopecky et al., 1978).

These observations could be explained only in terms of a perepoxide intermediate or transition state. The authors also concluded that conversion of perepoxide to product must have a very small activation energy, to the extent of being termed "perhaps concerted," from the fact that the isotope effect observed is quite small. Among the compounds that reacted to form allylic hydroperoxides, the yield was quantitative, indicating no dioxetane formation. For the compound in Figure 21-8, however, only dioxetane was formed. Comparison with the reaction of hydroxy (rather than peroxy) derivatives showed that epoxide formation from the alkoxide anion must occur at least four orders of magnitude faster than perepoxide formation from peroxy anion, as one would expect.

Singlet dioxygen, generated from hydrogen peroxide and hypochlorite, reacting with the parent olefin, gave the same product mixture as the perepoxide reaction but in much different ratios. This fact led to the contention that the perepoxide is not an intermediate in the primary singlet-dioxygen pathway. Nonetheless, several studies support the perepoxide intermediate or a species

Figure 21-6. Mechanisms of dioxetane formation.

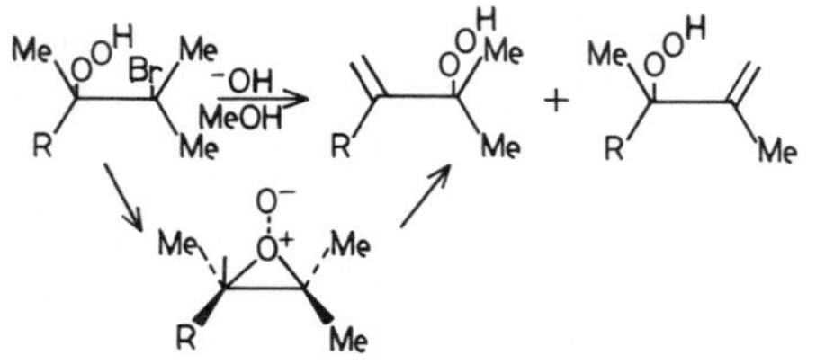

Figure 21-7. Migration of the hydroperoxyl group in the reaction of a β-halohydroperoxide with methanolic base as evidence for a perepoxide intermediate.

very similar to it in the formation of allylic peroxides (Frimer *et al.*, 1977; Grdina *et al.*, 1979).

Evidence for a perepoxide intermediate is found in the reaction of singlet dioxygen with adamontylidene adamantane in the presence of phenylmethylsulfoxide (Schaap *et al.*, 1983). The sulfoxide is oxidized to the sulfone and the olefin forms an epoxide, as would be expected from reduction of a perepoxide.

The isotopic rate effects of singlet dioxygen with partially deuterated tetramethyl ethylene (Grdina *et al.*, 1979; Stephenson *et al.*, 1980) depend on the position of deuteration, as shown in Figure 21-9. If the concerted mechanism were involved here, all isomers should show equal rate effects. Since they do not, one can infer that protons on *cis* vicinal carbons (A) are equivalent (i.e., competing in the reaction), but protons on *gem* (B) or *trans* (C) vicinal carbons are not. This equivalence is well explained by the perepoxide, given the assumptions that (1) its formation is irreversible and (2) there is a barrier to inversion of the epoxide ring. Compounds that have two forms of H–D competition possible show intermediate isotope effects. For example, compound in Figure 21-10 could have H–D competition with a "top" perepoxide (B), but no such competition with a "bottom" perepoxide (C). Intermediate isotope effects indicate that D or H substitution is not an important factor in whether or not the reaction of a given conformation proceeds, but it is a factor in which product is formed (*cis* competition).

21.5 Diradical Hypothesis

Harding and Goddard (1977) have performed *ab initio* calculations on simple olefins and have extended these calculations with thermochemical data to include

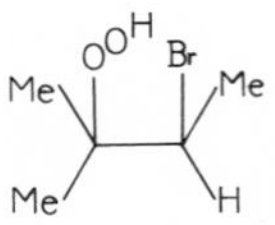

Figure 21-8. A β-halohydroperoxide that produces only dioxetane in methanolic base.

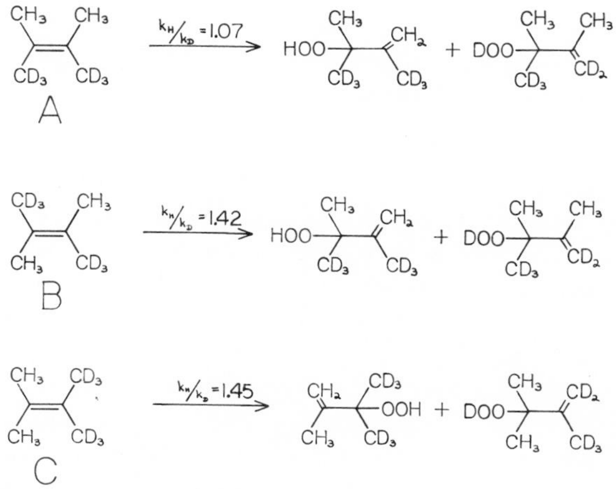

Figure 21-9. Isotopic rate effects in the reaction of tetramethylethylene with singlet dioxygen.

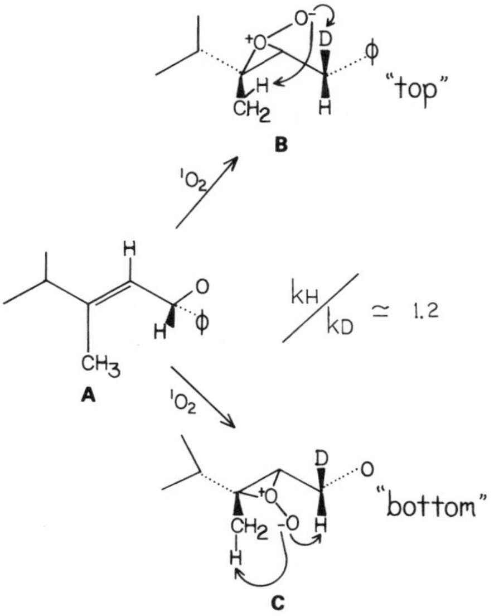

Figure 21-10. H–D competition as a function of perepoxide structure.

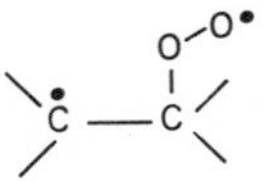

Figure 21-11. Hypothetical diradical intermediate in singlet dioxygen–olefin reactions.

larger olefins (Harding and Goddard, 1980) and their reactions with singlet dioxygen. They calculated heats of formation for dioxetanes, perepoxides, diradicals, and zwitterions, which have been suggested as reactive intermediates. Consistently, these calculations have shown the diradicals (Figure 21-11) to be the lowest-energy intermediates; perepoxides are energetically less favorable, and the zwitterions appear very high in energy.

There is no obvious reason why a perepoxide would be so unfavorable. Protonated epoxides are, of course, well known, as are epoxide–metal complexes. One would think that a negative ion attached to the positive epoxide would be, at worst, no less stable than the protonated epoxide. However, this is a classic charge-repulsion argument and does not take into account lone-pair orbital interaction. One can also argue that thermodynamic rationalizations are often incomplete; that is, reactions may not follow the thermodynamic path if another is quicker.

The perepoxide was found to be 53 kcal above the dioxetane. However, it was only 16.5 kcal above olefin and singlet dioxygen, while the diradical (Figure 21-11) is calculated to be only 9 kcal above reactants, in agreement with activation energies measured in the gas phase. The authors attributed the perepoxide instability to lack of delocalization of lone-pair charge density. They also estimate that in polar solvents, the zwitterion or perepoxide could be sufficiently stabilized to predominate over the diradical pathway.

The diradical energy calculated matches well the activation energies measured for this reaction in the gas phase, whereas the perepoxides and zwitterions are beyond the allowable error range. These activation energies (Ashford and Ogryzlo, 1975) are 3.2–8.3 kcal/mole in the gas phase and are expected to be smaller by approximately 3 kcal/mole in solution. The entropy change is about -23 e.u. and does not vary much with the olefin reacting.

Harding and Goddard (1978) have rationalized that the stereochemical results of several reactions are consistent with a diradical intermediate sterically controlled by the anomeric effect (Jeffrey *et al.*, 1972). For example, in the reactions shown in Figure 21-12, the hydrogen tends to be eliminated, i.e., *cis* to the methoxy group. Assuming the anomeric effect to dominate the initial O attack, two diradicals may be formed from each reactant. For the first (*trans*) reactant, one of the diradicals has the oxygen radical in the vicinity of the methyl group (A in Figure 21-13) and the other has the oxygen radical in an "unreactive" position (B). The reaction proceeds through diradical A by abstraction of a methyl hydrogen atom.

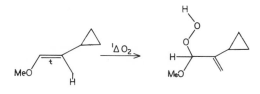

Figure 21-12. Reaction of singlet dioxygen with the *cis* and *trans* isomers of a vinyl ether.

The position of the oxygen radical is determined by the interaction of the lone pair on the methoxy oxygen with the CO_2 bond (an anomeric effect). The *cis* olefin will also produce an unreactive diradical and a reactive diradical. This reactive isomer will abstract a hydrogen from the cyclopropyl group. The barrier to conversion between the two radical stereoisomers (A and B above) is on the order of 3–5 kcal, which is larger than the expected activation energy for returning to reactants or for H abstraction leading to products.

While this is a perfectly logical argument, it has been shown that the reaction pathway favors H abstraction from the disubstituted side of the trisubstituted olefin even if the substituents have no lone-pair electrons; i.e., an anomeric effect is impossible (Orfanopoulos *et al.*, 1979).

Goddard and Harding also rationalized one of the main objections to a diradical intermediate: that the observed products lead to Markovnikoff orientation. Relative to carbonium ions, radicals are less sensitive to alkyl substitution at the radical center. Hence, the relative stability of a secondary vs. a tertiary radical (shown in Figure 21-14) depends on the relative bond strength at the alternative position. For a few highly electronegative radicals (including RO·), the tertiary carbon–OR bond strength is enough greater than the secondary carbon–OR bond strength that the equilibrium is shifted toward the secondary radical. Markovnikoff orientation is observed in gas-phase reactions.

The basis of the argument that a diradical is the first step in the reaction pathway is the rather surprising conclusion that singlet dioxygen is essentially

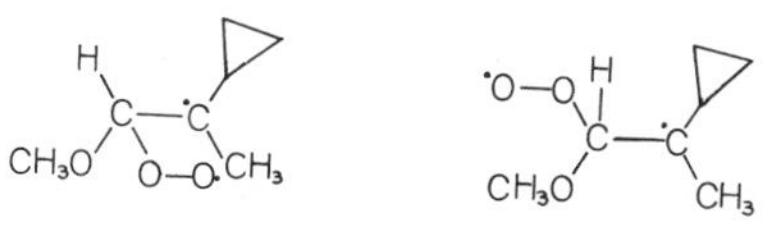

Figure 21-13. Possible diradical intermediates for the reaction shown in Figure 21-12.

A   B

Figure 21-14. Relative stabilities of tertiary and secondary radicals and carbonium ions as forces in rearrangements. A tertiary radical is not much more stable than a secondary, so that the equilibrium position depends on the relative bond strengths. A tetriary carbonium ion is much more stable than a secondary, so that equilibrium position is independent of bond strengths.

a diradical. Formation of a diradical from another diradical is expected to be quick and easy. Furthermore, abstraction of a hydrogen atom is similar to a bimolecular annihilation reaction, also expected to have a very low activation energy, as observed.

The products of the reaction of singlet dioxygen with 2-methoxy-*cis*- and *trans*-2-butenes (Figure 21-15) have been explained in terms of a diradical intermediate (Hammond, 1979). The reaction of singlet dioxygen with 2-methoxy-*trans*-2-butene (A) gives the expected allyl hydroperoxide (C), whereas the reaction of singlet dioxygen with 2-methoxy-*cis*-2-butene (F) gives the expected product (H) plus the same product (C) as obtained from the 2-methoxy-*trans*-2-butene. The enthalpies have been estimated from bond energies. Both diradicals (B and E) are relatively stable and lead to product (C). However, E must rotate to D before producing C, and this is competitive with the decomposition of E back to reactant F. Reaction through the less stable G does not require rotation

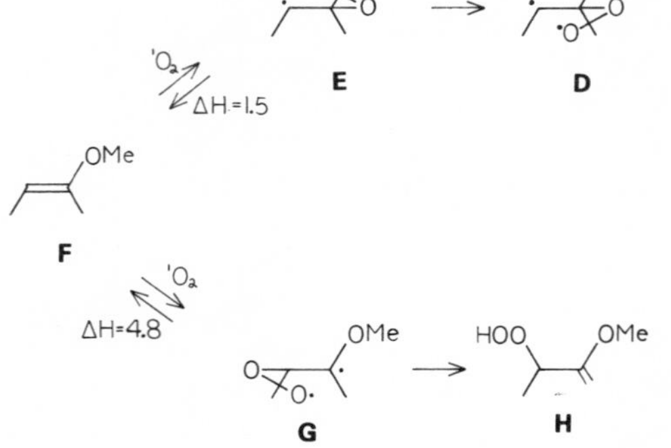

Figure 21-15. Reaction of singlet dioxygen with 2-methoxy-*cis*- and *trans*-butenes.

to give product H. Thus, a considerable amount of H should be formed, in agreement with observation.

Yamaguchi *et al.* (1981) have performed a further calculation on possible intermediates in this reaction. This calculation showed that semiempirical methods [(MINDO/3)] agree with *ab initio* results claiming that a diradical intermediate is considerably more stable than either a perepoxide or an open zwitterion if an unrestricted procedure is used. This unrestricted semiempirical method was then extended to trisubstituted olefins so that stereochemistry could be evaluated. The results showed that while diradicals are the lowest in energy of the suggested intermediates, they do not lead to the correct stereochemistry. That is, Markovnikoff directing effects are predicted for the diradical pathway, but not observed in solution. Also, the specificity that is observed experimentally in substituted olefins is predicted to be negligible for diradical intermediates. The authors concluded that diradicals are not involved in the reaction and that since the other intermediates are unreasonably high in energy, the reaction must be concerted.

The open-chain diradical is objectionable experimentally because the reaction is typified by a small number of products with systematic, if enigmatic, orientations. As Frimer (1979) points out, several observers have noted the lack of rearrangements and racemization, or inhibition by radical inhibitors, which are general characteristics of radical reactions. Also, the regiospecificity of the isotope effect cannot be explained by an intermediate in which a C–O single bond is clearly established (Stephenson *et al.,* 1980). For instance, compounds A and B in Figure 21-9 would give, respectively, intermediates A and B in Figure 21-16, which must have equal kinetic isotope effects. Since the observed isotope effect is different for the two olefins, this type of intermediate is probably not involved.

21.6 An Alternative Hypothesis for Allylic Hydroperoxide Formation

Jefford (1979) explained the propensity for H abstraction from the more hindered side of an olefin in terms of an open-chain zwitterion, and the explanation applies equally to an open-chain diradical. But the explanation does not account for regiospecific kinetic isotope effects.

The perepoxide geometry, on the other hand, explains the *cis*-only com-

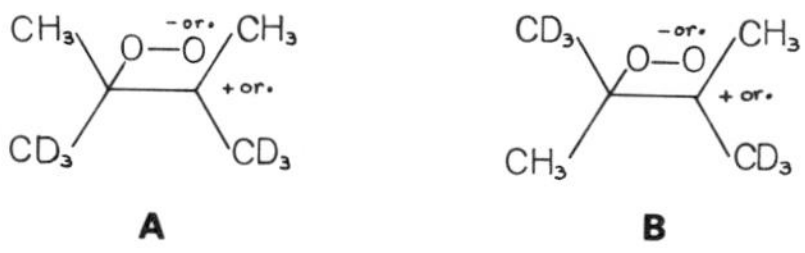

Figure 21-16. Possible intermediates in the reaction of compounds A and B in Figure 21-9 with singlet dioxygen.

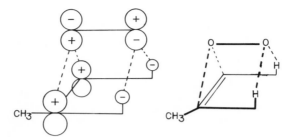

Figure 21-17. Transition state in perepoxide formation.

petition very well, but it does not explain the propensity for the more crowded side of the double bond. Stephenson *et al.* (1980) have provided an explanation of this enigma. They suggest that since the reaction has a very low activation energy, it is the initial interaction of electron acceptor (singlet dioxygen) and electron donor (olefin π bond) that determines the product. This initial interaction is determined by frontier orbital overlap. Since the olefin HOMO contains small coefficients on allylic hydrogens (hyperconjugation), the most favorable interaction has singlet dioxygen on the more highly substituted side of the olefin (Figure 21-17). This is analogous to a Diels–Alder reaction in which the sterically more hindered product predominates because of maximum frontier orbital overlap. After formation of the complex, the activation energy to continue the reaction is very low. This implies that the choice of whether the reaction will revert to reactants or continue to product is relatively independent of C–H bond strength, which, in turn, explains the negligible kinetic isotope effect for any hydrogens other than *cis*. Yet for *cis* hydrogens, the bond strength definitely would make a difference in determining the product, as observed.

This explanation also accounts for other observations of separation of rate-determining and product-determining steps without invoking a formal intermediate like the perepoxide. For example, the rate of the reaction is very little dependent on deuterium substitution (intermolecular competition), but the products formed can be isotope-dependent. Also, in the case of dimethyl sytrene, substituents on the phenyl ring affect the reaction rate markedly, but not the product distribution (Figure 21-18).

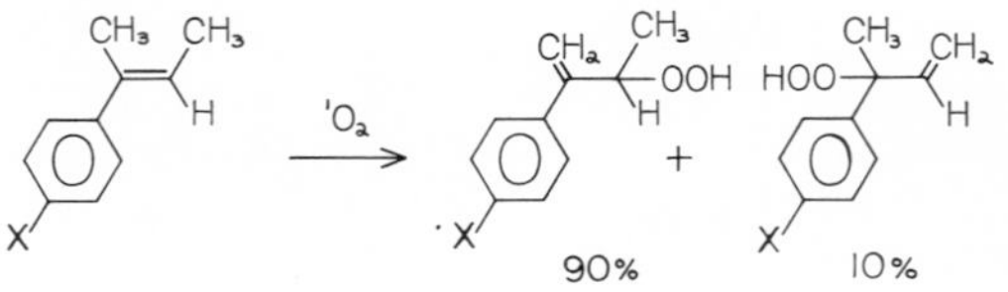

Figure 21-18. Product distribution in the reaction of singlet dioxygen with dimethyl styrenes. Relative rates: $X = OCH_3 \gg X = H > X = CF_3$.

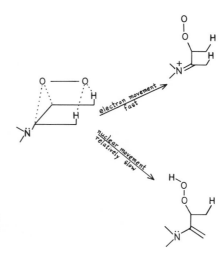

Figure 21-19. Importance of the Franck–Condon principle to the structure of the product in the singlet dioxygen–enamine reaction.

Thus, the frontier orbital complex explains many facets of the reaction more clearly than any of the previous suggestions. In accord with the Gorman *et al.* (1979) analysis of the indole reaction, it seems likely that the transition state for the carbonyl-forming reaction is also a Diels–Alder type complex that breaks down to a zwitterion. Especially in the case of enamines and enol ethers, formation of zwitterion intermediates should be faster than formation of the allylic hydroperoxide, according to the Franck–Condon principle (Figure 21-19).

From the standpoint of biochemical interest, it should be noted that (1) removal of hydrogen (proton or atom) does appear during or before the slowest step of the reaction, but the rate effect is small (however, one expects small effects with small activation energies); (2) there is a stereochemical inversion barrier in the crucial intermediate; and (3) singlet dioxygen does not generally form dioxetanes in concerted electrocyclic reactions; i.e., the Woodward–Hoffman rules appear to hold, even though there may be theoretical reasons that they would not in systems with several lone pairs not directly involved in the reaction.

21.7 Reaction of Singlet Dioxygen with α,β-Unsaturated Diketones

Another variation of the "ene" reaction has been observed (Figure 21-20) (Ensley *et al.*, 1980). α,β-Unsaturated ketones of correct (*s-cis*) configuration give allylic hydroperoxides with the double bond moved but still in conjugation with the carbonyl group. This reaction has been formulated as Diels–Alder (4 + 2) type because of the strict stereochemical requirement.

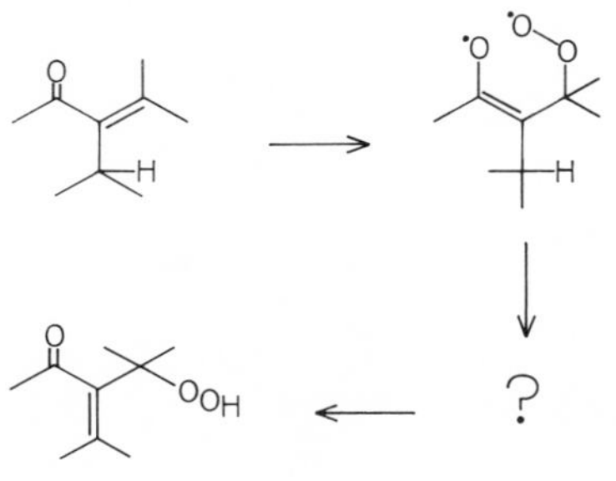

Figure 21-20. Reaction of singlet dioxygen with s-*cis*-unsaturated ketones.

An interesting variation is the singlet-dioxygen oxidation of 1-benzyl-3,4-dihydroisoquinoline to give 1-benzoyl-3,4-dihydroisoquinoline. The intermediate is again believed to be a zwitterionic peroxide that transfers a proton to form the hydroperoxide followed by decomposition to the product (Martin *et al.*, 1980). A dioxetane seems as unlikely intermediate because this would lead to carbon–carbon bond cleavage.

21.8   Reaction of Singlet Dioxygen with Strained Acetylene

Turro *et al.* (1976) have discovered a very interesting reaction (Figure 21-21) of singlet dioxygen with a highly strained aceylene (A). The fact that the C≡C − C bond angle in this compound is exceptionally small (146°) creates a nondegeneracy of the π_x and π_y orbitals. The π-orbitals in the approximate molecular plane (π_y) is destabilized relative to the perpendicular orbital.

Triplet and singlet dioxygen both react with A in Figure 21-21 to form B, the former reaction occurring at about 90°C and the latter as low as −90°C. The triplet-dioxygen reaction produces product B in an excited state, which is detectable by its fluorescence. Kinetic data were collected by enhancing the fluorescence by the addition of perylene (A in Figure 21-22), which has a more intense emission. The singlet-dioxygen reaction produces identical fluorescence if the reaction or product mixture is raised to −30°C. However, fluorescence

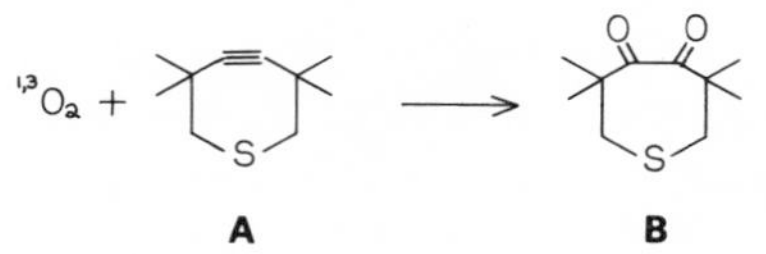

Figure 21-21. Reaction of singlet dioxygen with a highly strained acetylene.

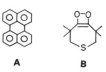

Figure 21-22. Perylene fluorescent enhancer (A) and the dioxetane intermediate (B) in the singlet-dioxygen reaction with compound A in Figure 21-21.

from the singlet-dioxygen reaction is thousands of times more intense than that from the triplet-dioxygen reaction. If held at $-90°C$, the product mixture is stable for days. Low-temperature infra-red studies ($-78°C$) showed very little dione present in the singlet-dioxygen product mixture.

These results indicate that the initial reaction product is an intermediate, believed to be dioxetene (B in Figure 21-22), which is stable at $-90°C$, but which decomposes to product A in Figure 21-21, in a symmetry-forbidden process at or above $-30°C$. The difference in activation energy between the two reactions is 21 kcal/mole, which, the authors suggest, is required to raise triplet dioxygen to singlet dioxygen thermally. Other examples of this thermal excitation are extremely rare because of the paucity of mechanisms for the required spin flip.

The authors explain that in the case of compound A in Figure 21-21, the destabilized (π_y) acetylene orbital can essentially donate an electron to a dioxygen π^* orbital. This interaction has the effect of isolating the two unpaired electrons in nearly degenerate p_x and p_y orbitals on the oxygen atom away from the acetylene, the ideal condition for rapid spin coupling. The result of this spin-flip and acetylene interaction would be a singlet zwitterionic species, either a perepoxide (A in Figure 21-23) or an "open chain" (B), that could rearrange to the dioxetane intermediate (B in Figure 21-22) or dissociate to form singlet dioxygen and compound A in Figure 21-21, which would then react very rapidly to form the dioxetane.

The dissociation to form singlet dioxygen agreed with the finding that various singlet-dioxygen traps and quenchers prevented the cperylene luminesence. This example supplies the possibility, at least, of a catalytic thermal generation of singlet dioxygen:

$$A + {}^3O_2 \xrightarrow[\text{Dark}]{\Delta} A \cdots O_2 \rightarrow A + {}^1O_2 \rightarrow B$$

Figure 21-23. Possible intermediates in the reaction of ground-state dioxygen with compound A in Figure 21-21.

References

Ashford, R. D., and Ogryzlo, E. A., 1975. Temperature dependence of some reactions of singlet oxygen with olefins in the gas phase, *J. Am. Chem. Soc.* 97:3604–3607.

Dewar, M. J. S., 1975. Computing calculated reactions, *Chem. Br.* 11:97–106.

Dmitrienko, G. I., Snieckus, V., and Viswanatha, T., 1977. On the mechanism of oxygen activation by tetrahydropterin and dihydroflavin-dependent monooxygenases, *Bioorg. Chem.* 6:421–429.

Ensley, H. E., Carr, R. V. C., Martin, R. S., and Pierce, T. E., 1980. Reaction of singlet oxygen with α,β unsaturated ketones and lactones, *J. Am. Chem. Soc.* 102:2836–2838.

Frimer, A. A., 1979. The reaction of singlet oxygen with olefins: The question of mechanism, *Chem. Rev.* 79:359–387.

Frimer, A. A., Bartlett, P. O., Bosching, A. F., and Jewett, J. C., 1977. Reaction of singlet oxygen with 4-methyl-2,3-dihydro-γ-pyrans, *J. Am. Chem. Soc.* 99:7977–7986.

Gollnick, K., and Schenck, G. O., 1964. Mechanism and stereoselectivity of photosensitized oxygen transfer reactions, *Pure Appl. Chem.* 9:507–525.

Gorman, A. A., Lovering, G., and Rogers, M. A. J., 1979. The entropy-controlled reactivity of singlet oxygen ($^1\Delta$g) toward furans and indoles in toluene: A variable-temperature study by pulse radiolysis, *J. Am. Chem. Soc.* 101:3050–3055.

Grdina, B., Orphanapoulos, M., and Stephenson, L. M., 1979. Stereochemical dependence of isotope effects in the singlet oxygen–olefin reaction, *J. Am. Chem. Soc.* 101:3111–3112.

Hammond, W. B., 1979. On the relative reactivity of $^1\Sigma$ versus $^1\Delta O_2$: Reactions of laser generated singlet oxygen with methoxy olefins, *Tetrahedron Lett.* 25:2309–2312.

Harding, L. B., and Goddard, W. A., 1977. Intermediates in the chemiluminescent reaction of singlet oxygen with ethylene: *Ab initio* studies, *J. Am. Chem. Soc.* 99:4520–4523.

Harding, L. B., and Goddard, W. A., 1978. Mechanistic implications of the stereochemistry of singlet oxygen–olefin reactions, *Tetrahedron Lett.* 1978:747–750.

Harding, L. B., and Goddard, W. A., III, 1980. The mechanism of the ene reaction of singlet oxygen with olefins, *J. Am. Chem. Soc.* 102:439–449.

Jefford, C. W., 1979. The role of zwitterion peroxides in controlling hydroperoxidation, *Tetrahedron Lett.* 1979:985–988.

Jeffrey, G. A., Pople, J. A., and Radon, L., 1972. The application of *ab initio* molecular orbital theory to the anomeric effect: A comparison of theoretical predictions and experimental data on conformations and bond lengths in some pyranoses and methyl pyranosides, *Carbohydr. Res.* 25:117–131.

Kopecky, K. R., and Mumford, C., 1969. Luminescence in the thermal decomposition of 3,3,4-trimethyl-1,2-dioxetane, *Can. J. Chem.* 47:709–711.

Kopecky, K. R., and van de Sande, J. H., 1972. Deuterium isotope effects in the oxidation of 2,3-dimethyl-2-butene via the bromohydroperoxide by singlet oxygen and by triphenyl phosphate ozonide, *Can. J. Chem.* 50:4034–4049.

Kopecky, K. R., Scott, W. A., Lockwood, P. A., and Mumford, C. A., 1978. Perepoxide intermediates in the conversion of some β-halohydroperoxides to allylic hydroperoxides by base, *Can. J. Chem.* 56:1114–1121.

Martin, N. H., Champion, S. L., and Belt, B. P., 1980. Regiospecific oxidation of substituted 1-benzyl-3,4-dihydroisoquinolines using singlet oxygen, *Tetrahedron Lett.* 21:2613–2616.

Nickon, A., and Bagli, J. F., 1961. Reactivity and geometry in allylic systems. I. Stereochemistry of photosensitized oxygenation of monolefins, *J. Am. Chem. Soc.* 83:1498–1508.

Nickon, A., and Mendelson, W. L., 1965. Reactivity and geometry in allylic systems. VI. Stereospecificity conversion of allylic alcohols to α,β-epoxy ketones by photosensitized oxygenation, *J. Am. Chem. Soc.* 87:3921–3928.

Orfanopoulos, M., Grdina, M. J., and Stephenson, L. M., 1979. Site specificity in the singlet oxygen trisubstituted olefin reaction, *J. Am. Chem. Soc.* 101:275–276.

Saito, I., Iouta, M., and Matsuura, T., 1972. Photoinduced reactions. LXV. Photosensitized oxygenation of 2-methyl indoles, *Chem. Soc. Jpn. Chem. Lett.* 1972:1173–1176.

Saito, I., Matsugo, S., and Matsuura, T., 1979. Mechanisms of indole–singlet oxygen reactions: Interception of zwitterionic intermediates and ene reaction, *J. Am. Chem. Soc.* 101:7332–7338.

Schaap, A. P., Recher, S. G., Faler, G. R., and Vallasenor, S. R., 1983. Evidence for a perepoxide intermediate in the 1,2-cycloaddition of singlet oxygen to adamantyl-ideneadamantane: Nucleophilic oxygen atom transfer to sulfoxides, *J. Am. Chem. Soc.* 105:1691–1693.

Schenck, G. O., Gollnick, K., Buckwald, G., Schoeter, S., and Ohloff, G., 1964. Zur chemischen und sterischen Selektivität der photosensibilisierten O_2: Ubertragung auf (+)-Limonen und (+)-Carvomenthen, *Annalen* 674:93–117.

Stephenson, L. M., Grdina, M. J., and Orfanopoulos, M., 1980. Mechanism of the ene reaction between singlet oxygen and olefins, *Acc. Chem. Res.* 13:419–425.

Turro, N. J., Ramamurthy, V., Kou-Chang Liu, Krebs, A., and Kemper, R., 1976. Reaction of strained acetylenes with molecular oxygen: A novel chemiluminescent reaction, evidence for a dioxetane, and a mechanism for thermal generation of singlet oxygen, *J. Am. Chem. Soc.* 98:6758–6761.

Yamaguchi, K., Yabushita, S., Fueno, T., and Houk, K. N., 1981. On the mechanism of photooxidation reactions: Computational evidence against the diradical mechanism of singlet oxygen ene reactions, *J. Am. Chem. Soc.* 103:5043–5046.

Index